Allan McLane Hamilton, George Fielding Blandford

Insanity and its Treatment

Allan McLane Hamilton, George Fielding Blandford

Insanity and its Treatment

ISBN/EAN: 9783337366803

Printed in Europe, USA, Canada, Australia, Japan

Cover: Foto ©berggeist007 / pixelio.de

More available books at **www.hansebooks.com**

INSANITY AND ITS TREATMENT

LECTURES

ON

THE TREATMENT, MEDICAL AND LEGAL, OF INSANE PATIENTS

BY

G. FIELDING BLANDFORD, M.D. Oxon.

FELLOW OF THE ROYAL COLLEGE OF PHYSICIANS IN LONDON; LATE LECTURER ON PSYCHOLOGICAL
MEDICINE AT THE SCHOOL OF ST. GEORGE'S HOSPITAL, LONDON

THIRD EDITION

TOGETHER WITH

TYPES OF INSANITY

AN ILLUSTRATED GUIDE IN THE PHYSICAL DIAGNOSIS OF MENTAL DISEASE

BY

ALLAN McLANE HAMILTON, M.D.

ONE OF THE CONSULTING PHYSICIANS TO THE INSANE ASYLUMS OF NEW YORK CITY, AND THE HUDSON
RIVER STATE HOSPITAL FOR THE INSANE, ETC.

NEW YORK

WILLIAM WOOD AND COMPANY

56 & 58 LAFAYETTE PLACE

1886

PREFACE TO THE THIRD EDITION.

THE discoveries and advances in the physiology and therapeutics of insanity, which have been made during the last few years, are not of so great importance as those recorded in the last issue of these Lectures. But time and experience enable us to estimate the value of the knowledge we possess, to test our remedies, and modify our treatment. By the light of such experience, I have revised this new edition, and trust I have added to its usefulness.

PREFACE TO THE FIRST EDITION.

THE following Lectures, in an abridged form, were delivered by me at the Schools of St. George's Hospital, and I now publish them with the hope that they may serve to some extent as a handy book concerning Insanity, on which subject there exist in our language few works of the character of a text-book.

Written for students, they make no claim to be a complete treatise on Psychology; neither have all questions connected with Insanity been discussed in them—such as the management of asylums, the problem of disposing of the crowds of our chronic patients, or of amending the present Lunacy law.

A difficulty which has ever attended the delivery of lectures on Insanity is not removed when they are printed for general use. It is the difficulty in speaking upon the subject with the precision and certainty demanded by those who are learning how to deal medically with a grave disorder. The lecturer on Medicine, when speaking of diseases of lung, heart, or kidney, can lay down the pathology in a way that to a psychologist is at present impossible. He can put his finger on the exact seat of the malady, can say what is going on amiss during life, and what will be discovered after death, and can speak with exactness of both the early and late symptoms. Far different is it when we have to teach the nature or treatment of that which we call Insanity—to speak not of respiration, circulation, or digestion, but of mind, of feelings, and ideas, in an abnormal or distorted state.

Nevertheless, I am convinced that the only method by which we shall attain an insight into the mysterious phenomena of unsound mind, is to keep ever before us the fact that disorder of the mind means disorder of the brain, and that the latter is an organ liable to disease and disturbance, like other organs of the body, to be investigated by the same methods, and subject to the same laws. In speaking of the pathology of insanity, I have endeavored to keep this in view.

For utterances too dogmatic, for iterations and recapitulations too numerous, and for the too frequent presence of the personal pronoun, the form and style of "lectures," which I have thought fit to preserve, must be my apology.

So, too, I have omitted references beyond a very few. To many authors, home and foreign, whose works I have consulted, I am under obligations not easy to acknowledge. Especially I would mention the now numerous volumes of the "Journal of Mental Science," which constitute a mine of information on this special subject, having, I believe, no equal. And I am not a little indebted for advice, suggestions, and corrections to many friends.

CONTENTS.

CONTENTS. ix

LECTURE XV.

INSANITY AND ITS TREATMENT.

LECTURE I.

Introductory—Impossibility of Avoiding Insanity—Why the Study of Insanity is a Branch of Medicine—The Organ of Mind—The Nerve Centres and Cells—The Nerve Fibres—Their Distribution—Chemical Composition of Brain—The Blood Supply of the Brain—Nerve Function—Method of Study Twofold.

GENTLEMEN:—If there be one branch of the great study of medicine which more than another deserves to be called an art and a mystery, it is the treatment and investigation of insanity. The treatment is an art, which, during the present century, has advanced in a degree not inferior to other arts, and in which by practice and example we may hope to attain skill, as in surgery or midwifery; but the disorder which we call insanity is a mystery not yet unraveled. Can we define it?

> "To define true madness
> What is't but to be nothing else than mad?"

In truth, its inscrutable appearance without assignable cause in a man hitherto sane, and its no less inscrutable departure, are things which we must confess are not yet explicable by human knowledge. Nevertheless, it is a branch of our art which is constantly forcing itself upon our attention. There are diseases described in the lectures you here attend which throughout your lives may never come before you: you may never see a patient die of hydrophobia; you may be so blessed with a healthy locality that you may never have to treat Asiatic cholera or ague. And other ailments due to locality or to special occupations you may never witness. But fortunate indeed will you be, if you spend many years in practice without being called upon to treat, or to pronounce an opinion upon, some case of unsoundness of mind. Your female patients after parturition will be attacked with insanity; their boys and girls as they come to the age of puberty will show symptoms of it; at the climacteric of life men and women will break down; and in old age insanity will merge into fatuity and dotage in the general decay of mind and body. From the cradle to the grave the mental no less than the bodily health of your patients must be your care. And not only will you have to treat them, you will have to send them away from home under legal restraint, to plead their irresponsibility in courts of law if in frenzy or folly they commit crime; and when they are dead, you will be called on to testify to their competency or incompetency to make the will they have left behind.

Now, if your attention has been drawn to insanity chiefly by notices of the grave disputes that arise in courts of law both upon the subject in general and also upon the sanity or insanity of individuals, you may shrink ·from the prospect I have set forth, and may determine to have nothing to do with such cases. You may resolve, like others I have known, never to sign a certificate. You may possibly carry out your resolution in London, but in the country you can no more escape from this duty than from other branches of ' our profession, which here we may hand over to special- ists; and an opinion you may have to give on the state of mind of any of your patients: therefore it behoves you to have some idea of what you may have to say on the subject, as well as of the treatment of the various kinds of insane persons. In my lectures I shall endeavor to direct your atten- tion to both of these points, that you may be able to refer to them as notes for future guidance in your medical, and also your medico-legal, capacity. About the mysterious psychological part of the question, I shall only say so much as will enable you to understand various doctrines and theories which have been put forth, and which may be propounded in opposition to your own by others, whether lawyers or doctors. You will be asked if you agree or disagree with the theory of Dr. A. concerning " moral insanity," or of Dr. B. as to " emotional insanity," or of Dr C. as to " volitional insanity." You will be asked whether delusions are essen- tial to constitute a man legally insane, and will have to deal with the doc- trine that insane people are responsible if they know right from wrong. Some amount of information as to mind in general is therefore necessary: this I shall try to make as brief as possible, in which I may succeed; and as little obscure as I can, in which my success may be less.

We will suppose that you have for the first time walked through the wards and inspected the patients in a lunatic asylum, sufficiently populous to have given you the opportunity of seeing all or nearly all the phases of this malady. You have seen some whose appearance, acts, and lan- guage showed at once that something was wrong with them, others whose manner betrayed nothing, but whose conversation immediately indicated that they were not like men in general: some, on the contrary, appeared rational both in conduct and conversation, while in a considerable number you noticed a negative, not a positive alienation; an absence of word and thought, of work or amusement, a mental blank. You have not, however, gained much in sight into the one thing, insanity, which you came to look for. You have failed to recognize one disorder from which all are suffer-' ing—to see how it is that these, who one and all are secluded in this abode, differ from the rest of mankind on the other side of the wall. Why they are here, and how they are to be cured so as to re-enter the world, is what I wish to show. But a great number never will be cured: they must be taken care of for the rest of their lives. The curability and incurability I shall point out, but into the details of the care and treatment of such incurable patients I shall not enter.

If you ask me in what particular the inmates of the asylum agree, I can only say that they are all persons of unsound mind. Some are congenital idiots; some are epileptic patients, whose minds, memory, and understanding have been destroyed by frequent fits; some have had paralytic strokes with a similar result; some are in the dotage of age; others are insane in the common meaning of the word, holding extraordinary opinions and doing outrageous acts, and incarcerated here because they cannot be taken care of in any other away, or because their chance of recovery is increased by the restraint and treatment they receive. They resemble one another only in this, that they are all of unsound mind, incapable of taking care of themselves or of managing their affairs. This, you see, is a mere legal or civil distinction, yet it is one we cannot overlook, for it is the link which binds them all together. On analyzing them, we find that in antecedents and history, in capacity, in peculiarities, in propensities, they are of infinite variety: no two are alike, yet all are confined here *nolentes volentes*, and are here upon the certificates of medical men. What, then, you ask, is meant by this legal term, unsoundness of mind? in what way does it become a branch of medical study and practice? by what application of the art and science of medicine is it to be removed?

The answer is, that unsoundness of mind is but another term for disorder of the human brain, or rather of that portion of nerve matter which has for its function that which we call mind and mental operation. For all that is contained within the cranium is not concerned with the operations of mind proper, but a portion only, so that the function of certain parts may be disordered, the mind remaining sound; and, conversely, the mind may be deranged and defective, while other nerve functions, as the senses, continue intact. The proper function of the mental portion of the brain is, moreover, not essential to existence, and patients may live in a state of unsound mind during the whole term of a natural life.

It is not within the province of my lectures to discuss the functions of the various parts of that which, as a whole, we call the brain. From your lecturer on physiology you will already have learned what is known about them, and will have heard that great doubt exists even now as to the duties which they discharge, and their relations one to another. I shall only mention such points with regard to them as are essential to my purpose; but it is to be remembered that there is within the cavity of the cranium a number of organs capable, more or less, of acting independently of each other, and more or less developed in different animals. These all agree, however, in certain particulars: all are compounded of nerve cells, having certain special properties, and connected with one another by means of nerve fibres, bundles of which conduct to other cell-groups or to the periphery. These groups of cells, which I shall call *nerve centres*, are supplied by blood-vessels, and on the supply depends mainly the proper performance of their special function. If this be inadequate or ill-regulated, or if the blood itself be impure, the function shows an equal disturbance. Here, then, we have four things to consider, the nerve centres, the nerve fibres, the supply of blood, and the function or property of the centres. These can be studied to some extent by experiment, as well as observed in health and disease. In this manner much has been learned by various experimental physiologists which can be advantageously applied to the consideration of the physiology and pathology of the special centres of human mind, the two great hemispheres of the cerebrum.

The nerve centres are aggregates of nerve cells which are embedded in a substance or *stroma* called the "neuroglia," which in the brain appears to be a "transparent, nucleated, homogeneous, non-fibrillated matrix, representing the fibrillated connective tissue of the spinal cord." [1] According to some, it is itself nerve structure or an element thereof. A more probable theory is that it is connective tissue, as held by Virchow and Kölliker, for in some diseased brains the so-called connective tissue cells can be observed. Throughout its substance are nuclei and nucleoli. Of this as yet little is known, but I believe that it will be found to play an important part in the pathology of the brain, if not as an important agent of the function we call mind.

The nerve cells or vesicles consist of a very fine membrane, containing granular matter of various colors. They differ in shape, being round, pyriform, pyramidal, oval, and irregular. They differ also in size, varying in man from $\frac{1}{18000}$th to the $\frac{1}{300}$th of an inch in diameter. They do not appear to possess a distinct cell wall, but are contained in the surrounding stroma, and present the appearance of a finely granular body with a nucleus and nucleolus. They are arranged in the convolutions of the brain, according to Dr. Lockhart Clarke, in six layers, which are alternately paler and darker from the circumference to the centre. In these layers the cells vary in shape, size, and number. The superficial layer contains only a small number of round, oval, fusiform, and angular cells. The second, a darker layer, is densely crowded. The third is paler, and contains, like the fourth, narrow and elongated groups of small cells, and nuclei radiating at right angles to its plane. The fifth is pale, with but few cells, while the sixth abounds with examples of all those above named, and also some rather larger. This arrangement is most noticeable at the extremity of the posterior lobe. The other convolutions, however, differ from those of the posterior lobe, not only by the comparative faintness of their several layers, but also by the appearance of some of their cells. Proceeding forwards, the convolutions contain a number of cells of a much larger kind, triangular, oval, and pyramidal, the latter being quadrangular

---

[1] Drs. Tuke and Rutherford on Morbid Appearances in the Brains of the Insane, Edinburgh Medical Journal, Oct., 1869.

at the base, giving off four or more processes which frequently subdivide into minute branches, the opposite end of the cell tapering gradually into a straight process which runs directly to the surface of the convolution.

In other convolutions, modifications of the vesicular structure occur. In the surface convolution of the great longitudinal fissure on a level with the anterior extremity of the corpus callosum, all the three inner layers of gray substance are thronged with pyramidal, triangular, and oval cells of considerable size, the inner orbital convolution situated on the outer side of the olfactory bulb, containing a vast multitude of pyriform, pyramidal, and triangular cells, but none so large as those at the vertex. In the island of Reil, which overlies the extra-ventricular portion of the corpus striatum, a great number of the cells are somewhat larger, and the general aspect of the tissue different. A further variety is presented by the temporo-sphenoidal lobe which covers the *insula*, and is continuous with it; for while in the superficial and deep layers the cells are small, the middle layer is crowded by pyramidal and oval cells of considerable and rather uniform size. Even different parts of the same convolution may vary with regard to the arrangement or relative size of the cells.[1]

The nerve fibres, or nerve tubes, as they are often called, which conduct to or from the nerve centres, also vary in size from the millionth to the $\frac{1}{1800}$th of an inch in diameter. They form the white or commissural substance of the nervous system, being gathered into fasciculi, comprising a large number of these fibres. Each fibre consists of a very delicate membrane, forming a tube, and containing a viscid, albuminous, and fatty pulp. Enclosed in this tube is a minute fibre, or "axis cylinder," which is in truth the real nerve fibre, and which exists in places without its investing sheath. At the peripheral extremity it parts with the latter, and is divided and subdivided as it permeates and blends with the structure in which it is distributed, while at the other end it becomes continuous with the contents of the nerve cells, also parting with its sheath at this point. That the nerve fibre is protected from disturbance by the investing sheath and pulp there can be no doubt, and that the latter plays a part in the preservation of the healthy working of the whole system seems probable from the changes observable in it in certain diseases. You will not forget that a nerve fibre possesses a power of its own, even when severed from its connection with a nerve centre, and that this power, if exhausted by stimulation, may by rest be again recovered; and it appears to me likely that the nerve elements surrounding the fibre may contribute materially to the restoration of this power.

Without going at length into the history of the development of the nervous system, I may tell you that the earliest nerve cells are small,

---

[1] Dr. Lockhart Clarke, quoted in Dr. Maudsley's work on the Physiology and Pathology of the Mind, 2nd ed.

round, and devoid of processes, that afterwards they become larger in size, of various shapes, sending out one, two, or more fibres to communicate with other cells: in like manner we find at first extremely fine fibrils, which are packed together in larger fibres, and these in turn are covered and protected by a sheath. And as the cells and fibres become more and more developed, so by disease they may be more and more reduced to their primitive condition, till we return to the simple round cell,· till the fibres disappear, both sheath and axis cylinder, and only connective tissue remains.[1]

Out of these component parts—the nerve cells and nerve fibres—are built up the structures, which, as a whole, we term the nervous system. How is it constructed? What are the relations of the constituents? We turn to anatomy for information on these points, and are not disappointed for here, as elsewhere, anatomy points out to us the physiology of the organ. Throughout the entire nervous system, from the spinal cord up to the cerebral hemispheres, anatomy indicates the relations between the white fibres and the gray matter of the nerve centres. As, however, it is with the brain itself that you, as students of psychological physiology, are chiefly concerned, I shall say but little of the spinal cord. Now, the actual mode of connection between the spinal cord and cerebellum on the one hand, and the central ganglia and cerebral hemispheres on the other, cannot be said to be definitely known. The motor tract from the hemisphere is in the anterior part of the internal capsule, so called, and is in special relation with the corpus striatum, and the sensory tract is in its posterior part and in relation with the thalamus. It is, however, uncertain whether the cerebral cortex is directly connected with the medulla and cord by the fibres of the internal capsule, or indirectly by means of the central ganglia. There is strong evidence in favor of a continuous tract from the hemispheres to the cord, but not sufficient to overthrow the view of the intermediate functions of the corpus striatum and thalamus. According to this, which, you will bear in mind, is hypothesis only, the fibres of the motor tract of the cord and medulla end in the corpus striatum; and although it is not universally admitted, yet it may be assumed that the sensory tract finds its terminus in the optic thalamus, a connection being maintained between the latter and the corpus striatum by means of fibres issuing from the thalamus, and entering both divisions of the corpus striatum. The corpus striatum, then, is the motor ganglion for the entire opposite half of the body. It translates volitions into actions, or puts in execution the command of the intellect; it selects, so to speak,

---

[1] Dr. Ross, Diseases of the Nervous System, i. 70.

[2] For the following account of the construction of the brain I am indebted to Dr. Broadbent, *vid.* Journal of Mental Science, April, 1870, and British Medical Journal, April, 1870.

the motor nerve nuclei in the medulla and cord appropriate for the performance of the desired action, and sends down the impulses which set them in motion. These impulses are transmitted through fibres, and the fibres must start from cell processes in the corpus striatum. A given movement, therefore, must be represented in the corpus striatum by a group or groups of cells giving off downward processes, which become fibres of the motor tract of the cord. When the movement is simple, or when the co-ordination required can be effected by the cord, as in walking, the cell group will be small, and the descending fibres few. When the movement is complex and delicate, and guided by vision or by conscious attention, as in writing or drawing, the cell groups will be large and definite, and the descending fibres numerous. There will not be a separate group of cells for each movement; but the same cells may be differently combined, just as different combinations of carbon, hydrogen, oxygen, and nitrogen form the basis of all organic substances. Words which require for their utterance the simultaneous co-operation of muscles of the chest, larynx, tongue, lips, etc., and the exquisite and rapid adjustment of their movements concerned in phonation and articulation, must be represented in the corpus striatum by very large groups of cells; and not in that of one side only, but of both.

The function of the thalamus as the sensory ganglion will be to translate an impression arriving from the cord into a crude sensation; it is a stepping-stone between impressions and perceptions; and automatic actions may be performed by the thalamus and corpus striatum just as the cord may perform such automatic actions as standing and walking.

Nerve fibres proceed from the thalamus and corpus striatum to the hemisphere; but beside these, other fibres also proceed thither, which pass by these ganglia. They come from the crura cerebri, both crusta and tegmentum, and are probably in direct or indirect relation with the cerebellum. Wherever fibres of crus go, thither go also fibres of thalamus and corpus striatum. These fibres of crus, thalamus, and corpus striatum may be called, briefly, radiating fibres. Also, where these radiating fibres go, thither go also fibres of the corpus callosum, though not necessarily in the same proportion. Their chief distribution is as a commissure between corresponding convolutions of the two hemispheres; accordingly, those convolutions in which radiating fibres terminate are also bilaterally associated. These radiating and callosal fibers are not, as has generally been stated, distributed impartially to all the convolutions. On the contrary, many convolutions do not receive a single fibre from crus, thalamus, corpus striatum, or corpus callosum, but have only an indirect communication with the central ganglia or great commissure by means of looped fibres which pass from them to other convolutions, which are supplied with central and callosal fibres.

The convolutions to which the radiating and callosal fibres go are chiefly those along the margin of the hemisphere—the margin of the great longitudinal fissure on the one hand, the margins, superior and inferior, of the Sylvian fissure on the other—continued forward by the inferior frontal, backward by the inferior occipital gyri to the frontal and occipital extremity of the hemispheres respectively, which are well supplied; the free margin, again, formed by the hippocampus major. To these must be added the ascending convolutions on each side of the sulcus of Rolando, named ascending frontal and parietal; and perhaps the second frontal-callosal fibres pass more abundantly to the margin of the longitudinal fissure; radiating fibres to the Sylvian border of the hemisphere and the ascending gyri.

The convolutions which receive no fibres from central ganglia or corpus callosum are all those on the flat internal surface of the hemisphere, those on the inferior aspect of the temporo-sphenoidal lobe and orbital lobule, the convolutions of the island of Reil, and those on the convexity of the occipital and parietal lobes, not near either margin, as far forward as the ascending convolution which lies behind the sulcus of Rolando. It may seem less strange that there are convolutions without central or callosal fibres, if we recollect that nowhere do these fibres pass to the gray matter within the sulci, but only to the crests of the gyri, so that by far the greater part of the cortex is without them.

The functional mechanism of the cerebral hemispheres is at once suggested by the facts of anatomy, and so obviously that it may, perhaps, be well to say that the anatomical investigations preceded entirely the physiological deductions. The convolutions which have no direct connection with the crus and central ganglia, and which are thus withdrawn from immediate relation with the outer world, must clearly be those concerned in the purely intellectual operations which can be carried on independently of existing sensations, and without at once prompting movements. They will, on the one hand, receive the raw material of thought from convolutions in relation with the sensory ganglia and tracts, and on the other will employ convolutions in communication with the motor ganglia and tracts, to transmit to the muscles the volitions which are the product or outcome of thought. Now, the convolutions pointed out by structural arrangement as the seat of the higher intellectual operations are those which are gradually superadded to the fundamental convolutions in the progress of development, which is of itself a strong reason for the same conclusion. They are, moreover, those which distinguish the human from the simian brain. The motor centres in the convolutions, again, identified by the experimental researches of Hitzig and Ferrier, and Ferrier's recently discovered perception centres, are in situations in communication by central fibres with the sensory and motor ganglia and tracts, while the intermediate convolutions are not irritable.

The following, then, is in outline the theory of the employment of the structural arrangements of the brain in the evolution and expression of thought. In the thalamus, as has been already stated, we assume that impressions undergo the first development into crude sensations. These sensations will be transmitted to the hemispheres by fibres radiating from the thalamus to some part of the marginal convolutions, in which a further development is effected translating or transmuting sensations into perceptions. Perception is the conscious recognition of the external cause of a given sensation—is objective, and not merely subjective—and it is a necessary postulate that each kind of perception will have its own special seat. And Dr. Ferrier has recently identified the position of the chief "perceptive centres." The visual perceptive centre is found in the angular gyrus round the end of the fissure of Sylvius; the auditory is situated near the apex of the temporo-sphenoidal lobe; the tactile centre is to be found in the hippocampal region; and the centres of smell and taste are located together at the lower parts of the temporo-sphenoidal lobe. Now, ideas result from the combination of perceptions derived from vision, touch, smell, taste; and the structural counterpart of this operation will be the convergence of fibres from the different perceptive centres to a common spot, which will be in one of the superadded convolutions. But an idea is something beyond the mere combination of perceptions; there is an intellectual elaboration of the association of a name, which is thenceforward the symbol of the idea recalling more or less vividly the whole group of perceptions it represents. Names or ideas then are stored up as the subject of thought, or material for intellectual operations.

The motor centres through which the volitions resulting from mental operations are transmitted by means of the corpus striatum to the motor nerve nuclei in the medulla oblongata and cord, are situated in the ascending convolutions bounding the sulcus of Rolando. Those in relation with the foot and leg are in the upper and posterior portions of these gyri, those for the upper extremity occcupy the central region, while those for the face and tongue are at or near the lower and anterior part abutting on the fissure of Sylvius. It is a remarkable but well ascertained fact that the motor centre for speech, or "way out for words," is in the left hemisphere only, and is located in the lower end of the ascending frontal and the adjacent part of the third or inferior frontal gyrus continuous with it.

If we turn to the chemical composition of the gray and the white matter, we find that in many respects they resemble one another. Each is a colloid substance capable of transformation, or of propagating transformation, each containing a large quantity of water, each is supplied by blood-vessels, each is influenced by chemical changes, by the salts, alkaloids, and other substances conveyed by the blood, by the water in which it is bathed, and the pressure rested upon it. But their composition also points to differences of function. The gray matter contains a larger quantity of water than the white, and the presence of water indicates a facility of decomposition and transformation. The rapidity with which the brain decomposes is familiar to every student of anatomy, and this facility of decomposition occurs not only after death, but also in life, and specially pertains to the gray matter. The latter also contains a far larger supply of blood—a supply which is brought into immediate contact with the nerve cells. The axis-cylinder or nerve fibre, which is the essential part of the white matter, is protected by its tube or sheath throughout its length, and does not come into immediate contact with the blood like the nerve cells, but communicating by its extremities with other parts, it conveys motion to the centres, and sets up in them the decomposition, the proneness to which is so plainly indicated by their chemical composition.

The next point we have to consider is the blood supply of the brain; and this is of so great importance, that if we could comprehend it beyond all manner of dispute, we should go far toward explaining most of the phenomena of brain function and disorder. One might, in truth, write a volume in discussing the brain circulation, as has been done ere now, and the most than I can do within the limits of a lecture is to direct your attention to some of the most striking facts connected with it.

The arterial system, whether in its trunks, branches, or capillaries, is peculiar; not less so is the venous. The arteries supplying the entire encephalon are four in number—the two internal carotids, and the two vertebrals coming from the subclavian and uniting to form the single basilar. Now, if you look at the arrangement of these arteries within the cranium, you will see that the encephalon is not supplied with blood, as the kidney, or the liver, or the spleen, by means of a large artery entering the substance of the organ and subdividing into smaller ones; but, on the contrary, all the larger vessels are arranged on the outside, and only their fine prolongations enter it. And in the case of the cerebral hemispheres, which are the parts with which we are chiefly concerned, we find that the blood is conveyed to the gray matter entirely from the outside, by means of very minute vessels dipping into it from the branches ramifying in the pia mater. The white matter is also supplied from the pia mater by larger vessels. These are not so numerous, however, as those which supply the gray matter, the latter receiving five times as much blood as the white.

Most of these arteries have great facilities for rapid contraction and dilatation after they have once passed through the bony canals by which they enter the cranium. The internal carotids pass through the cavernous sinus, and are protected from pressure by the blood there; and then, like the vertebrals and basilar, they are surrounded by the subarachnoid fluid at the base of the brain, and the vessels in the pia mater have facilities for contraction and dilatation which they could not have were they embedded in the interior of the organ.

I need not here describe the course of these arteries within the cranium —how the two vertebral join to form the basilar, and afterward divide into the posterior cerebral; how these are connected by means of the posterior-communicating with the internal carotids, which are themselves connected by the anterior-communicating, forming in this manner the so-called circle of Willis. We are not to suppose that this "circle" implies a circulation, or that under ordinary circumstances there is any admixture of the blood coming from these various sources. The posterior-communicating arteries are the merest twigs, and in the rapid and constant contraction and dilatation of the cerebral vessels their function must ordinarily be *nil;* consequently we may practically affirm that the whole posterior portion of the encephalon is supplied by the vertebrals; viz., the posterior cerebral lobes, the back part of the corpus callosum, the fornix, optic thalami, corpora quadrigemina, pons, cerebellum, and medulla oblongata.

We find, then, the anterior brain supplied by the carotids, viz., the anterior and middle lobes, the anterior and greater portion of the corpus callosum, the corpora striata, the olfactory lobes, and the front part of the optic tract. That free anastomosis does not exist between the vessels supplying these parts and those of the vertebrals, and even between the two carotid arteries, is proved by the results of many an operation upon the carotid. That compression of the carotids in man produces stupor and collapse, was known before the time of Galen, and during this century it has been proposed as a remedial measure. Tying the common carotid on one side is generally followed by some amount of cerebral symptoms, and is occasionally the cause of paralysis and shrinking of one hemisphere, showing that even the comparatively large anterior-communicating artery is not sufficient in all cases to provide a due supply. You know, of course, that after deligation of an important vessel in other parts of the body, some time must elapse before the collateral circulation is established, and the patient is in danger till this happens. But there is greater danger to the brain than to structures less highly organized, and an interruption of brain function may occur when one or other of these arterial streams is checked, not stopped, by causes other than deligation.

Now the contraction and dilatation of arteries are under the control of what we call the vaso-motor system of nerves, that system which is now so greatly attracting the attention of physiologists, about which, however, there are so many doubts and difficulties. The trunks and branches within the cranium are largely supplied with vaso-motor nerves, which may be seen running along them so far as the pia mater. In the gray cerebral substance the smaller branches and capillaries are without nerves, and so it needs must be that the contraction and dilatation of these vessels must depend on that of others, and on causes external to the gray substance. It depends, in fact, upon the influence of the vaso-motor nerves, which influence may be derived from various vaso-motor centres. For it would appear from late researches that vaso-motor centres are disseminated throughout the nervous system; that there are perivascular centres, and centres in the gray matter of the cerebral convolutions, besides the centre in the floor of the fourth ventricle and those of the sympathetic ganglia and spinal cord. But much doubt and mystery still envelop all the subject. When this is removed, the pathology of insanity will stand revealed in a clearer light.

All these considerations assist us in our endeavors to form some idea as to the functions of various parts of the encephalon. They confirm the probability of various vascular areas, as Dr. Hughlings Jackson suggests. They indicate how MM. Prevost and Cotard's experiments on animals produced their effects. These gentlemen injected extremely fine seeds into the carotids of animals, which, being carried into the brain substance, caused softening, showing that no anastomosis came to the rescue. They assist us, moreover, in understanding many of the phenomena of sleep, dreams, and somnambulism, and these it is incumbent on every student of insanity closely to watch and analyze.

Although the ultimate terminations of the blood-vessels which ramify in the gray matter may be said to bring the blood into immediate contact with the nerve cells, yet a prolongation of the pia mater does, according to some writers, surround these vessels, forming at first a fibrous membrane, which gradually becomes translucent and hyaline. This extension inward of the pia mater is, according to Dr. Batty Tuke,[1] of considerable importance; for, as he believes, the space between this sheath and the outer fibrous coat of the vessel is that by which the waste products of the brain are removed, and constitutes the cerebral lymphatic system. These spaces are not, however, to be confounded with the rings or so-called perivascular canals, which are the result of hyperæmic dilatation during life. That the brain requires to be relieved of the products of the excessive decomposition and waste always going on seems almost certain. That the blocking or obstruction of any such system of excreting organs would be productive of immediate mischief seems also certain, but on these points we still require much investigation and observation.

Passing from the consideration of these various organs, the nerve centres, the connecting fibres, and the blood-vessels which supply them with life, we have next to inquire into this life, into their uses and properties, into what is commonly called their function. We must ascertain if there by any functions common to all of them, or, where they differ, in what respect they differ, and what their relations are to one another. By such researches we shall be better able to understand how these functions may be disturbed and altered; in other words, what is the pathology of the nervous system.

---

[1] Edinburgh Medical Journal, Dec., 1874.

If we go down to the lowest forms of life, we shall find that nerve force or function is in itself a specialization of something lower, which we must call vital force. We have no other name for it. We see animals leading lives and manifesting the phenomena of sensibility and motion, in whom we can discover no nervous system at all; and as we ascend the scale, at first the same organs appear to subserve many purposes, and only at last do we find special organs for special functions. Nay, it is even yet a question whether in man the sensory and motor nerves may not occasionally interchange functions, and motor nerves become sensory, and sensory motor.[1]

The method by which we approach the study of nerve function is two-fold. On the one hand, it may be contemplated as it is exhibited by all beings—men, children, lunatics, idiots, animals of every grade; on the other, we may examine ourselves. The former is called the *objective*, the latter the *subjective* mode of inquiry. By the first is revealed to us the nature of the nerve functions of other men and animals, as derived from the operations thereof, from the movements, acts, and speech. By the other is to be studied our own feeling and our own consciousness, that knowledge of ourselves, and what goes on within us, which to some has appeared the only true knowledge, and the only real subject of contemplation and study. Now, it is quite true that we can only judge of the feelings of others by inference; that which is really known to us is our own feeling, and nothing beyond; our own consciousness limits in a sense our *knowledge*. But it is clear that there is much beyond our consciousness which it is necessary to examine. There is the whole range of mental phenomena exhibited by others, whether normal or abnormal. This we must study objectively, as we see it exhibited. But we may also bring to our assistance that knowledge which we derive from the contemplation of our own feelings and existence, always keeping in mind, that although this helps us much in the examination of the minds and consciousness of others, such examination must ever be imperfect, and the wider the difference between ourselves and the individual we are studying, the less correctly shall we be able to analyze the feelings of the latter by the light of our own. Pre-eminently is this the case in the consideration of the insane mind. People cannot comprehend how an insane patient can do this or say that, because they judge of his disordered *feelings* by their own. And if he does or says certain things as they would do or say them, they argue that his mind must be in all respects similar to their own.

By experiment and observation, by post-mortem examination and vivisection, we connect the movements and actions of other beings with the functions of the various organs of the nervous system, and we infer from such examinations that there are in ourselves also a brain and nerve organs,

---

[1] Carpenter's Physiology, 6th edition, p. 451.

a fact of which consciousness reveals nothing. Examining experimentally the actions of other living creatures, we arrive at the conclusion that they consist of contractions of various muscles or groups of muscles brought about by nerves emanating from nerve centres, which centres are stimulated by impressions conveyed to them by other nerves coming from the periphery. Thus, it appears that every act is the result of a stimulation of a nerve centre, which in turn gives rise to a movement. The stimulus may act directly from without—as light, or sound, or touch; or it may be central and mental; it may be conscious, in which case some present feeling, either by itself or united to the experiences of the past, gives rise to conscious action; or it may be unconscious, as what we call automatic or reflex action: yet the law is the same, that from the conscious or unconscious stimulation of some nerve centre the act arises. The movement is all that we experimentally can observe and know; the feeling that accompanies the stimulation we can only conjecture; and although we can learn much from the information afforded us by other men, by comparison of our own feelings under similar circumstances, and by the evidences of pain or pleasure displayed by children and animals on the application of various stimuli, yet there comes a point when we can only surmise, and that doubtfully, concerning the presence or absence of feeling and consciousness in beings other than ourselves. Hence arise the different opinions of authors upon the presence of "mind" in certain animals. We can observe in the lower animals the progressive development of movements, and the increasing specialization of their organs and acts; but when we try to determine their feeling, or consciousness, or mind, we are driven to conjectures and arguments from analogy, without having any facts or foundations to rest our theories upon. One writer makes "mind" coextensive with nervous action, and sees "mind" in the movements of a headless frog; another sees in this not "mind," but "consciousness;" another, "reflex action." Now, it is plainly a mere affair of words to discuss the question whether such movements imply mind or consciousness. And similarly with regard to animals low in the scale of creation, one person says that those which have no cerebral hemispheres have not mind, but mere sensation, their acts being sensori-motor. Another attributes to them mind, thinking he sees purpose and deliberation in their acts. On this point I would say that, doubtless, they have mind—a mind not specialized, as is that of higher beings, or our own, but one suited to their life and surroundings, and endowed with a capacity of feeling in accordance with the excitations of their daily life. But to construct a comparative psychology guided merely by the analogy of our own mind and consciousness, will only lead us further and further from the truth, and blind us to that which we can really see and determine objectively.

LECTURE II.

The Phenomena of Mind—The Growth of Mind—The Divisions of Mind—Ideas—
Feelings or Emotions—Emotions.correspond to Ideas—Feelings vary accord-
ing to the Condition of the Centres—Will—Conditions necessary for the Right
Operation of Mind—A Healthy Blood-Flow—Food—A Normal Temperature—
Light—Sleep.

IN considering the objective or physiological aspect of mind, we speak
of nerve centres and nerve fibres, of stimuli conveyed by the latter to the
former, resulting in movements visible to the eye, but we must now consider such mental phenomena as feelings and ideas. In treatises on insanity, you will see, as I have already said, such expressions as "emotional"
or "volitional" insanity. If you turn to works on the subject, you will
see that the mind is divided into the intellect, the emotions, and the will;
and therefore I am bound to speak of these subjects. But what I have
to say will be as brief as possible, my object being to point out what
questions are essential for you to examine previously to entering upon the
study of insanity and what, in fact, are the component parts of that which
we call mind, and which may be at times disordered, if the conditions of
its healthy working are not fulfilled.

The first remark that I shall make with regard to mind is, that it
varies greatly at different periods of life; that we are not born with it,
but that it is developed by slow degrees and by the aid of our external
surroundings. Our study of it must be carried on not only in adults of
fully-developed and mature mind; we must also observe all its immature
or imperfect manifestations and developments in children, idiots and the
savage dwellers of uncivilized lands, and compare these with minds diseased
and disordered, and with the still less developed mental functions of the
lower animals.

Let us take an infant and see what its brain functions are, and how
they grow into the full intelligence of manhood. You have probably
heard of the controversy as to innate ideas, a sense of duty born with us,
and the like. But infants are not born with ideas and knowledge. They
acquire ideas, but they are born with a brain, and the power of developing the function of it, and of acquiring knowledge. The brain they
inherit, but their acquisitions must depend on its healthy working, and
on their surroundings and opportunities of receiving ideas. We shall find
no intelligence at first; an infant's nerve phenomena are those of bodily
feeling and sensation rather than of mind. It passes a considerable
portion of its time in sleep: when hungry or cold, it wakes and cries; when
fed and warm, it sleeps again. It sees nothing in our sense of the word;

the rays of light strike on its eye, and, according to the intensity of the luminosity, it receives an excitation, pleasurable or painful, of the organ of vision. So we find that, when exposed to a very strong light, it cries, but in the dark it is often pacified by being brought in view of a lighted candle. Thus, although the infant cannot be said to have any true perception of objects, we infer from its movements that it experiences certain *feelings* of pleasure or pain from the excitation of its faculty of vision, and the changes produced in the vision-centres. As time goes on, it habitually sees certain things, some more agreeable than others, as its mother and nurse. Not only is the sight pleasant at the moment, but the object remains fixed in the memory, and a pleasant feeling is associated with it. The child *recollects* the mother, and the sight of her arrests its tears, even before the wished-for nourishment is afforded. Here it experiences pleasure from seeing a well-known face,—a great advance from the time when it was pleased by a certain amount of light; the pleasure being, however, the stimulation of a portion of the brain through the eye.

Similarly, if we examine the sense of hearing, we find that at a very early age it can hardly be said to exist. The infant's slumbers are not disturbed by a noise that would wake adults. Yet, when awake, a violent, sudden, or harsh sound will cause it discomfort; a soft, rhythmical, or musical one will please, especially if it come from some familiar person, as the singing of its mother or nurse. Then the sound becomes associated with the individual, and it testifies delight at the voice, even when the speaker is out of sight. When first it begins not merely to recognize sounds, but to recollect names, it associates the latter with certain concrete objects—with a horse, a dog, or a cat; and as each of these objects when seen causes a feeling of pleasure or the reverse, so does the memory of it cause the same, when laid up in the mind under the name associated with it.

There can be no question that the great majority of objects which are laid up in the memory are taken into the mind by the avenues of the senses of sight and hearing: yet the experiences of taste, smell, and touch are registered in the same way, though they are not so multitudinous and varied. Those who are so unfortunate as to have been born without one or both of the former senses, bring the latter three to their aid in a degree which others who have no such need of them can hardly appreciate. However they enter, whether by sight or hearing, taste, touch, or smell, all impressions are conveyed by the senses to the brain, and are there stored away and associated together, and so become food for thought and reflection, and thus we attain to a thinking mind. Till this comes to pass, the young infant is much in the condition of the animal whose hemispheres have been removed. It passes most of its time in sleep, cries when in pain, follows a light with its eyes, and executes such movements as sucking.

All these ideal feelings stored up in the child's memory have, as I have said, come to it from without, and have for the most part been associated with some feeling either of pleasure or pain at the time they were perceived. The excitation of vision representing the nurse is associated with the pleasurable sensation of appeased hunger, and so the two are connected afterwards, and the sight of the nurse gives pleasure, and also recalls the gratification of the hungry appetite. The association of such ideas is plain enough, and it must be that certain portions of the brain corresponding to these are also associated, and give rise to associated action.

A child a twelvemonth old shakes his head when I ask him to come to me: he must have learned by his eye to discriminate persons, and to know mine to be an unfamiliar face; also he must have learnt that unfamiliar faces, i.e., strangers, are productive of less gratification to him than are nurses and friends. So he dissents by shaking his head; but he does not cry as he would have done at an earlier age, or as he would now, were I to take him by force: he has learned that shaking his head indicates unwillingness, and averts the evil. Here is a variety of ideas arising out of sight and sound, which must have passed through a number of associated brain-centres, and, culminating in a deliberate act of volition, finally result in setting in motion the muscles that move the head. We see decided will, the outcome of a feeling of dislike or distrust, which the sight of a stranger has roused, but there is very little that deserves the name of intellect. Some degree of memory and association we see connected with the stimulation of the organs of sense; but the intellectual powers of a child of this age are below those of an intelligent dog.

If we contemplate the same child at the age of three, we see that he has made a vast stride. He has gone far beyond canine intelligence and canine powers. Supposing him to be of fair average capacity, mental and bodily, we see in him in a certain stage all the capabilities of adult mind. He can convey to others his ideas in intelligible speech; he can commit to memory what he hears; he can reason and perceive the consequences of his acts, and can abstain from what he would fain do if permitted. He has, it may be, a determined will, earning for himself the character of being a " willful " child. His intellect and will have greatly developed. But what of his emotional phenomena? We find this portion of his organization more developed still. His whole day is devoted to enjoyment, to gratifying his bodily sense and appetite, to running about, eating, singing, shouting, and amusing his mind by pictures, sights, and play. As the infant passes its days in sleeping and feeding, in looking at the light, and kicking its limbs in the delight of muscular exertion, so does the three-year-old child live in the perpetual indulgence of his pleasures and his self-feeling. If we observe the manifestations of emotion, we shall see that he exhibits anger, fear, jealousy, hatred, and love, wonder combined with pleasure, or with pain which becomes fear, also

2

self-importance, and a desire to be first and to have precedence in all
pleasant things. These all grow out of the mere feelings of pleasure or
pain, which at first constituted the whole of the infant's emotional state.

In this microcosm of humanity, a strong and healthy child, you may
study without fear of mistake the development of mind. You will find
no better field elsewhere, and he who has learned what mind is from an
analysis of his own internal consciousness, and has not studied it in the
manner above mentioned, has only half learned his lesson.

In the works of modern writers upon mind, there is a general agreement among most of them to divide it into three,—the intellect, the emotions, and the will; and these are described separately, and spoken of as
having an independent function. Hence it has happened that physiological writers and inquirers, hearing from metaphysicians that there are
three divisions of mind, have thought it necessary to have three portions
of the brain corresponding to the three parts of mind; and we shall hereafter find that classifications of insanity have been laid down in accordance
with this threefold division; and it has been supposed that one such
portion of the brain might become unsound, the other remaining sound.
The separate existence of these faculties, however, is more apparent than
real. That for which an independence has been claimed, more than for
any, is the will. On this are supposed to hinge the questions of free will,
necessity, responsibility, and the like; yet, after the examination of the
acutest reasoners of all ages, what is there laid down concerning it on
which men are agreed? Looking at these divisions from the phenome-
nalist point of view, and considering them as they appear in their developing stage in childhood, we may, I think, come to some practical conclusions without much difficulty.

I have already spoken of the storing up in the brain of the combinations
of perceptions derived from the external world by the various senses—
sight, hearing, taste, touch, or smell. We call these perceptions, when
they are thus stored up, ideas—*ιδέαι*, the images of the original perceptions. The brain receives these by means of its machinery of cells and
connecting fibres, and deals with them, associating them into groups, so
that the idea of one thing calls up another, which is habitually associated
with it. The idea of form calls up color; the sound of a trumpet calls
up the form of one. And when we have thus filled our brain with ideas,
we unconsciously compare them, discover the *simile inter dissimilia*, and
the *dissimile inter similia;* we advance from the *simplex apprehensio* of
the logicians, the mere reception of things, to *judicium* and *discursus*,
the forming judgments, and proceeding from certain judgments to another founded upon them. For all this it is necessary that the organs
involved be in sound working condition. The child that sees, collects
ideas inaccessible to one that is blind; the latter's store is so much the
less; and ideas of certain kinds, as of colors, are absolutely unknown to

him, though he obtains wonderfully complete ideas of objects by means of his more carefully exercised and more highly developed sense of touch. It is necessary that the brain-cells in which the stores of ideas are laid up, and the fibres by which they communicate, should be perfect, so that they may all be brought to bear upon the formation of a given judgment. The man whose memory is good, who has all his knowledge available, and can concentrate quickly the whole of it on a single point, will form a sound judgment, and is popularly said to have all his wits about him. The various operations of the intellect, call them what we will, may be reduced to those of retaining or remembering, discriminating or sorting, and reproducing or creating new ideas or judgments out of our previous stock.

For all this we must have *mens sana in corpore sano.* Perfect intelligence, pure intellect, has been conceived as existing without the drawback of corporeity, because we cannot but feel that the condition of the body constantly interferes with perfect mental action. How this comes about, how man's intelligence and sound mind are marred by his bodily imperfection, it is the chief object of these lectures to show.

We have seen that the child, besides showing signs of a developing intellect, indicates that it possesses emotions and will—that is, it lets it be known that it likes or dislikes persons and things, and executes various movements in accordance. This brings me to the consideration of what is meant by emotion and will in children and in adults. They are constantly spoken of, together with intellect, as being divisions of the mind. It will, however, be seen that they are something very different.

What we generally call emotion, as rage, terror, or joy, is the feeling of the higher brain, the pleasure or pain of the mind, as ordinary pain or its opposite is of the body. It is a physical condition or state depending on a certain excitation of a nerve centre or centres, varying in its character according to the centres excited; consequently, as the latter are developed in complexity and specialty up to the highest point of refined and educated adult life, so will the emotions be complex and special in their character. A feeling of pleasure or pain is essential, we may almost say, to life itself. The humblest animals, far below those in which we first find cerebral hemispheres, indicate by their movements that their well-being is promoted or retarded. The youngest infant testifies to pain. If we use the term "feeling" instead of "emotion," we shall better understand how much there is in common between the bodily and mental feelings, how they are exalted or depressed by similar causes. An infant cries if it experiences bodily pain, or the discomfort caused by cold or hunger. Its bodily well-being is arrested, and it testifies this by appropriate acts, which are in accordance with its nerve development at the time. If this development advances no further, and the child remains in the condition of the lowest idiot, it may go through life without indicating any higher feeling; but if it progresses normally, we find in a short time signs of mental feeling in addition to bodily. Besides crying, ceasing to cry, and going to sleep, which at first is almost all that we can observe in addition to the movements of the extremities, we notice that it smiles when it sees an accustomed face, and shows in a short time by vocal signs its pleasure as well as its pain. In all this we see that an excitation of its centres is followed by appropriate movements, which may at first be called voluntary or involuntary, so closely do they follow the excitation. If, however, we try to discover why a child is pleased or displeased, we find that this depends either on the character of the excitation, or on its physical condition at the particular time. A strong light, too loud a noise, a nauseous taste, a prick of a pin, a wound or blow, produce pain at once; but, besides these, we constantly see that when the child is ill everything causes pain, and nothing brings pleasure. It cries even with those it loves best, and will not be comforted by its toys, or by any of those things that pleased it when in health. We judge of a child's health by its emotional state, by its being fretful and cross, or gay and hilarious. As it grows in mind and brain, and lays up a store of ideas of all kinds, passing from mere concrete objects to abstractions, from particulars to generalizations, we find that excitation produces feelings equally varying and advancing in complexity and specialty. Herein we shall see the difference between one child and another, as between one man and another. The educated child, the descendant of a line of educated ancestors, becomes a highly specialized man, with feelings and emotions of a refined and complex nature. The savage child—the child, that is, of savages—becomes a savage men, and his ideas and feelings are, compared with those of civilized man, childlike throughout life. He is, like a child, easily moved to joy, terror, or rage; but he is incapable of comprehending abstract ideas, as truth, justice, honor, or of feeling the complex emotions that belong to such ideas. And as his mental manifestations are simple, so we find also that the convolutions of his brain are simple, alike in both hemispheres, and more resembling the brain of the apes than does the complicated and convoluted brain of an educated European. His brain is undeveloped, and his mind is incapable of development. Little by little in successive generations this brain may increase in complexity and specialization, but in the individual savage this cannot take place. And as one savage inherits the simple and imperfect brain of former savages, so we shall find that amongst the educated nations a child may inherit the imperfect brain of ancestors who have retrograded from the development of civilization, or who have, from disease, overwork, debauchery, or drink, become degenerate and fallen. Like the savage child, this one will be incapable of attaining the perfection of intellectual and emotional life. Either he is so stunted and blighted that he is an imbecile and an idiot from the beginning, or when he enters upon life he is unequal to contend with the chances of fortune, or his organization is so unstable that every ordinary illness disturbs his reason. And even if he does not fall into a sudden and marked state of insanity, he nevertheless is unlike other people. His notions are warped and eccentric; he is destitute of the sense of duty and of right possessed by others of his country and social status. He becomes a criminal, if not a lunatic, and hands on to his descendants, if he has any, the inheritance of criminality or insanity, swelling the ranks of the criminal or the lunatic class. Each of these is a degenerate and degraded section of the community, which might be reclaimed through several generations, if we could select the healthiest specimens, and leave the worst to die out, as in fact they often do, in a state of sterility.

In ordinary health, excitations of our nerve centres produce certain feelings, which, though often very complex, may yet be all resolved into pleasure or pain. The result of the excitation, if it be powerful, is action of some kind—verbal, facial, or bodily; or desire for action, which we may repress. In the child or the savage this repression is not exercised, and there is an immediate display of muscular action demonstrating the feeling experienced. More civilized men, from other ideas habitual to them, repress these signs if they are able. But this they cannot always do, the pent-up storm of rage or grief finds vent in words or action.

To a centre in its ordinary state any stimulation may be at once pleasant or painful; or it may be at first pleasant, and may afterwards be so prolonged as to cause pain. Familiar instances of the latter occur to every one. The exercise in which we at first take keen delight becomes irksome and painful if continued so as to produce great fatigue. We are said to " get tired " in time of almost anything—of music, of conversation, of our amusements—and hence we see that the particular feeling aroused by such things depends on the condition of the centres at a given moment; and as the act follows the feeling, this also will be regulated by the condition, whatever it may be.

A dog let loose from its kennel, a horse turned out of its stable into a field, a young child fresh from its rest, feels the highest delight in exercising its limbs in jumping and running; its centres are full of energy almost spontaneously discharged in motion. If, on the contrary, either of them does not move at all, or crawls along languidly and dejectedly, we say that it is not well with it. Similarly, if accustomed pleasures fail to delight a man, and he is gloomy and melancholy without any cause, we know that something is amiss. He may have been subject to stimulation of a very painful character, to some grief, or loss, or pressing anxiety, which has exhausted him, has robbed him of sleep, and brought him to this condition. He may have encountered some event which has caused so sudden a shock that he may have fallen as in a fit, or which has excited him to rage or terror, with corresponding action and consequent exhaustion. Even pleasurable emotion becomes exhausting, if prolonged.

Laughter may turn to sobbing, and men may faint from excess of joy. On the well-being of the nerve centres will depend our recovery from the effects of these excitations. They must needs violently disturb the balance of the brain circulation, which is roused to supply the force expended. If the circulation fall, and sleep return, all is reduced to its former level, and we are said "to get over it." But if not, a permanent disturbance may be established.

Here, then, we may lay down the component parts of that which, for our purpose, we describe as mind. We see that it is evolved out of feelings, under which name we group the sensations derived from the excitation of our senses, and the emotions attending the excitation of ideas or past feelings laid up in memory, which being recalled to consciousness, are united to the feeling of the moment, from whatever source this may have arisen. This union of the past and the present is effected by various processes of reasoning and judgment, and the carrying out of the action consequent upon this emanates from what we call will, about which I must say a few words.

Will is not one of the primary divisions of mind. Our mind is composed of feelings, present and past, that have been produced by the various stimulations brought to the nerve centres or cells by the conducting nerve fibres. Will is only a process of energizing, which these structures possess when in a healthy and normal state—a process which intervenes between the stimulation of the centres and the motion which is the ultimate result. If will does not intervene, the act is said to be automatic; if will intervenes, it is voluntary. And when we say that will does intervene, we mean that in our mind we form a judgment concerning something which we wish to carry out, and then regulate our movements so as to accomplish this end. This judgment may be slow and deliberate, or so rapid as scarcely to appear in consciousness, as in the case of habitual voluntary actions. In the latter, however, the mental operation of judging must have been gone through in every case at some time previous to their becoming habitual. Will is concerned with action of some sort, mental or bodily; it does not exist as a metaphysical entity. There is no such thing as will apart from something willed. And if we examine the acts, mental or bodily, which are the result of deliberate will, we shall find that a vast number of our actions are not comprised in this category, and that those which really deserve the name are the result of the whole collected knowledge of our mind. Being what we are, we cannot help acting as we do. Involuntarily we avoid that which is painful, and seek that which is pleasant. In sudden self-defense or danger, we do that which truly is called involuntary. Many things are done unconsciously by habit and custom, and if we court danger, or choose the painful rather than the pleasant, it is because the ideas and knowledge stored up in our brain teach us that it is better for some reason or other so to do. People differ in that which they choose to do, because of the difference in the general constitution and furnishing of their minds. One man can practice great self-denial which another cannot. He is enabled to do this, not because one part of his brain, inhabited by a function called his will, is larger than that of the other man, but because, in the first place, his whole mind is stored with feelings and experiences which counteract the impulse to gratify a present desire—a desire not resisted by a man less endowed—and, secondly, because he has the power of concentrating these feelings and experiences on the question to be decided, this power depending on the healthy condition and working of his brain. Our criminal law can only be enforced by supposing that all men are alike, even to the point of all being acquainted with every law that is made; but, practically, we make great differences and allowances for individuals according to their opportunities, their rearing, education, and past history. Were an educated gentleman to steal a watch, we should affix to the act a stigma very different from that which accompanies the theft committed by one who has been reared in vice and crime.

If we consider what can be done by dint of our will, we shall find that willing can do very little *per se*. We learn to walk, we learn to ride, to write, to dance. By long and laborious practice we acquire the power of executing such movements, and no effort of will can enable us to perform them till we have learned the method. When this is acquired, consciousness is so slightly involved that such things are regarded as being done unconsciously. If we apply our will to mental operations, frequently we cannot fix our attention on one subject for ten minutes at a time, or, do what we will, we cannot exclude an idea from our thoughts. We cannot by our will recall a name or a circumstance; and when we are conscious of exercising a choice, and of deliberately resolving to do this or that, it is because our reason, judging by the aid of experience of the past, and the probabilities of the future based on such experience, indicates that which will best satisfy our predominant feeling. We say emphatically of a man when we wish to assert that he did something freely and voluntarily, that he *deliberately* did the thing; *i.e.*, that he, after due reflection and consideration, proceeded to act. Therefore *volitional* insanity must imply an insane reason and judgment, and an insane emotional condition, not only an insane will, and is no more a separate and special form than are ideational and emotional insanity, which are supposed severally to represent an insane intellect and insane emotions. We cannot consider intellect as having a separate existence apart from emotions.

I discard the will, then, as a third component of mind, and retain only feelings and ideas, which in truth are not two, but one, as they arise from present stimulation, are stored away in memory, and in new combinations come again into consciousness upon fresh excitations. Under the term feelings we may range the bodily sensations of pain or pleasure, the sensations of the special senses, as the eye or ear, and the emotions which are but the feelings of the highest centres of the brain concerned with the intellectual, the æsthetic or the religious. For the due operation of our feelings and resulting ideas, we require the healthy working of the nerve centres, with the system of nerve fibres and adequate supply of blood, which I have described to you. When this goes on aright, our minds are healthy; when anything interferes with the proper working, we have the evidence of it in an irregular or abnormal manifestation of mental action. And before I come to the subject of insanity as generally understood, it may be as well to glance for a moment at a few of the conditions of the healthy working of brain and nerve.

We may sum up in a few words these several conditions. Given a healthy apparatus, free from defect, we require for its working a due amount of material in the shape of food, to be converted, through the agency of the digestive and circulatory system, into healthy blood, supplying the waste in the brain cells. This blood must be in all respects fit for its purpose, rich in oxygen and all necessary ingredients, and free from all impurities, as urea, bile, carbonic acid, or other poisons. Secondly, we require for the due discharge of mental action a certain amount of heat. Thirdly, we must at stated intervals have a period of rest and cessation, which in man is given by sleep. Failing any of these, mental action becomes disordered, and finally ceases.

The mere amount of blood circulating through the brain must of necessity influence to a material degree its power of acting. Mechanically, I mean, the pressure of an undue quantity, or conversely, the removal of the accustomed pressure, must affect the relations and the functions of such delicate structures as the nerve cells and nerve fibres. That pressure can be exerted upon the brain cells by increased blood-supply is a fact which I believe, may now be considered fully established, though formerly some held that this could not be the case. We may concede, however, that the amount of blood sent to the brain is, compared with that which may be injected into other organs, limited by the conditions of the arteries; nevertheless, it is certain that enough may be sent there to interfere with the healthy state, for after death we have traces of active hyperæmia plainly apparent. Another probable result of excessive hyperæmia and pressure is stasis of the blood in the vessels and capillaries—stasis both of the red corpuscles and the white—with blocking of the minute vessels, and consequent delirium or stupor. Upon this point I shall have more to say hereafter; but I wish only here to remind you of the writings of Mr. Lister on the phenomenon of stasis in inflammation, and of Dr. Charlton Bastian's paper, in which he describes this blocking as discovered by himself, in a case of erysipelas of the head with delirium, narrated in the "British Medical Journal" of January, 1869.

Into the varieties of food necessary or adequate to the proper discharge of brain function I shall not here enter. You will have heard of them elsewhere. It is a fact of observation that the dwellers in northern regions eat quantities of animal food and fat and grease of all kinds, which could only be consumed by those who live under such climatic conditions, while those who inhabit tropic lands may pass through an active life without eating anything save a vegetable diet. And this brings me to another head. For the due discharge of brain function it is necessary that the individual should live in a certain temperature, not too hot nor too cold. Life—the life, that is, of man—can only exist in a certain temperature; and the first mode in which the invasion of cold is evidenced is in the effect produced upon the nervous system. An overwhelming desire to sleep comes over a man exposed for a long period to extreme cold; and, as you know, to sleep under such circumstances is fatal, unless some one is at hand to wake the sleeper. If this be in the case, the sleep is beneficial, and recruits the exhausted powers, showing plainly that nervous exhaustion is the condition which the cold produces. In his most interesting book of arctic travel, Dr. Kane relates how he and his companions were once nearly lost in the cold: "Our halts multiplied, and we fell half-sleeping in the snow. I could not prevent it. Strange to say, it refreshed us. I ventured upon the experiment myself, making Riley wake me at the end of three minutes, and I felt so much benefited by it that I timed the men in the same way. They sat on the runners of the sledge, fell asleep instantly, and were forced to wakefulness when their three minutes were out." [1]

The blood must be pure: it must contain no deleterious substance which may interfere with the healthy nutrition of the brain, or may actually poison it, and set up therein that inflammation and stasis which I have alluded to. It must not be vitiated by poisons introduced from without, as alcohol, opium, lead; neither ought it to contain those poisonous matters which, generated within the body, and in a healthy individual excreted thence, are occasionally retained, and give rise to symptoms of brain disorder, such as delirium, coma, or convulsions.

However we may explain the metamorphosis of other forms of motion or energy into mind, it is a fact of experience that an adequate supply of food is required for the wants of the nervous system, and that a failure of food results in a corresponding diminution of nerve-energy, and often in nervous disease. I shall have to return to this again and again, believing, as I do, that a plentiful supply of food is of all things the most efficacious in restoring exhausted nervous power, and in removing nervous disorder. If we read the accounts of shipwrecked sailors and others who have been compelled to live for some time upon a scanty supply of food, we see how weakness was the prominent symptom experienced—weakness rather than

[1] Arctic Explorations, vol. i. p. 198.

hunger, weakness not to be acccounted for by the diminution of the muscular tissues, but rather by the want of nervous power. Not merely for
the nourishment of the brain and other portions of the nervous system is
the food required. It is demanded not only that the brain may live, but
that it may duly discharge its function in a normal and healthy manner.
The brain may live, and the owner may live for years in a state of the
most abject fatuity.

The animals that have to pass the winter in countries where the cold
is very severe, do so, many of them, in the state of hybernation. So
little is expended in this condition of sleep, that often without food they
exist for months, with all the functions of life reduced to a minimum, the
amount of nervous energy required for these being derived from the
blood, which is in turn renovated from the stores of their bodies, or by
occasional meals from the supply of winter food laid up by some of them.
But hybernation is not possible for us or for the more highly organized
and developed animals, and cold deprives us of our nervous power even
when we are supplied with adequate nourishment. Too great heat also
incapacitates us from properly discharging our brain function, and causes
that disorder called *coup de soleil*, which has its seat in the cerebral organs.

Not only is the warmth of the sun beneficial to the health and energy
of the human mind, the light of it is also essential to its well-being. The
protracted darkness of northern countries has been observed to bring about
insanity, especially melancholia. Dr. Lauder Lindsay draws attention to
this in a paper on "Insanity in Arctic Countries,"[1] to which I refer you.
And Dr. Kane, from whose book I have just quoted, mentions the depressing effect of darkness, which affected, he says, even the dogs, though
they were born within the arctic circle. A disease, which he considered
clearly mental, affected them to such a degree that they were doctored
and nursed like babies. They ate and slept well, and were strong, but
an epileptic attack was followed by true lunacy. Of course, we cannot,
in speaking of arctic countries, eliminate the joint effect of cold, fatigue,
and want of fresh food; but in other countries, in cities, dungeons, and
elsewhere, the depression caused by prolonged darkness has been felt and
noticed.

For the due discharge of its function, our brain requires rest, which
rest it takes in sleep. Only in complete sleep does it throughly recruit
itself, and lay up stores of energy to be expended in the waking hours.
In sleep all work ceases save the processes of organic life, and these are
reduced to the lowest point. There is no expenditure, but, on the contrary, there is constant renewal of nerve power. According to the exhaustion and previous waste of power will be the demand for sleep and
the continuance of it. Men become so worn out that the strongest im-

---

[1] British and Foreign Medico-Chirurgical Review, January, 1870.

pressions, the loudest noises, or the most exciting news, cannot avert the sleep that comes over them; but the amount that will refresh them varies greatly in different individuals. Some require much sleep, some little; at times a brief snatch, even of a few minutes, will greatly recruit the wearied man. As you study insanity and observe insane patients, you will have to be constantly watching the phenomena of sleep and sleeplessness. You will see the consequences of the entire loss of sleep, of the partial loss. You will meet with some whose minds break down because their brain is constantly and habitually overworked by day, and not sufficiently renovated by sleep at night. Either their work pursues them into the night, and haunts their couch and disturbs their sleep by harassing thoughts and grave responsibilities, or they allow themselves an amount of sleep far short of the proportion demanded by their daily task.

Such are the subjects to which I wish you to direct your attention before you enter on the study of the disorder termed Insanity, or Unsoundness of Mind. I have brought them under your notice roughly, not with the accuracy and perfect delineation of a photograph, but as men draw diagrams on a black board, giving in broad outline just so much as will illustrate what they have to describe. You must bear in mind that you have to consider the nerve centres and the nerve fibres which connect them to each other and to other parts, the blood which furnishes them with life and energy, the vessels that bring blood to them and take it away, and the nervous system that regulates its supply; the nature of the resulting operations, whether of mind or motility, and the conditions under which they are carried on. Such are the data. When these organs all work harmoniously and healthily, sound mind is the result. When the mind is unsound, we must discover the defective spot in one or other of these parts or processes.

LECTURE III.

The Pathology of Insanity— Sound Mind—Unsoundness of Mind—Insanity with Depression—Insanity with Excitement—Ætiological Pathology—Ætiological Varities—Insanity from Anæmia—Insanity with Tuberculosis—Climacteric Insanity—Senile—Insanity of Pregnancy—of Lactation—Puerperal Insanity— Insanity of Puberty—of Masturbation—Insanity of Alcohol—Insanity from other Poisons—from Lead Poisoning.

HAVING thus glanced at the phenomena of healthy mind, we are in a position to study those of unhealthy or disordered mind; having laid down the physiology, we proceed to the pathology.

Now, by daily intercourse with sane people we know very well what is meant when in ordinary phraseology we speak of a man as being " right in his mind." It is assumed that, being of full age, he has stored his brain with a reasonable amount of knowledge and facts of experience; that he can recall a fair proportion of what he has seen and heard during the past years, and can act upon this experience in an intelligent manner, giving good reasons for so acting; that he can understand what is said to him upon subjects within his comprehension and knowledge, and display judgment in what concerns him personally. Such is what lawyers call a man " of sound mind, memory, and understanding." Conversely, we say a man is " unsound of mind," when he forgets most of what happens to him; can form no judgment from what he has seen or learned; when his acts are outrageous, and he can give no good reasons for them; when his ideas concerning himself are palpably false, i.e., delusions; or when he is quite unconscious of what he is doing.

According to the nature of the defect, we say that he is idiotic or imbecile, insane or delirious. These forms of unsoundness of mind may vary in degree, and no less in duration, lasting from hours to years. Yet certain is it that they depend on alteration or defective action of those organs I spoke of in my first lecture, the nerve centres and conducting fibres, the blood, the blood-vessels, and the system by which the supply of blood is regulated. Unsoundness of mind may exist by itself, the bodily functions being apparently intact; it may be coupled with epilepsy, apoplexy, and other cerebral affections, or arise in the course of such diseases as measles, pneumonia, acute rheumatism, and fevers of all kinds; or may be traced to blood-poisons, to alcohol, haschisch, or opium. Whether we call it delirium, coma, wandering, or idiocy, mania, melan- cholia, or dementia, it depends on some pathological condition of the nerve centres, and implies a total or partial mental alteration or defect.

Having thus widened out the subject to the full, I must proceed to consider some of the details, for it is not within the scope of these Lectures to examine *seriatim* every one of the conditions just enumerated. I pass over idiocy and congenital idiots, whose undeveloped organs and faculties are incapable of receiving the data of experience, and of forming out of them judgments. I leave to other teachers the condition of coma, which is rarely seen by those who observe the insane, except when it is the forerunner of death. That to which I chiefly wish to direct your attention, and which is involved in the greatest obscurity, is the alteration that takes place in the mind of a man previously sane, perhaps in a longer or shorter time to pass away, leaving him sane as before. The alteration may be so transient, and the restoration so complete, that it is impossible to believe the pathological change can be anything more than what is usually called "functional."

If we carefully consider the symptoms noticeable in an individual whose insanity is of recent origin, we find in the majority of cases that they may be classed under one of two heads: either they are symptoms of depressed feelings, or of angry, exalted, or hilarious excitement. Examining them closely, we learn that the depressed patient gradually became dull and dejected, so gradually that friends are unable to point to the actual commencement, which may date back a considerable period of time. He is less capable of work, cannot efficiently perform his daily duties, takes a gloomy view of everything, thinks that all he is engaged in will fail, becomes penurious, sits indoors all day, probably loses appetite, almost always becomes thinner, and looks wan, yellow, and ill. If he has hitherto been an active man, taking much exercise and occupying himself with various amusements, he will give it all up, and will either sit brooding and gloomy, or will walk the room in an automatic and purposeless way.

This is the stage so often spoken of as one of emotional "alteration." There are no delusions, and the alteration, if it amounts to insanity, is described as "moral" insanity, or insanity without intellectual defect.

What is the pathological condition here? Beyond question there is a lack of nerve force, a failing genesis of nerve power which is unable to supply and permeate the centres and fibres of the brain.[1] But although this failure of nerve force produces the feeling of depresssion and inadequate power in the higher centres, it may not do more than this. A man's consciousness may be at first but little affected. He is aware of the depressed feeling, but does not assign it to any cause beyond physical weakness or ill health. His judgment concerning things in general is unimpaired, and he is able to converse and argue rationally; nay, at certain times of the day, under influence of meat and drink, the stimulus of society, and pleasurable surroundings, his brain may be so permeated by

---

[1] *Vide* Spencer's Psychology, i. 586–603.

the nervous fluid,[1] and so incited into energy, that his gloom is thrown off and he is himself again; to relapse, probably, into greater gloom afterwards. This depressed condition often continues for a long period without going further. Some are subject to it periodically. They have their fits of gloom, which pass away again and again without ever reaching the stage ordinarily called insanity. The intensity of the depression may even drive them to self-destruction without their having displayed any marked intellectual aberration.

When depression, however, advances to the melancholia of insanity, delusions characterized by the prevailing feeling soon manifest themselves. The self-feeling of the individual is shown in fears for his eternal salvation, because this is one of the channels of constant thought which is easily permeated by the sluggish current of his nervous fluid;[1] or he thinks he is ruined, reflections concerning his wealth and worldly position being his usual line of thought; or his relations to others make him think that all are looking at or regarding him, or are conspiring to do him injury and hurt. His highest cerebral centres are not only pervaded with the feeling of depression, but their force is so reduced that they are unable to bring together the centres which must unite in order to test such ideas, and prove them to be phantoms having no real existence. In the case of an apparition, we test its reality by the assistance of some other sense, as touch. The brain centres are unable to combine to prove the falsity of the delusions, and until they can do so the delusion remains. When by rest, food, sleep, iron, and such things, the nerve force is raised to its proper level, the delusions vanish without argument or demonstration.

The brain, thus reduced by lack of nerve force, manifests its condition in the majority of cases by the well-known delusions of melancholia, a considerable amount of intellectual power being often retained. Although the patient cannot get rid of his gloomy feeling and ideas, he is able to converse rationally on many topics, and is what is termed *partially* insane. But the disorder may advance in one of two directions. The lack of nerve force may be so complete that he sits motionless and lost in the state termed *melancholia cum stupore*. Here the action both of the highest and lowest centres is arrested by the failure of nerve force, and the powers of both mind and body seem in abeyance. This form is closely allied to another, which it nearly resembles both in symptoms and pathology, known as *acute dementia*. In each there is the same lack of nerve power, but there are differences which will hereafter be described. The latter may be, indeed considered a more advanced form than the melancholia,

---

[1] "Nervous fluid," "nerve force," "nerve power" is here spoken of metaphorically, and it may be objected that, even speaking metaphorically, I have treated it too much as if it were an entity. As one draws diagrams, however, not with absolute truth of representation, but to illustrate one's meaning for some particular purpose, so have I used these expressions throughout these Lectures.

but it occurs at a different time of life, and is more curable. The disorder sometimes, however, advances in another way. We find not a mere lack of force leading to stupor, torpor, and gloom, but acute and active symptoms, with violent resistance, incessant motion, and delusions of terror and horror, often with hallucinations of various kinds.

This is clearly an advanced stage of the disorder. The action of the higher centres of the brain is completely suspended, and the lower or automatic centres run riot for want of the control of those superior to them.[1] There is no power of conversation on any subject, no power of listening to advice, no power of keeping quiet. The centres of automatic motion are almost as uncontrolled as in an epileptic attack; the centres of automatic thought, uncontrolled by those of reason and judgment, and such lower ideas as self, self-preservation, and fear of death, predominate. The patient is in a state which is termed acute melancholia,—a state not far removed from that known as acute mania, or acute delirium. Such symptoms are generally found in people broken and weak in health, coming on occasionally at the close of other disorders, and often being merely the precursors of death.

Turn we now to the consideration of the opposite condition. Instead of gloom and inaction there is excitement, irritability, or hilarity, at first not amounting to insanity, but betokening a change in the individual, growing into gay or noisy mania, or into incoherent and violent delirium.[2] This form of disorder differs from the former not only in the symptoms but in several other respects. First, it may come on much more rapidly. Instead of a patient slowly drifting into melancholia, acute symptoms of mania may appear after an indisposition of a few days, or even hours.

---

[1] Dr. Hughlings Jackson has pointed out in two articles on the *Comparative Study of Drunkenness* (*British Medical Journal*, 16th and 23d May, 1874), that by alcohol, epilepsy, or insanity, the patient is "reduced" to a more automatic condition of mind, just as in a case of aphasia he is reduced to a more automatic condition of language, and to a more automatic condition of movement in hemiplegia. We find overaction of lower centres from the removal of the influence of the higher centres; not merely action of those processes which are more automatic, but increased action; and he enumerates four factors, according to which the mental automatism will vary. 1. It will vary according to the *depth* of the reduction. The "shallower" this is, the higher and more special is the automatic mental action permitted. 2. The *rapidity* with which the reduction is effected will influence the extent. 3. The insanity of the person whose brain is reduced will vary according to the normal peculiarities of the individual, no two persons being alike. 4. It will vary according to the influence of external circumstances and internal bodily states.

[2] Dr. Tigges tried to measure the amount of electric excitability in the insane by a galvanometer, using the constant current. He found the greatest excitability of the nerves and muscles in mania. The excitability in melancholia, in general paralysis, and in dementia, is less than in ordinary health, and occurs in the order stated.—*Zeitschrift für Psychiatrie*, 1874, Band xxxi. Heft. 2.

Then the sufferers are young rather than old, men and women often in the spring or heyday of life, so that to this insanity has been given the name of "sthenic," to distinguish it from the "asthenic" variety I have been describing. Frequently we are unable to assign any physical cause for it; the patients are not wasted by disease, broken in health, prematurely old, or enfeebled by long-continued care and toil. It seems to come on without a cause, and hence is sometimes termed "idiopathic." And as it began almost suddenly, so it may end; and a very violent attack may rise and subside in a few days, and thus it has been called *transitory* mania.

An excessive discharge of nerve force, and no defect, is what characterizes an attack of acute mania or delirium. There is a certain tendency in the nerve centres of many persons to what has been termed "instability" or "explosiveness." This tendency is frequently inherited, and so it comes to pass that in young people who apparently have had no worry, or illness, or other cause of insanity, we see violent "sthenic" and "idiopathic" mania lasting a short time, and perhaps recurring with a certain periodicity. In all this we are reminded of the explosive attacks of epilepsy. The brain-cells in the latter disorder are rendered unstable by some cause, or by inherited tendency, and "explode, or liberate much energy, or, in other words, discharge excessively,"[1] in the convulsions with which you are familiar. The centres affected are not the same in acute mania as in epilepsy, but I have not unfrequently seen violent and acute mania go on till it resulted in an epileptic fit, just as we know that after epileptic attacks a patient may exhibit all the symptoms of ordinary mania. The instability of the one set of nerve centres is communicated to the other.

Can we lay down anything more definite with regard to the pathology of these cases? A man is unduly elated, spends money recklessly, thinks he is worth more than he is, does things he would not have done formerly, but is not manifestly insane, only reckless and excited, and defends argumentatively all he does. His sleep is in defect, and he cannot remain quiet. In talkativeness and expansiveness he much resembles a man who has had a little too much wine, and in the latter case we say that there is arterial relaxation, owing to the alcohol. In both it is certain that the highest cerebral centres are those in fault; they have lost the power of control, and the whole function and capacity of the mind is impaired thereby. In both it seems probable that there is arterial relaxation, producing a hyperæmia of these highest centres, and consequent defective action. The problem to be solved is, on what does this increased arterial action depend? In the case of the wine-drinker the alcohol may be assumed to be the cause; it acts as a narcotic, causing a temporary paralysis.

---

[1] Dr. Hughlings Jackson, Brain, iii. 436.

But in a case of idiopathic insanity beginning suddenly, without previous loss of sleep or any mental cause, what is the pathological condition? Are the symptoms to be ascribed to disorder of the brain-cells, or to the apparatus of circulation?

It is commonly noted that almost all recent insanity is marked by want of sleep; the more acute the insanity, and the more urgent the symptoms, the less is the sleep; when the stage is reached of constant restlessness and overwhelming delusions, sleep is well-nigh absent, as in acute delirium. Now the phenomena of sleep have a close relation to those of insanity, and the study of them ought to throw some light on this disorder. We know that a reduced blood-supply is, if not the cause, at any rate the condition, of sleep. The circulation in the brain is diminished in force and volume. In sleeplessness the opposite condition prevails. The flushing, heat of head, and pain noticeable in mania indicate an increased vascularity, and after death signs of hyperæmia may be revealed by post-mortem examination. Are these symptoms due to disorder of the apparatus of circulation, the vaso-motor centres, as they are called, or to that of the supreme cerebral centres, the centres of mind? I fear that at the present time we cannot give a certain answer to this question. Much has been written about the vaso-motor centres, but little is definitely known. If we consider the causes of ordinary sleeplessness, we see that they are sometimes mental, sometimes bodily. Mental anxiety or worry may increase the cerebral activity and blood-flow, and banish sleep, but in a healthy individual this will only last for a certain time. Fatigue will exhaust the brain-cells of their force; they will cease to be stimulated, and sleep will ensue. But in insanity the instability and tendency to discharge are such that fatigue does not bring healthy sleep; the activity of the brain centres, or some of them, continues in an abnormal and irregular way, and a large blood-flow is directed to them. Now, in a healthy individual who cannot sleep, we may procure sleep by drugs, such as chloral. In the insane we may also procure sleep of greater or less duration by the same means. How does such a drug act? Does it produce its effect by reducing the circulation and increasing arterial tonicity, or by some direct action upon the brain-cells and centres, whereby it renders them less susceptible to stimulation, less unstable, and less prone to " discharge." When we observe the action of bromide of potassium or chloral in checking the attacks of epilepsy, it seems probable that the effect is produced directly upon the nerve centres. On the other hand, chloral is said to depress the heart's action and circulation. The sedative action of morphia, however, does not appear to be produced in this way, and upon the whole, there is reason to think that these medicines act directly upon the nerve cells, whether those of the supreme centres or those of the vaso-motor centres, of which we know but little.

Deficiency of nerve-force is the outcome of various physical conditions,

3

and may be brought to light, so to speak, by the various events of life, whether ordinary or extraordinary. For instance, we are constantly told that a man's melancholia "cannot be accounted for;" the cause is "unknown." In other words, he has had no shock or trial, but his nerveforce is unequal to the demand made upon it by the everyday routine of his usual life, and he sinks into despondency. On the other hand, he may have experienced some great loss or sorrow, or have been exposed to long harass and worry, and this has exhausted his nervous power either by its own depressing influence, or because his physical strength quickly fails, owing to some bodily condition or infirmity. Among the latter we may class such conditions as anæmia, various chronic or wasting diseases, lactation, the climacteric period of life, dyspepsia, or unhealthy work, and lack of hygiene. In the same way, the instability of the centres and increased arterial action may be due to an over-stimulation caused by external circumstances, as shock, protracted worry, or over-work, or to such causes as masturbation, alcohol, epilepsy, or sunstroke, or the brain-oragnization may be specially unstable under the influences of puberty, pregnancy, or child-birth.

But besides the "positive" symptoms of insanity which indicate a change in the cerebral brain-life, whether in the direction of depression or the reverse, there is constantly to be seen a weakening and deterioration of mind—"negative"—symptoms which betoken some brain disease or degeneration. Such are seen in connection with the progressive disease known as general paralysis, with tumors, blows on the head, syphilis, lead-poisoning, chronic alcoholism, lastly with old age.

There is a factor, morever, of which we must not lose sight, which may operate in conjunction with any of the causes we are about to consider: this factor is hereditary taint. It is never to be forgotten that insanity is in the majority of cases the result of a number of causes, not of one. There is a conjunction of a moral and a physical, an exciting and a predisposing cause. And again and again, we find that the chief, perhaps the only assignable, reason for a man's becoming insane, is that his parents were insane before him. Probably there is always some exciting cause, but it is often hard-to find: the latent tendency is kindled and brought to light by something so trivial that it escapes observation. The insanity is assigned to a "cause unknown," and friends and relatives, ignoring the family history, profess their astonishment at its occurrence. Yet people with this inherited taint will pass through life without becoming insane. The tendency may not be strong, and external circumstances may befriend them. Exempt from pecuniary troubles or harass or ill-health, they pursue the journey of life to its end, and are called sane to the last. But the evil which they have escaped may appear in their offspring if adverse circumstances call it out. The strength of the tendency will depend on the other parent's nervous system, and on the healthiness and mode of bringing up of the children. The insane temperament of a parent may have a baneful influence on the education and training of a son, who may thus be deprived of the wholesome discipline he so much needs.

I propose now to consider a number of ætiological varieties singly. It will be seen that many of them belong to the classification of the late Dr. Skae; a classification which, with all its faults, will probably, for practical purposes, endure longer than any other, and will serve as the basis of all others. Of this I shall have again to speak in considering the subject of classification.[1] I wish now to examine the various conditions which may reduce the full working power of the brain.

When we speak of anæmia causing insanity, the term is to be understood to comprise all those conditions of bodily weakness which render a man, in the common acceptation of the term, anæmic. A man may be anæmic from loss of blood, or poverty of blood: a woman may become anæmic from repeated menorrhagic attacks, or from an inability to form healthy blood from the food she eats. Anæmia may proceed from actual want of food, and hallucinations or desponding and suicidal thoughts may be the result of prolonged starvation, as probably happens not unfrequently in the case of fasting ascetics and religious fanatics of all kinds. Patients continually come before us who, from one reason or other, take an insufficient quantity of food. Some think that too much food is carnal, and a lust of the flesh; others say they cannot work if they eat much; young ladies abstain, because they have a horror of becoming fat, and numbers fancy that they suffer from dyspepsia, and abstain from food for their stomach's sake. So that when the mental symptoms are first noticed, the patient is already reduced in health and strength, and then depression is the prevailing character of the disorder in the majority of cases. Unhealthy habitations or unwholesome air may also undermine the constitution, and bring about nervous depression. In short, any cause which operates hurtfully upon the general health may have insanity for its result and outcome.

There is a variety characterized by melancholic symptoms, which is usually ascribed by the friends of the patient to dyspepsia, or disorder of the "stomach" or "liver." This is closely allied to the milder form to which we give the name of hypochondria, and the symptoms of the latter often grow into those of the former. Instead of the sufferer thinking that he has some ordinary bodily ailment, for which he seeks relief in vain by visiting all the doctors within his means, he imagines that he has some incurable disease beyond the reach of medicine, or that his inside is completely gone, or that the drugs he has already so freely taken have done him some irremediable harm. Conversely, a patient may, as he is getting

[1] Lecture V.

better, improve from melancholia into hypochondria. He may give up
delusions about his soul, and concentrate his attention on his body.   He
may give up the delusions first mentioned, and merely think, as hypochondriacs do, that his liver is grievously out of order, or his digestion entirely
gone.   You will constantly find that these people will, under the plea
that they cannot digest this or that, take a very insufficient quantity of
food—often a very innutritious food.   Many—I might say most—of them
suffer from constipation, and this is made an excuse for not eating.   Many
are of opinion that if the bowels do not act freely, little food should be
taken; of course the constipation increases, and for this reason they diminish the food.   From the unwholesome diet such people live on, it frequently happens that they suffer from chronic diarrhœa.   The ideas they
entertain concerning what is proper or improper to be eaten amount
almost to delusions.   In one way or other, they become half-starved,
emaciated, and weak, and then they may drift into melancholia.   These
patients are curable if we can compel them to take what is necessary, but
we shall probably find that nothing short of forcible feeding will do this.
    In the course of a wasting illness, or during the subsequent convales-
cence, great mental depression, or even temporary dementia, may occur.
With returning strength the mind generally recovers.   There is an insanity, however, which coexists with one chronic disease, viz., tubercle,
which has given rise to discussion.   In tubercular patients may be noticed,
according to Dr. Clouston,[1] a special form of insanity characterized by suspicion, irritability, unsociableness, and disinclination to exert the mind
or body.   I must refer you to his paper on the subject, for I have met
with so few cases of insanity and tuberculosis conjoined, that my observation of them is worth little.   In an asylum population ranging from fifty
to seventy patients of the upper classes, and a large number of insane
patients not in asylums, who have been under observation during the last
twenty-five years, I have only known four who exhibited symptoms of
pulmonary phthisis.   Three were males, one a female: one, a male, was
a congenital idiot.   Two, also males, were examples of recurring insanity,
mania alternating with melancholia, and in one the lung symptoms made
their appearance after a long period of refusal of food and great emaciation.   The lady also suffered from periodical mania.   The two persons
whose lungs post-mortem examination showed to be the most tubercular
died of general paralysis, which ran its usual course.   Dr. Clouston says
that tubercle is found in the bodies of those dying insane much more frequently than in the sane; in the former, it is to be found in 60 per cent.;
in the latter, in about 25 per cent.   My own experience does not bear this
out, and I think it useless to compare the chronic insane dying in large
asylums with the sane people who are not in asylums.   I have found, how-

[1] Journal of Mental Science. April, 1863.

ever, that phthisis and insanity do frequently coexist in the same family, but that some members will be afflicted with insanity, while others suffer from phthisis, and that this is the rule rather than that both disorders coexist in greater or less degree in all the members of the family. The subject is discussed in the first volume of the West Riding Asylum Reports by Drs. Nicol and Dove, and the figures they adduce differ considerably from those of Dr. Clouston, which are based mainly on the records of the Royal Edinburgh Asylum. In the latter, phthisis was the assigned cause of death in something over 29 out of every 100 deaths, while in the eight principal towns of Scotland, in the year 1861, out of every 100 deaths, above five years of age, 20 were from this disease. In the West Riding Asylum there was a proportion of 15.6 deaths from phthisis to every 100 deaths in the Asylum, while over the whole kingdom, out of every 100 deaths, over five years of age, 17 are ascribed to phthisis. The experience of Dr. Boyd, at the Somerset Asylum, is equally unfavorable to Dr. Clouston's view.

The varieties of insanity hitherto discussed are marked by symptoms of depression rather than noisy, angry, or gay excitement. They belong to the asthenic rather than the sthenic order of cases. I now pass on to the consideration of another group of patients, some of whom present the one set of symptoms, some the other. Their insanity would be comprised, according to Dr. Skae, under the heads of climacteric insanity, the insanity of pregnancy, lactation, and child-bearing and the insanity of pubescence; and in this order I will consider them, because the first three varieties are characterized in the majority of instances by depression, the last two by the opposite. And all specially illustrate what I have said about insanity being the outcome of many causes rather than one. In almost all such patients you will find a strong inherited tendency, which lights up the disease when some perhaps trivial circumstance causes mental worry, or when such events as parturition or pregnancy occur.

At the time when women undergo what is called the change of life, and at an analogous period, from the age of fifty to sixty years in men, insanity of a melancholic type is common. It may have been preceded by worry or ill health, or may make its appearance without apparent cause. It is noticeable, however, that there is no immediate connection between the cessation of the catamenia and this disorder. Frequently it does not make its appearance till some time after menstruation has ceased, and it is as common in men as in women. It may be the result of weakness caused by the menorrhagia which so often accompanies the cessation of the menses; it may be due to weakness caused in other ways; but it is not to be looked upon as in any way a uterine disorder. It is, in fact, due to the weakness of approaching old age; and whether the immediate exciting cause is mental worry or not, the physical condition is one of failing strength, conjoined in many cases with hereditary taint. If the weakness yields to food and remedies, the mind is restored, and this, fortunately, is a common result: if the bodily strength is too much impaired for renovation, the patient declines, and death ends the scene; or the mind and brain may degenerate, and the depression may give place to the dementia and fatuity of age, the "second childishness, and mere oblivion." The earliest symptom of this is loss of memory—a sure sign that the end has begun. Of this symptom, bad at any time, the cure is hopeless when it is noticed in a person of advanced years.

Midway, as it were, between the insanity I have termed "climacteric," which generally presents the features of curable melancholia, and the senile dementia which is the termination of all mind, there may sometimes be observed a variety which may be termed senile insanity. My lamented friend Dr. Anstie has described this. It is characterized by great restlessness, suspicion, and irritability, a phase inexpressibly trying to the patient, and still more to all those who are about him. It may be said to consist in a peculiar perversity, a tendency to offer vexatious and frivolous delay and opposition to everything which is suggested by others, however important the occasion. It may also manifest itself in changed habits and tastes, a "moral insanity" of a peculiarly painful kind, accompanied by extravagance and immorality very difficult to deal with, as the acts themselves may indicate vice rather than insanity.

Let us now consider the pathology of another group of cases where insanity shows itself during pregnancy or after labor. When we see the thousands of women who go through these periods with perfect immunity from all such symptoms, it is clear that the pregnant or parturient condition is only one of a number of causes, which must be sought and investigated. It must be connected with the insanity by means of a series of links which are often very hard to find. As I suppose not one woman in a thousand becomes insane after her confinement, we must search among the so-called predisposing causes for some reason of the mind-disorder in those who thus break down, and we must very closely investigate the physical condition at the time the mental symptoms first appear. Those of you who attend cases of midwifery will see such symptoms at an earlier period than I do, and you may be able then to arrest them. In insanity, as in so many other diseases, prevention is better than cure, and it is to this that modern science must direct its efforts.

And first of the insanity of pregnancy. This is almost always melancholia, indicating an asthenic condition of the bodily strength. In my own experience I have found it most frequently in women who have had many children rapidly. It has occurred in the later pregnancies, not in the earlier; has lasted up to the birth of the child, and then gradually

[1] On Certain Nervous Affections of Old Persons, by F. E. Anstie, Journal of Mental Science, xvi. 31.

passed away. Whether, in fact, it occurs in the first pregnancy or after several, it is almost always found in women who are not very young, and thus forms another illustration of what has been laid down so often, viz., that depression is a disease attacking not the young, but the old, whether the latter be old in years or have become prematurely aged in bodily strength.

Insanity may appear after confinement at various periods, and it has been found that the symptoms will vary according to the time that has elapsed since parturition took place. The insanity of lactation, as it has been called, makes its appearance after an interval of some months, and is due to the anæmia brought about by prolonged suckling, or by the mother making undue efforts to nurse, and so overtaxing her strength. And, as may be expected, the insanity is, like the last, marked by depression in the large majority of cases. Like the last, too, it is curable. Where we have to deal with a weakened bodily state, we can by food and tonics restore the strength, and with it the mind. Consequently, the prognosis in all these cases of asthenic insanity is much more favorable than when we meet with symptoms of what is called mania in a person who is strong and well nourished.

Symptoms of insanity may first be noticed during labor or immediately after, within a week, within a month, two months, six months—in fact, at almost any period within a twelvemonth. After this we should not attribute them to the confinement; at any rate, it would be a much more remote cause. They may appear in or after a perfectly easy and natural labor, or after one very difficult, and attended possibly with great exhaustion or hæmorrhage. They may come on in a woman who has not suckled at all, or has long ceased to do so. Now, according to the interval between the birth and the first manifestation of the insanity, so is the character of the latter. That which commences in the course of a week or fortnight is almost always attended with violence, sleeplessness, and great excitement, amounting to acute mania. We have a great and rapid discharge, all the symptoms of cerebral hyperæmia, entire want of sleep, and a disorder which may run its course to death in a few days. If the insanity does not begin till a month or more has elapsed, the symptoms are almost always those of melancholia with suicidal tendencies, less acute and less rapid in its course, and less formidable to deal with. In the first-mentioned class of cases, the brain is prone to rapid discharge; the sufferers are almost all persons of an excitable and irritable temperament with marked hereditary taint. Some trifling circumstance, some piece of news, the excitement of visitors, a few sleepless nights, or the mere "sympathy" of the uterine organs, light up the mischief in the unstable cerebral cells, and the train of evil commences. But when a longer time has elapsed, and depression appears, we find not an irritable and unstable brain, but one whose nerve-power has been weakened by childbearing, by imperfect bodily nutrition, or by the exhaustion of anxiety or worry. Failing strength of body and depression of mind here as elsewhere are linked together; here as elsewhere renovation of the one brings about the removal of the other.

In the next variety of insanity we find, as we should expect, no depression. It is the insanity of puberty, of boys and girls who are just passing into manhood and womanhood. At this time of life there is no failing or declining strength; the cares and anxieties of life are unknown, the brain has not been racked by deep thought or intellectual toil. What, then, do we observe? First, that this period brings more dangers to girls than to boys; that more girls between the ages of twelve and eighteen become insane than boys. This we should expect. The period of pubescence causes a greater functional change in a girl than in a boy, with an increased risk of functional disturbance. A boy grows into a man imperceptibly, as it were. His development is marked by a capacity of procreation; but this is something very different from the establishment of the menstrual function, which so often is attended by great general disturbance. We may assume that at this time every girl is in a condition peculiarly susceptible of nervous irritation; therefore in one who inherits from her ancestors an unstable organization, two conditions exist very favorable for the production of mental disorder. These may be of themselves sufficient to originate it; but there may be superadded either a mental cause, as a fright or loss, something startling, harassing, or afflicting; a dreary, companionless, and cheerless life; a physical disorder, as menorrhagia, or an illness of an acute character.

You will find that such a conjunction of causes does not always produce insanity in a girl or boy; it may produce chorea: instead of the highest mind-centres being affected, there is disturbance of the motor centres, resulting in that irregular and spasmodic extrication of force which is recognized in all the protean forms of chorea. So prone are the motor centres to be affected in early life, that even when the mind is upset, and genuine insanity is recognizable, it is most frequently accompanied by noisy and violent action. Irregular movements or cataleptic rigidity, with more or less of a choreic tendency, are constantly to be seen; in fact, the insanity is more often shown in violence and powerful demonstrations of emotional activity than in delusions and disorder of the general intellect. Here, then, is a pathological condition to be marked off from others,—a nervous temperament undergoing the change which accompanies puberty, and unstable in consequence—upset, it may be, by some mental shock or debilitating illness, and characterized by disturbance of both mind and body, generally of a violent and spasmodic character.

Insanity accompanied by masturbation is not to be invariably assigned to masturbation as its cause. It is a different disorder—different in its oncoming, different in its course and character, and rendering different the prognosis to be made concerning it. Insanity caused by masturbation is, generally speaking, gradual in its approach, not attended with any sudden or acute symptoms, but manifested in unpleasant conceit and exalted self-feeling, with delusions in accordance; this gradually increases, nor is there any great hope of cure, for the brain seems to have undergone permanent damage from the constant irritation to which it has been exposed by the practice of the habit. Such a state of things altogether differs from the violent, but often transient, outbursts of hysterical or nymphomanical insanity, which are by no means incurable; and it points to a different pathological condition. In the one we have brain disturbance coming on suddenly, possibly from some sympathetic uterine or ovarian irritation, which causes great disorder of the cerebral circulation, and an acute attack of insanity, from which the patient may recover. In the other the brain is gradually altered and impaired by the unceasing demands made upon it, and the constant excitation caused by the act of masturbation. We may compare the latter, though less in degree, to frequently-renewed attacks of epilepsy or of alcoholic intoxication, which produce mental disorder by their constant recurrence through a series of years.

The insanity which masturbation produces is for the most, part seen in young persons, but there is another form found in those of middle age, often the result of sexual excess or masturbation, which is known under the name of general paralysis. I shall hereafter describe this at length. Suffice it to say, that I believe a frequent cause is sexual excess, whether in married or single life. Like the insanity produced in the young by masturbation, it is characterized by intense self-exaltation, by ideas of grandeur and importance and a feeling of the most perfect health and strength; and it would seem to be lighted up in the first instance by the constant irritation to which the brain is exposed by the frequent repetition of the sexual act. We see other effects of sexual excess every day in ordinary practice, such as lassitude, dyspepsia, giddiness, dimness of sight. These are the results of expenditure and exhaustion of nerve power; but here, if the cause is removed, the effect ceases, and the patient recovers. Once, however, the insanity called general paralysis is set going, there is at present no cure for it known. This, of course, points to a pathological condition entirely differing from that of any curable or transient form of insanity.

As in some persons mere nervous exhaustion and bodily disorder are produced by masturbation and sexual excess, while in others genuine insanity is the result; so other causes, as epilepsy and alcohol, give rise to insanity in some, to decay of mind and body in others. The original constitution and pathological condition of the individuals being different, the result is different, though the cause is the same. We find in practice that patients in various ways, and through various stages, arrive at legal unsoundness of mind. In the eye of the law they are all alike, all in-capable of taking care of themselves or their affairs; but to the patholo-gist they present infinite diversities, and the points of difference will be multiplied more and more as our means of scientific research are extended.

Let us take the various pathological conditions produced by alcohol: there can be no better illustration of what I have just said. First of all, a man may be drunk, paralyzed in speech and ideas by alcohol, furious with drink, or wholly insensible. Another may suffer from *delirium tremens*, even after he has ceased drinking, perhaps for days. Besides these, we may notice a third state; instead of delirium tremens, which runs its course in a week or so to recovery or death, an attack of ordinary insanity with delusions may come on in a person accustomed to drink. From this he may recover after a considerable period. And, besides, the habitual drinker, man or woman, may lapse into dementia, into utter obliteration of memory and mental power, into premature old age, from which he never will emerge again. In each of these states there is for the time unsoundness of mind from drink, but how great is the difference in the pathology of them! In the man who is drunk we see the effect of the alcohol circulating in the blood, and conveyed in it to the brain cells and fibres, in fact, to the whole nervous system: mind centres and motor centres, afferent and efferent fibres, are all affected, and their functions more and more impaired, till absolute insensibility and paralysis, nay, death itself ensue. Can we say which of our nerve organs are implicated in these various states? In intoxication, as I have said, all seem affected. There is a disordered cerebral circulation, after even a moderate amount of wine or spirits, shown in flushing of the face and excitement in talking and manner. Very soon the movements of the tongue and lips are affected. There is some loss of control. The words are somewhat clipped, are not enunciated in a measured and even manner. This may occur in some before there is any confusion or impairment of mind, and appears due to an affection of the motor centres or fibres, or of the com-missures which co-ordinate and focus the movements. I hold that these phenomena are due to the presence of alcohol, and not to mere alteration in the blood-supply of the part, because we notice them in some who are very drowsy after taking wine—and by drowsy I do not mean comatose from drink—while others who are noisy and talkative exhibit the same. The mental symptoms correspond to the motor. There is at first a want of co-ordination of thought, an inability to recall just what is wanted at the moment, and this after a very small amount of wine; and yet there may be an entire absence of excitement, or anything denoting any great difference in the cerebral circulation. The presence of the alcohol, then, in the blood, is the main pathological fact in this condition. When this is eliminated, the man is well again; but if it be present in large quantity, he will die, and the experiments of Dr. Anstie on animals show that the mode of death is paralysis. The good effects of a glass of wine are possibly due to an increased circulation, brought about by the influence exercised over the vaso-motor system. When a man is intensely sleepy after taking strong drink, it would seem that his brain is deadened to ordinary stimuli. Owing to impure blood, all changes therein cease, as in narcotization from choloroform; only violent shaking or shouting can then rouse him.

But in delirium tremens we have a widely different state of things. This is but little, if at all, removed pathologically from the acute delirium of the insane, though it is shorter in duration. It is not the actual presence of alcohol which causes the symptoms, for none may have been taken for days, so that it is often asserted that the withdrawal of drink is the cause, and we are told that we must not fail to give the accustomed stimulus when treating the disease. The truth, however, seems to be that by constant drinking—and, I may add, by loss of food, which is almost always the concomitant—the nerve centres are reduced to so unstable a condition, that the slightest thing, an accident, or grief, or anxiety, or any mental shock, upsets the balance; and then ensues that indication of a lack of power exhibited by muscular tremor, together with the incessant talking, sleeplessness, and mental disturbance, with which you are so familiar, denoting great discharge and rapid molecular decomposition in the brain-centres. We may see nearly the same attack in patients brought to a like unstable condition by exhaustion or want of food. But we shall not in these see the peculiar muscular tremor, though there may be convulsive or cataleptic phenomena of other kinds.

In delirium tremens it is evident that there is a disturbance of the brain to such an extent, that unless it subsides the patient is liable to die of exhaustion of his nervous power, which has no chance of renewal. There is an incessant discharge and emission, and the renewal of the exhausted force is prevented by the absence of sleep, and frequently by the difficulty in administering food; only by means of food and sleep can the exhaustion be remedied, and the sickness so often present may render it very difficult to give a sufficient amount of food. With this we may compare the epileptic attacks which are so often the outcome of continual drinking. In one person the alcohol produces the variety of malnutrition and instability which leads to the discharge of delirium tremens; in another, we find the discharge of epilepsy; and a comparison of the two will show us how nearly akin they are.

Not rarely does insanity, mania, or melancholia, make its appearance in people who have for years led lives of intemperance. Here we meet with a somewhat different pathological condition. There may never have been an attack of delirium tremens. The patients either have not taken enough, or they have escaped the accidents which lead up to it, or their constitution does not expose them to this particular form of disorder. So, instead of the busy wakeful delirium and incoherent wandering, we find the delusions, suicidal tendency, and outrageous acts of ordinary insanity. Patients recover from these attacks, return to intemperate habits, break down again, and may recover again and again, but, rarely giving up drinking, they for the most part die in confirmed insanity. A few consent to remain under some sort of control or surveillance, and so escape. If we consider the probable pathology, we must conclude that the repeated states of intoxication, like constantly renewed anxiety or incessant labor of brain, bring about in time an irregularity of circulation and of function, a certain instability and defect which do not amount to delirium, but may be manifested in delusion and violent and maniacal conduct, with irregular and imperfect sleep. There may even be loss of memory and temporary paralysis, which may pass off in apparently very hopeless cases. Abstinence, and the quiet of a life under control, may enable the brain to recover its balance; but I have known two years elapse before the cure was effected.

There is one more pathological state connected with the habitual use of alcohol. After years of habitual drinking, drinking which may hardly have amounted to intoxication, far less to delirium tremens, we may perceive the mind weakening, memory failing, and the dotage of premature old age coming on; and not unfrequently with this decrepitude of mental power, we notice some amount of bodily paralysis, which slowly advances at the same time. This is the manner in which women often show the effect of drink, and for them there is no hope. We may reasonably infer that, from the long-continued poisoning, irritation, or whatever we like to call it, the nerve centres and nerve cells and fibres are degenerate, and have lost their structural perfection and efficiency. These cases are very curious and interesting to watch. Quite suddenly, without illness, sleeplessness, or excitement, memory gives way. The patient talks quite rationally and calmly, but does not distinguish yesterday from last week, thinks friends long dead are alive, and when set right, makes the same mistake five minutes afterward. Here is commencing dementia, a pathological condition resulting from many antecedent events, of which drinking is one. Such people closely resemble those who have completely lost their memory from old age. I have known some recover to a considerable extent, but in all there was left some amount of weakness, mental or bodily. They did not regain their former state as a man does after delirium tremens.

There are some substances besides alcohol which, by constant use, may bring about insanity. Notably, this is the effect of bhang or Indian hemp, which sends many patients to asylums in India, producing a form of excitable mania with delusions, from which they for the most part recover. You will not meet with these cases in our own country, but in certain parts of the East they are common. In Europe there is a preparation of *absinthium* which is also said to produce mental symptoms, if largely taken. There is much controversy, however, concerning it, many asserting that the harm done by *absinthe* depends on the alcohol taken, and not on any peculiar properties of the herb. I am not aware that any but an alcoholic preparation is ever drunk, so that it is difficult to come to any decision on the subject. That opium produces curious phenomena and trains of ideas out of the control of the will, must, I think, be admitted by all. I cannot say, however, that in my experience it has been often found to produce insanity. The anomalous symptoms are due to the actual presence of opium, to a poisoning going on at the time; but if it is withdrawn these vanish. The brain and nervous system are affected, as are those of a man breathing nitrous oxide gas; but when the cause is removed the effect ceases, and the pathological condition is not one which deserves the name of insanity, any more than that of a man who is drunk with alcohol, or delirious under the influence of cholroform or nitrous oxide. Practically, we do not find that opium eating or smoking swells the population of the lunatic asylums of this or other countries. There is an immense difference between the results of the continual use of opium and alcohol. Dr. Christison even thinks that opium-eating does not necessarily shorten life. Certain it is, that our two most noted opium-eaters, De Quincey and S. T. Coleridge, lived, the former to the age of seventy-four, the latter to that of sixty-one years.

The poison of belladonna sometimes gives rise to mental symptoms. A single large dose has produced insanity in a man into whose eye atropine was introduced. The symptoms resembled to some extent those of delirium tremens, and there were hallucinations of sight of a corresponding nature.[1]

Cases of insanity from lead poisoning are not unfrequently met with, and several have been recorded by various observers. The mental symptoms vary considerably, according as the lead intoxication has been rapid or protracted. In many the symptoms simulate, to a certain extent, those of general paralysis. Hallucinations of sight and hearing are common, and here may also be noticed the blue line on the gums, and a certain degree of paralysis, which passes off under treatment. The few cases I have myself seen were characterized by melancholia, but in more advanced cases it is more common to find the contentment and elation which may cause them to be mistaken for general paralysis. In the majority of such patients, the mind remains damaged to some extent, but a few recover.[2] Lead poisoning may also bring about epilepsy.

---

[1] Journal of Mental Science, xxvi. 430.
[2] Insanity from Lead Poisoning, Journal of Mental Science, xxvi. 222.

LECTURE IV.

The Pathology of Insanity, continued—Insanity in Acute Diseases—with Rheumatism or Gout—with Disorder of the Sexual Organs—Peripheral Insanity—Insanity with Diseases of the Head—Liver—Heart—Kidneys—Stomach and Bowels—Insanity with other Neuroses—Myxœdema—Insanity from Mental Shock—Long-continued Worry—Insanity tending to Dementia—with Epilepsy—Syphilis—From a Blow—Sunstroke—Do the Mental Symptoms correspond with the Variety?—Concluding Remarks on Pathology.

THERE is a series of pathological conditions which I shall briefly review, where we find a combination of bodily disease, such as may exist alone, and mental disturbance giving rise to what we call insanity. The latter may present itself in patients who are suffering from some acute inflammatory disease, as pneumonia or fever, or from disorder of the uterus or some other organ, as the heart, liver, or kidneys. It may accompany epilepsy, or come on quite suddenly in the course of acute disorders, measles, pneumonia, fevers, and the like. The consideration of all these conditions involves some most obscure and intricate questions of pathology; but I think that a careful examination of the phenomena throws some light on the nature of the disorder of the brain. In all these it is clear that we have a combination of conditions making up one pathological whole. As thousands of patients suffer from the above-mentioned diseases without any signs of insanity, we must look for other concomitant causes besides those I have mentioned.

The first class of such cases of which I shall speak is that of insanity occurring in the course or at the decline of acute disorders. And here, to shorten what I have to say, I may direct you to an interesting paper on the subject by Dr. Hermann Weber, in the 48th volume of the "Medico-Chirurgical Transactions." In this he gives the particulars of seven cases of measles, scarlatina, erysipelas, pneumonia, and typhoid fever, where, toward the decline of the disorder, maniacal delirium came on, with delusions of an anxious nature, and hallucinations of the senses, especially of hearing, but also of sight. The duration of the derangement was short, extending from less than eight to forty-eight hours. The outbreak was sudden, the times was in general the early morning. *Almost always the commencement was stated to have occurred immediately after waking.*

These cases are called by Dr. Weber the "delirium of collapse"—a name, however, which throws but little light on their pathology. The points to be borne in mind are the transitory nature of the attack, its sudden oncoming, especially on waking from sleep, and the partial nature of the delirium, this taking the form of delusions, and not being the general incoherent wandering of the ordinary delirium of fever. We may suppose that the combination of conditions is something of this kind. First, the patient is by inheritance " excitable;" his nerve-balance is easily upset. "Excitable," "naturally anxious and excitable," " over-conscientious," are terms applied by Dr. Weber to several of his patients. By the acute disorder and the accompanying fever this condition of instability is greatly augmented. During the heightened temperature, Dr. Weber tells us, there was no delirium, but the circulation and temperature fall ; less blood is carried to the exhausted brain centres; sleep comes on and the brain nutrition is lowered to its minimum, and on waking suddenly from a short sleep, the instability is at its height and delusions and delirium manifest themselves; food and opiates procure a long sleep, and the cerebral force having been restored the mania vanishes. The sudden outbreak of insanity on waking from sleep is familiar to all who have the care of the insane. In acute mania, patients who sleep half an hour or an hour often wake in a furious paroxysm. Many a crime is committed by others subject to transient attacks of insanity when just awakened from an insufficient sleep. Suddenly to wake even a sleeping dog is a thing most persons look upon as attended with risk. It is a matter of common observation also, that melancholic patients are almost invariably more dejected at first waking. The lack of nerve force is shown in them, not by instability leading to excessive discharge, but in the absence of force, the symptoms of which are depression and gloom. It is also to be noted that epilepsy in many occurs chiefly during sleep, at which time the centres appear to reach their highest degree of instability.

It is to be noted that Dr. Weber is speaking of cases which occurred in a general hospital, and in all recovery from the mental symptoms took place in a few days. In the experience, however, of those who see cases of the kind, where the insanity has not passed away quickly, the prognosis is much more unfavorable. Dr. Clouston [1] gives the particulars of ten cases of " post-febrile" insanity following scarlet fever, smallpox, typhus, typhoid, and intermittent fever; of these only two recovered. The rest were incurable, six being hopelessly demented, and two helplessly melancholic. " Post-febrile insanity may be said to be generally characterized by sub-acute symptoms, to result clearly from the brain being poisoned by zymotic poison and exhausted by fever, not to require an hereditary tendency for its development, and to be a most incurable form of insanity from the beginning." The cases seen by Dr. Weber may be likened to the short delirium of acute alcholism, while those of Dr. Clouston resemble the

---

[1] Morisonian Lectures, Journal of Mental Science, Ap., 1875.

dementia which marks the condition of patients who have been long poisoned by alcoholic excesses.

The connection between insanity and rheumatism or gout seems to be of a most special kind. It is a commonly recognized complication, and we not unfrequently find stated in books, and observe in practice, that the symptoms of the one disorder vanish as those of the other appear, and *vice versa*. I myself have seen phenomena of very acute insanity disappear quite suddenly, to be followed by great pain and swelling of hands or feet. This has been called *metastatic* insanity, and it has been supposed that the seat of the inflammation, caused by a blood poison, is transferred from the brain or its membranes to the joints. We may, at any rate, remark, that there is a marked analogy between the way that the acute symptoms of transient mania often entirely subside and vanish, and that in which we see the pain, heat, and redness of a joint disappear within a few hours, leaving nothing behind but some stiffness and tenderness as relics of what appeared most acute inflammation. It is not for me to tell you what is the exact pathology of acute rheumatism or of gout; the differences of opinion which still exist among observers of these common maladies may at any rate be some set off against the doubts and difficulties that beset me in trying to lay down as facts the pathological conditions of insanity. But when we see how the connective tissue is attacked by these disorders, how acute are the symptoms, and how rapid and complete their disappearances, we may suspect that the connective tissue of the brain may occasionally participate; and as the vessels going to the gray substance are meshed in this, it may well be that brain circulation is in this manner greatly disturbed.

The connection between rheumatism and insanity, rheumatism and chorea, and rheumatism and other head symptoms, has been dwelt upon by many physicians—Griesinger, Trousseau, Sander, Hughlings Jackson—and a most interesting account of two cases in which rheumatism, chorea, and insanity coexisted, is given by Dr. Clouston in the "Journal of Mental Science," July, 1870. Here the rheumatism was the first symptom. In one case, a woman aged twenty-four, the rheumatism existed for two months before the insanity. It suddenly abated, and mental and chooreic symptoms commenced.. There was loss of power in all the limbs, the legs were quite paralyzed, and reflex action destroyed. The mind was in a state of stupor and depression. Recovery took place in about three months. The second patient, a lad of nineteen, had already had chorea at the age of seven and thirteen. After an acute attack of rheumatism of two weeks, choreic symptoms commenced, with strangeness of manner, restlessness, and inattention. His mind then became greatly confused, with hallucinations of vision, delusions, and occasionally refusal of food. Recovery took place in about three months. Dr. Clouston remarks that the symptoms were very similar in both these cases; in both the rheumatism was the commencement of the attack ; in both there were choreic symptoms, paralysis of motor power, a deadening of the reflex action of the legs, similar mental phenomena, *high temperature*, increased at night, and a tendency to improvement in all the symptoms coincident with the lowering of the temperature, showing that the same lesion existed in both. All this, he thinks, indicates "a serious but transitory interference with the functions of the nerve cells and fibres in the spinal cord, such as might be produced by slight rheumatic inflammation, and infiltration of the connective tissue of the cord causing pressure on the nerve elements."

The pathology of chorea is a much argued question of the present day, and its complications with rheumatism on the one hand and insanity on the other, and occasionally, as Dr. Clouston has shown, with both, ought to throw some light on the pathology of that disease with which we are specially concerned. There is a theory that it is due to the obstruction by means of emboli of the vessels supplying the corpus striatum and surrounding parts; and Dr. Hughlings Jackson, who supports this view, holds the condition of this portion of the brain to be one of hyperæmia. These emboli are said to be derived from the heart in those cases where rheumatism coexists. But, on the other hand, there are many cases of chorea where rheumatic symptoms are altogether absent, and where the disease is clearly due to a mental cause, as a fright or other strong emotion. It may, however, be conjectured that the hyperæmia observed by Dr. Jackson may be the condition in chorea whatever be the exciting cause, that fright may cause hyperæmia and chorea, as it undoubtedly causes hyperæmia and insanity, that rheumatism may also cause hyperæmia and chorea in one case, hyperæmia and insanity in another, the rheumatic inflammation attacking the connective tissue of the brain, either *metastatically*, leaving the structures of the joints and going to the head, or affecting both simultaneously. And it may produce an hyperæmia which will lead to blocking of the vessels, and so bring about those appearances noticed by the advocates of the embolism theory. Dr. Lockhart Clarke has noticed this metastatic action in another allied disorder, viz., delirium tremens. In a case related by him,[1] the patient, a great drinker, was first attacked by acute and severe articular rheumatism, from which he recovered, and was again attacked a second, third, and fourth time. "Some months afterward," he says, "I was again requested to see him, and found him suffering, not in the least degree from his usual attack of acute rheumatism, but from the most violent delirium tremens. Every means employed failed to procure sleep or allay the severity of the symptom. Recollecting that his habits of intemperance had always been followed by articular rheumatism, I thought his only chance of relief

---

[1] British Medical Journal, 6th April, 1872.

depended on the reappearance of his usual attack.   I immediately ordered
the application of mustard plasters to his ankles, knees, and elbows.   On
the following morning the toes on the right foot were found to be swollen and red.   Then the knee of the same side and elbow of the opposite
side assumed a similar appearance, with marked alleviation of the cerebral
symptoms.   Soon after the patient fell asleep, and when he awoke there
was scarcely a trace of delirium."   Here the inflammatory action was conveyed to the joints—a fact which indicates the nature of the cerebral
affection in delirium tremens and the acute hyperæmia which is its condition.

The presence of marked sexual and erotic symptoms in many insane
patients, particularly women, and various irregularities in the uterine functions, have led writers to constitute a class of insanity to which they have
given such names as satyriasis, nymphomania, amenorrhœal insanity, and
the like.   That a connection exists between the brain and sexual organs
needs no demonstration; everyday experience shows it, but shows that
these organs are affected from the head downward quite as often as the
reverse, e. g., the sight or smell of the female excites the male and the male
organs, hitherto quiescent.   When we see violent sexual excitement in the
insane we must not always assume that the origin is in the sexual organs,
for I am convinced that it may be propagated from the excited brain to
them.   Here we must search through the analogies of other diseases and
the general teachings of pathology, and not be led into rash conclusions
drawn from phenomena which but too readily present themselves.   In a
great number of young females, the subjects of acute delirium, there is a
furious erotomania often accompanied by a continual desire to masturbate.
But when the acute attack subsides, all these symptoms vanish without
special treatment of any kind.   It is extremely common to find in recent
cases of insanity, where the patient is debilitated and out of health, that
the catamenia are altogether absent, and the cessation accordingly is
returned as the " cause " of the insanity.   When the patient is recovering,
and bodily and mental health returning, the catamenia reappear, and are
said to have " cured " the brain disorder.   On the other hand, a patient
may suffer from menorrhagia, and being greatly debilitated thereby, may
fall into melancholia; and in a case I lately saw, there were delusions of
pregnancy.   But the whole history of the case forbade its being called one
of uterine insanity.   If we look at the general constitution and antecedents
of the majority of these patients, what do we learn?   That she has always
been what is popularly called nervous, or peculiar, or hysterical, prone to
emotional displays, to bursts of temper, to hysterical crying, exaggerated,
it may be, at the menstrual period.   Most likely other members of the
family are nervous, possibly some are insane.   Thus the old story is
repeated—hereditary predisposition to nervous disorder, a proneness to be
upset by trifling occurrences, to be greatly upset by serious trouble, an

unstable cerebral condition, much influenced by sympathetic disturbance of the sexual organs and functions, and in turn reacting upon and violently exciting them. At first all these disturbances are for the most part temporary, very variable, and often periodical; later, they may become fixed and incurable. We sometimes find insanity caused or conditioned by a prolapsed or displaced uterus. In such a case the delusions often have reference to the sexual organs, and cure of the disorder will cure the insanity. But such delusions may exist where there is no local ailment, and examination may do far more harm than good.

The brain seems to be occasionally stimulated into active insanity by an irritation arising in some distant portion of the periphery. Thus cases are on record where an attack has passed off on the expulsion of a tapeworm, upon the removal of a piece of glass from the sole of a boy's foot, or the cure of a displaced uterus. Here a central disturbance is caused by an irritation of a peripheral nerve, akin to the reflex paralysis which often follows injuries of the periphery. Investigation would probably show that such patients were of an extremely unstable nervous temperament, and highly prone to nervous discharge. They, moreover, show the necessity of closely examining the bodily state of all patients, for by the discovery of such matters we may cure an insanity which otherwise might have become confirmed.

Insanity is found coexisting with diseases of the head, lungs, heart, liver, and kidneys. I must pass in review each of these organs, but this much may be said with reference to them all. In estimating the connection which exists between insanity and diseases of the above-mentioned organs, it is of no use to examine the statistics of post-mortem examinations conducted in large lunatic asylums, unless we tabulate the results in a manner much more precise than that usually met with. The commonest source of fallacy is this: A large number of the patients who die in asylums have been insane for years; it is therefore impossible to say whether the disease of lungs, heart, or liver, found after death, was, or was not, a pathological condition coexisting with the commencement of the insanity. When we discover tubercles in the lungs of a patient who has for perhaps twenty years undergone confinement in an asylum, with all the concomitants and depression arising out of such a life, we may well believe that the lung disease is long subsequent to the insanity. When another patient has been for an equal time constantly subject to violent excitement, we need not be surprised to find his heart in an altered state.

Let us take the head: if you examine the records of the post-mortem examinations made in any large asylum, you will probably be struck with the absence of disease in the brain sufficient to account for death; and if you compare them with the case-books, you will learn that patients die comatose after the exhaustion of acute mania, or sink from exhaustion with-

out coma, or die of apnœa or syncope, but will comparatively seldom find
the diseases of the brain met with in the post-mortem theatres of our
general hospitals, such as tumor, abscess, hydatids, inflammation of an
acute character of the brain or membranes, or tubercular meningitis.
From the latter causes unsoundness of mind may undoubtedly spring, and
legally we may have to deal with patients brought by one of them to this
condition, but they chiefly produce feebleness of intellect, and wandering
and dementia, or delirium or coma, such as you see in the wards here,
and not insanity proper, such as I am more especially speaking of, which
is treated in special asylums and institutions for the insane. Nevertheless,
the presence of a tumor in any part within the cranial cavity may give
rise to insanity, and in our autopsies of the insane we occasionally find
such. We must conjecture that the derangement is caused by the sympathetic irritation of such foreign bodies rather than by their actual invasion
of the parts concerned in mental operations. When the latter happens,
we should expect dullness and imbecility, not insanity. We may also conjecture that in a patient in whom insanity is caused by a tumor, other
pathological conditions exist favorable to the development; thus, of a
woman who died at St. Luke's Hospital, in whose right cerebral hemisphere was found a tumor of the size of a pullet's egg, it is recorded that
her sister was insane. In this case mania existed for two months, but
before death there were noted drowsiness, ptosis of the left eyelid, and loss
of memory. Dr. Clouston has recorded the particulars of six cases of
tumor,[1] and the conclusions he draws are that the brain structure is
affected in three different ways. They may cause an irritation of the
neighboring substance tending to softening. They may cause pressure on
distant parts, and so produce alteration in the nutrition. And, thirdly,
they may set up in certain parts *progressive* disease and degeneration, the
true nature of which is not yet well known. The results pathologically
are an increase of the connective-tissue in the form of granules and
enlargement and thickening of the coats of the vessels; but all these seem
to be secondary changes. Dr. Batty Tuke says that carcinoma produces
the greatest perversion of psychical function, this being probably due to
the great rapidity of its production. As Dr. Hughlings Jackson[2] would
say, the brain is more rapidly "reduced" by a quick-growing than by a
slow-growing tumor. Dr. Clouston says that the symptoms appear to be
irritability, loss of self-control, and gradually increasing dementia, rather
than the delusions of ordinary insanity, and in this they resemble those of
another variety, hereafter to be mentioned—insanity after a blow on the
head. There is often a strong resemblance to the symptoms of general
paralysis. In assigning a pathological value as "causes" of insanity to
the various morbid products found after death, we must consider well

---

[1] Journ. Ment. Sci., xviii. 153.
[2] *Vide* p. 47, *note.*

whether they may not be the results of the excessive action which has gone on in the head. The various effusions, extravasations, and layers of exudation found among the cerebral membranes may be given as instances, with other appearances hereafter to be mentioned. Important questions may hinge upon our interpretation of these, as, for example, the duration of mental disease estimated by post-mortem examination—a point not unfrequently raised in the Court of Probate.

An attack of apoplexy with hemiplegia may be followed by insanity, and this may pass away, or continue and become chronic even when the paralytic symptoms have disappeared. Here the mischief caused by the apoplexy may be various in kind. It may cause an irritation and alteration in the blood supply, and upon the absorption of the clot, or by the surrounding structures becoming habituated, so to speak, to its presence, recovery may take place; or the alteration in the organs of mind may become permanent and the insanity remain, as we see happen in other cases even after the exciting cause is removed.

We may conclude that the ordinary forms of insanity, such as we term mania, monomania, or melancholia, are rarely caused by morbid growths within the head, or by acute inflammation of the brain or membranes; that where they are thus produced, there is usually a strong predisposition depending on other pathological conditions coexisting in the individual ; and this insanity may disappear for a time and then recur, being lighted up anew by the irremovable cause. We may learn from this what a functional disturbance insanity is, even when springing from an organized cause, and how strong the tendency of the individual life of the brain-cells is toward recovery, so long as it is not seriously interfered with by conflicting morbid conditions.

In spite of the ancient theories of insanity being caused by black bile, I believe the liver has little to do with the pathological condition of a patient who has recently become insane. In the general disorder of the system existing at such a time, the liver may participate, but it is not the cause of all the mischief, as so many, especially the friends, try to establish. The liver is the best abused organ of the body; whatever be the matter, nine people out of ten ascribe the blame to the liver, and cannot be convinced of the contrary. Little is to be learned from post-mortem appearances: there is nothing more than may be observed in the livers of those dying of ordinary diseases in a general hospital. Congestion, a softened and friable state, enlargement rather than shrinking, are what we chiefly see.

In the post-mortem theatre the hearts of insane patients very frequently present morbid appearances, but we are not to connect these with the outbreak of the insanity. Their nature indicates that they are the result of the long-continued, violent, and irregular action of the organ during many years. That which we shall most commonly find in chronic cases is that the right side is very much thinner than usual, and probably dilated, and that the left ventricle is thickened. In quite recent cases the whole may be healthy, or possibly there is a softened, friable, greasy condition without loss of substance. The valves are not diseased more frequently than usual, but the right auriculo-ventricular opening is often very large. Dr. Sutherland states in his Croonian Lectures that of forty-two patients examined at St. Luke's Hospital in the years 1853–56, the heart was healthy in eight cases only.   .

There is little to be said concerning the kidneys. In the pathology of commencing insanity they play a very unimportant part, and even after death they are not often found diseased. Acute renal disease with albuminuria and dropsy is decidely rare amongst the insane.

Dyspepsia and functional derangement of the stomach are common, but are not to be looked upon as primary disorders, but rather as caused by the disturbance of the nervous system. Vomiting is not common even amongst patients who refuse their food from alleged dislike and want of appetite, and the whole of the phenomena of dyspepsia are to be studied most carefully from an objective point of view. We are not to be too much influenced by the patient's own subjective sensations: numberless are the delusions connected with the stomach and intestines.

Similarly, functional derangement of the bowels and the uterine organs may exist without actual disease. Obstinate constipation is a common symptom also to be referred to the nervous condition, and varying with it. Much has been made of an occasional displacement of the large intestine met with after death. The transverse colon descends and lies across the hypogastric region or in the pelvis. I think it possible that violent struggling and straining, especially under mechanical restraint, have something to do with this. We find tumors of the womb or ovaries, and are able to connect these with the insanity, and even with the delusions; but whether the insanity is to be attributed to them and whether they do more than color it, so to speak, is, in my opinion, a very open question.

In speaking on the subject of hereditary transmission, I shall have occasion to point out to you that amongst the children of nervous or insane parents one may be hysterical, one hypochondriacal, a third a drunkard, a fourth epileptic, or a fifth insane. But you may also find two of these neuroses combined in one individual, or alternating in him or her. Thus, hypochondria may drift into insanity, and, as the insanity passes off, the symptoms of hypochondria will again become prominent. So with hysteria. Besides the hysterical mania so often witnessed, insanity, mania, or melancholia may take the place of the former hysterical attacks. And it has often been remarked, and I myself have several times met with instances, that neuralgia may be followed by insanity, the pain vanishing during the mental disturbance, and reappearing as the latter passed away. In one case, neuralgia of the spine, for which a gentleman was kept to the sofa for some months, was followed by deep melancholia. In another, very acute mania followed neuralgic pain, which a young lady had endured for a long period. Sir B. Brodie narrates similar cases in his work on "Local Nervous Affections." Now, that hysteria and hypochondria are mental affections is all but universally acknowledged, and we merely see an alteration in the symptoms of the neurosis, depending upon some altered condition of the patient, just as mania alternates with melancholia in the so-called *folie circulaire*. And in the transition of a neuralgia we may well believe that the neurotic affection is merely changed from one nerve centre to another, from the centres of sensation to those of mind. The prognosis, so far as the passing away of the first attack of insanity is concerned, is favorable in all these cases, but the whole state of the patient is one of evil omen; for it is clear that the nervous system is highly unstable, and not likely to be free from disorder of some kind or other, while, as years go by, the form of this will probably be more and more serious and incurable.

Dr. Clouston has recorded[1] two cases only in which insanity was associated with diabetes mellitus, and he terms it Diabetic Insanity. The symptoms in both these patients were those of melancholia; and inasmuch as mental excitement and worry are among the recognized causes of diabetes, we might expect to find the two diseases co-existing in not a few cases. I have never myself seen ordinary diabetes in an insane patient, but I have known sugar to be found in the urine of a gentleman when convalescent after melancholia. This passed away in a short time, without giving any trouble. I think it possible that sugar may be found in the urine of insane patients without the presence of ordinary diabetes.

The co-existence of exophthalmic goître and insanity is not so common as might be expected,when we reflect that the former disease is a neurosis, and frequently follows violent nervous excitement. I had under my care a lady who suffered for some years from this disorder, but just as the symptoms were abating she became pregnant, and after parturition insanity supervened, with suspicions and fear of her husband. She was for upwards of a year in an asylum, and although now at home, she is not free from delusions, and there still remains some protrusion of the eyeballs. Dr. Clouston[2] mentions a woman affected with exophthalmic goître and acute mania. After the subsidence of the mania fainting seizures began, with loss of power on the left side, and she died about twelve months after admission. An examination showed that the right hemisphere was extensively diseased. The pia mater, on removal, dragged away in places the whole depth of the convolutions, the white matter was pink and

---

[1] Clinical Lectures on Mental Diseases, p. 592.
[2] *Ibid.*, p. 604.

mottled, the cortical matter universally soft and red, and in many places quite disorganized.

Dr. Savage[1] records three cases of exophthalmic goître, the patients being all young women with strongly-marked hereditary insanity, the symptoms being melancholia passing into violent mania. Two died. *Post-mortem* examination revealed no disease of the cervical sympathetic, to which some have ascribed the disorder. In one there was disease of the supra-renal capsules. The brain presented the appearance we should expect after acute mania.

If we read in works on medicine the descriptions of diabetes or exophthalmic goître, we shall not find mental disorder mentioned as a constant or even frequent complication. We cannot, therefore, constitute special varieties of insanity out of such a conjunction. But I now come to a disease where mental defect is always present; and inasmuch as we have just been considering the goîtrous affection of the thyroid gland, I may fitly bring it before you in this place. It has been called "myxœdema," from the œdema or infiltration of the tissues by a mucus-yielding, jelly-like dropsy occurring sometimes in men, but more frequently in women of adult age. The skin is thickened, dry, and rough; the face and hands lose their shape and expression; the hair falls off, and the teeth decay. And along with these symptoms there is to be noticed a loss of power in the nervous system, a slowness and feebleness of movement, and a corresponding languor and dullness of mind. This disorder was first brought to the notice of the profession by Sir W. Gull in 1873, who termed it "a cretinoid state supervening in adult life in women." Since then many have observed it, and all have noted the mental alteration as well as the bodily phenomena. At a meeting of the Clinical Society in November, 1883, a remarkable communication was made of the experience of Professor Kocher of Berne, who had partially or totally extirpated the goîtrous thyroid gland in 101 cases. Examining as many of these patients as he could, some time after the operation, he found that where partial extirpation had been performed the result was excellent; but where the gland had been entirely removed in 18 patients, all except two showed more or less deterioration of health, according as a long or short time had elapsed—the two exceptions being patients in one of whom a small accessory thyroid gland had undergone an hypertrophic change, while in the other a recurrence of the goître had taken place. The deterioration of health corresponded in every respect with that described as myxœdema. Dr. Ord had previously noticed atrophy of the thyroid body in those affected by this disorder, and these remarkable facts lend great interest to it, though its pathology can hardly yet be looked on as definitely settled. A lady came under my care in 1880, who had been

---

[1] Guy's Hospital Reports, xli. 31.

suffering from myxœdema for some years. The swelling had greatly diminished, and the baggy, pendulous skin showed what distention had formerly existed. Many of the appearances coincided with the published account. There were no teeth, and very little hair. The face was flabby and expressionless; the skin dry and rough. She looked 70 instead of 52, which was her age. Her mind was greatly affected. Before coming under my care she had remained constantly in bed, pretending she could neither stand nor walk. She could do both, though feebly; could converse rationally, and write a sensible letter if she chose, though generally she looked as if she were in a state of hopeless dementia.

With any of the preceding pathological conditions may be conjoined that which arises from mental sources, or the latter may exist alone and may by its overwhelming effect unhinge the mind. Some observers ascribe to "moral causes" the chief rôle in the causation of insanity; others would deny to them any *status* in this regard and reduce all to physical causes and conditions. And with some reason; for in almost all the cases where insanity follows a mental trial, we shall detect in the physical constitution of the individual a predisposition, an instability of nerve constitution, which enables the mental cause to overthrow him. We must consider, not the events of the preceding month or year, but the history of the individual from his birth, and that of his parents before him. There are men who, with the misfortunes of a Job, or the anxieties of a Damocles, are nevertheless calm, equable, and active, ever ready to catch the turn of the tide, prepared for everything, good or bad; and, on the other hand, some fall down before the slightest buffet of fortune, and are said to be driven mad by grief, or loss, or worry. A man may have griefs and anxieties so severe as scarcely to fall short of disease; but the moral causes of the insanity of many are of so slight a nature, when looked at objectively, that it is at once clear that other causes and conditions must be sought. I have said already, and shall have to repeat it again and again, that insanity is the result of a number of events and conditions which go to make up one pathological state.

There is one group of patients in whom disorder of the brain is produced by a mental shock, with sleeplessness, heat of head, depression, or painful excitement, which may be accompanied by confusion of thought or delusions. Sudden shock may be followed by a variety of grave consequences, by death if there be heart disease, by apoplectic diffusion of blood, by epilepsy, by chorea in young persons, and by various forms of insanity, from acute mania to what is termed acute dementia.

Most men know what it is to have been kept awake for a considerable portion of the night by something which causes them anxiety or grief. There is a strong stimulation of the mind, causing continued thought, and followed by an excited brain circulation. Even the anticipation of pleasure may produce the same result. In the young this may more frequently be the disturber of sleep than care or sorrow, which comes to them seldom and sits on them lightly. From one cause or other almost every one knows what it is to be rendered sleepless. Now, if we observe the succession of phenomena in persons of ordinary health, what do we find? The reception of a mental shock causes immediate activity of brain, rapid molecular change in the centres, and in consequence a determination of arterial blood to the head. Even muscular structures may be set in motion, and this involuntarily. Very likely there will be trembling, sobbing, or crying. The sufferer may pace the room, or rock himself, or wring his hands. All such acts imply a continued change going on in the centres, mental and motor, and they also imply a want of controlling power. The weaker the individual the more violent will be these manifestations. Take the first, trembling. This indicates a lack of force. If we hold out a weight with extended arm, a weight too heavy for our strength, our hand trembles more and more, till it falls exhausted. The muscles do not balance one another, and the mission of force is jerky and uneven; the organs are not co-ordinated. The same thing may be said of convulsions, which so freqently occur after hæmorrhage or other exhausting causes. As the higher centres become exhausted, the loss of control is manifested by the irregular and spasmodic action of the muscles which are set free from one another, and act separately and automatically. A common sequel of shock is chorea, which is another variety of . this spasmodic action.

In ordinary health sleep comes to the disturbed brain after a longer or shorter period, the sufferer wakes refreshed, and on the next day his mental disquiet wears a very different aspect. But in a man doomed to insanity the rapid molecular changes do not cease, sleep does not come, and soon there is evident emotional change, with confusion and want of co-ordination of ideas.

The form of the insanity will vary according to the condition of the whole nerve centres of the individual; in a young and vigorous subject it will most likely take the form of mania, with violent ebullitions of anger, and much muscular action; in the old or weakly, melancholia more commonly appears; while in one whose nerve power is for the time utterly prostrated, there will be some such variety as acute dementia, or *melancholia cum stupore*, where mind is a blank, and voluntary action altogether gone. That the patient presents this or that kind of mental disorder is, it is true, in one sense an accident. The exciting cause may act as a depressant or a stimulant, and produce in one man melancholia, in another mania; but I hold that the condition called melancholia is very different pathologically from that of mania. An important pathological condition, then, for your consideration and study is that of insanity caused by sudden mental shock, a rapid molecular change being set up in the brain, giving rise to accelerated circulation, heat, and want of sleep, with the phenomena of emotional disturbance and confusion of thought and idea, the symptoms varying according to the general constitutional energy of the individual. Here, to use an old expression, we may say that the circulation is disturbed by a *vis a fronte*, or, using another formula, *ubi stimulus, ibi fluxus*, we indicate that the changes which bring about the increased blood-flow have commenced in the organs of mind under the influence of undue stimulation from without.

I now come to another pathological condition, where not a sudden mental shock but long-continued mental worry or anxiety, or long and laborious mental application and work, has overset the reason. We see the result of this very frequently in brain diseases not involving insanity, in so-called softening and disorganization of structure, produced by years of overwork. Here the machine gradually wears out; but in another, whose brain function is more liable to disturbance, there may be at an earlier period signs of disorder rather than decay. The carking cares of poverty, and the lack of means to support a family, the chronic torment of a bad husband or wife, or of prodigal sons or profligate daughters, constant harass and anxiety in businesses of a speculative character, or a perpetual craving ambition perpetually disappointed,—these and a thousand other miseries of human life are the things which upset reason and fill asylums. And what is the pathology here? We do not hear of a sudden mental shock causing an overwhelming emotional excitement, and bringing about almost at once sleeplessness and acute insanity. But we know that there must have been a constant strain upon the brain, with increased emotion and increased expenditure going on for years. The brain circulation during all this time has been disturbed, and the nerve centres exposed to a greater demand and a greater amount of change than they are able to bear. Some men may endure this, may work early and late, and retain their faculties unharmed; but others, who are by nature more prone to change, who easily display emotional excitement, and do not easily subside into their normal calm, are one day excited beyond recovery, and insanity is manifested. Owing to the length of time that the stimulation has existed, and the consequent weakening and exhaustion which the whole of the machinery must have undergone, we find that this form of insanity does not usually subside rapidly, and too often see along with it signs of irremediable disorganization of the brain or the nutrient blood-vessels.

It follows that these patients are not very youthful, and as long-continued and exhausting anxiety is the main cause, we should expect melancholia to be the form the insanity will take, or monomania, with suspicion and ill-temper, rather than violent sthenic mania.

The late Dr. Skae applied to every case of insanity caused by mental or moral causes only the term *idiopathic*, if he was unable to bring it under one or other of his varieties. Dr. Clouston says,[1] "that contrary to the common belief that insanity is *usually* the result of severe disappointments, anxieties, afflictions, and distresses of life, we find as a matter of fact, that only about one-fourth of all the cases are so caused, and of these about two-thirds can be usually referred to some of Dr. Skae's varieties, leaving about one-tenth or one-twelfth of the total number of the insane that are really idiopathic."

Could we accurately examine this tenth or twelfth part of the insane we should probably find that in almost all the idiopathy means hereditary disorder; the patient becomes insane, possibly from some very trifling moral cause, because he inherits the constitution which has already shown its unstable nature in the insanity of his father or mother, or grandfather or grandmother. He has an inherited tendency to become insane, which is what we mean by an inherited taint. If his surroundings are favorable he may go through life sane. If circumstances favor the development of insanity, whether it be a moral cause, as shock or trouble, or such events as pregnancy or parturition, he or she succumbs to the conjunction of causes and becomes insane. But the insanity is no more idiopathic in the former group of cases than in the latter. We do not expect insanity in every parturient woman, and when we meet with it after a perfectly healthy and normal labor, the mental disorder must be just as idiopathic, *i.e.*, special to the individual, as that which is caused by overwork or worry, which is endured by thousands without insanity supervening.

There remain some varieties in which the mental unsoundness tends to hopeless and incurable dementia, even though at first the symptoms may be those of ordinary insanity, such as we term melancholia or mania. In some there may be no insanity in the ordinary sense of the word, but from the first may be noticed a mental defect, with loss of memory and failure of intellect and power, which slowly progresses till the mind is obliterated. The first that I shall consider is the insanity caused by epilepsy; as we said that there was a form of insanity, as well as imbecility and chronic dementia, brought about by alcohol, so after an epileptic attack, or series of attacks, we may have a transient outburst of delirium, or an invasion of insanity with delusions, lasting from days to weeks, and then passing off, and leaving the patient sound in mind until a fresh seizure occurs. This is a condition quite distinct from the loss of memory and general dementia, which, if the fits are at all frequent, gradually encroach upon the mind and almost without exception terminate the career of every epileptic, even when there has been no insanity properly so called throughout the entire illness. Can we in any way account for an attack of delirium or of mania supervening in one person after epileptic attacks, while another comes out of them comparatively unscathed? As there is every degree in epilepsy, from the slightest *petit mal* with scarcely any impairment of consciousness up to the severest fit, and as fits may occur singly or in larger numbers, so may there be every kind of effect, from a brief arrest of attention to the wildest mania or the most profound coma. Of the discharges of unstable gray matter which constitute epilepsy I have already spoken.[1] According to the degree of the discharge and the extent to which the cerebral centres are involved, will be the effect and the duration of it. A considerable period may be required before the brain force is restored to its normal state, and the various centres are in a state of equilibrium. The more numerous the attacks the more likely is mania to supervene. I have known a patient who, after three or four fits, woke sane; but if they were more numerous, if he had twelve or thirteen, he was for days in a state of mania, with many hallucinations and delusions —a state which was not cured at once by a sleep, but which gradually subsided like an attack of ordinary insanity, and was, in fact, in all respects the counterpart of it. The point to notice is that so many patients suffer from epileptic fits without undergoing this general disturbance; though their brain is so disordered that we see first the fit, and then possibly a prolonged condition of comatose sleep, yet here the disturbance terminates, and the sufferer wakes sane.

The dementia attending upon chronic epilepsy is at any rate more easy to comprehend; but we must not too hastily set this down to structural change, for it is marvelous how some long lost in speechless idiocy recover intelligence and power, if by some chance or medicine the fits are arrested. Exhaustion of force is here clearly to be noticed. The constant fits prevent it from ever accumulating in adequate quantity; but if they cease, it again shows its presence in the centres. For in violent and frequent convulsions the whole force of the entire cerebral system is discharged. When the attack is slight, consciousness is not lost for more than a moment.

A case is related by Mr. Whitcombe[2] of a female who was admitted into the Birmingham Borough Asylum in the year 1860. Her state on admission was described as one of " total imbecility, unconsciousness of. everything and every person about her." She was then thirty years of age, and had had epileptic fits from the age of nine, becoming insane at the age of twenty-three. She remained epileptic and insane till 1872, when she suffered from an attack of acute rheumatism, which was followed by chronic rheumatic arthritis, necessitating in 1874 the amputation of the left leg. From June, 1872, she had no fits, and in November, 1875, she was bright, cheerful, and happy, rational in conversation, and constantly employed. With the cessation of the fits mind and memory had returned, even after this long time.

---

[1] Those who desire information with regard to this, I would refer to Dr. Hughlings Jackson's valuable papers in the West-Riding Asylum Reports, in the Medical Press and Circular 1875-6, and in other journals.

[2] Journal of Mental Science, xxi. 588.

Besides the mania which results from a number of fits and the dementia which is the termination of years of epilepsy, there is another form of
mental affection which it is most important you should recognize. This
is *nocturnal epilepsy*.· After some hours' sleep a patient may have an epileptic attack with scarcely any convulsion, or with the latter limited to
the face, but with frightful dreams and hallucinations which may impel
him to acts of insanity and even homicide.   Such acts will be done in an
unconscious state which may last for a very considerable time.   If he is
roused during the attack, or wakes in the morning at his usual hour, he
may have a confused and vague remembrance of dreadful dreams, but he
will be absolutely unconscious of what he has done.   This form has been
called larvated or masked epilepsy—*epilepsie larvée*.   It has been confounded with somnambulism, but the latter is not usually accompanied by
acts of violence, and the pathognomonic signs of epilepsy, such as nocturnal incontinence of urine, the bitten tongue, and petechial eruption on
face and neck, are absent in the somnambulist.   It is very important that
you should rightly diagnose these patients, for you may be called upon to
give an opinion as to their responsibility for some act of violence.   Responsible they are not, but they constitute a most dangerous class, requiring restraint in an asylum.   It occasionally happens that an attack of violence of this nature is terminated by a fit of epilepsy of the ordinary kind.
Whether there is at the commencement any slight convulsion or *petit mal*
is a question very difficult to decide.[1]

The connection existing between syphilis and insanity and other nervous disorders is one which is attracting much attention at the present time.
Insanity is found depending upon syphilitic intra-cranial diseases, on
syphilomatous growths, periostitis, and other evidences of coarse disease
in the brain membranes or vessels.   The mental affection corresponds,
and is a gradually advancing dementia, usually with some amount of
hemiplegia or other paralysis, epileptiform convulsions, and general motor
feebleness, sometimes simulating general paralysis.   There may be also
acute cranial pain at night, with impairment of the senses of sight and
hearing.   The prognosis, I need hardly say, in these cases is extremely
unfavorable, yet some patients recover.   But it not unfrequently happens
that syphilis and insanity may co-exist in an individual, the mental symptoms being those of ordinary mania or melancholia, and the syphilitic
symptoms being those of secondary syphilis, without any evidence of intra-
cranial mischief.   The question arises, how far such insanity deserves to
be called syphilitic.   It would, in my opinion, be going much too far were
we to set down the insanity to the syphilis in every case where they coexist.   We find cases where special treatment does not benefit the insanity,
nay, where such treatment is manifestly hurtful, and where recovery

---

[1] Echeverria on Nocturnal Epilepsy, Journal of Mental Science, xxiv. 568.

takes place without it. That syphilis may cause a deterioration of the blood, and hence a malnutrition of the brain, is possible, but can we create a variety of insanity owing to this circumstance? Should we not rather place it in the category of the insanity due to anæmia? There is no uniformity in the symptoms found in such patients, neither are we told whether such insanity arises at a short or a long interval after the syphilitic infection. Probably, in the majority of such cases there exists a predisposition to insanity, and an exciting cause other than the syphilis, which occupies the same position in the ætiology as any other debilitating disorder. The very fear and distress caused by such contagion doubtless contribute in many cases to bring about the mental disturbance. We should do well, I think, to restrict the name syphilitic insanity to cases where either the blood-vessels, membranes, or other tissues are invaded by the disease.[1] The head symptoms here will vary indefinitely. They may be those of insanity with hallcuinations and delusions, of paralysis, of a gradually advancing dementia, even of general paralysis of the insane. Among these unfavorable cases we find here and there one where recovery takes place under special treatment: none are likely to be cured without it. Even, those which most closely resemble general paralysis are by it occasionally cured, and give rise to the belief that the latter is a curable disease.

Turn we now to a different pathological state, to that of a patient in whom insanity has appeared consequent upon a blow on the head. So numerous are the cerebral symptoms which at varying intervals follow blows, that it can cause no surprise if among these we occasionally meet with insanity. Imbecility, paralysis of limbs, a gradually advancing decay of mental and bodily power, is perhaps a more frequent sequel than insanity, strictly so called; yet this is to be found. When after a blow a man or a woman develops insanity in the form of mania or melancholia, or gradually becomes altered in character, or subject to fixed delusions, we may conclude that the injury the brain has undergone is not of the coarse character described in works on surgery. There is no traumatic inflammation of the brain-substance or the membranes, such as we are accustomed to see in the post-mortem theatre of our hospitals. We have to deal with a much more minute and molecular change—a change commencing, it may be, in the contusion of the gray matter caused by a blow or fall, and producing an alteration in the nourishment and growth of the part, in the blood supply, or in the nerves presiding over it. That condition which I have vaguely called instability of nerve function and force may be in this manner set up, as it comes to others by inheritance; and so it frequently happens that men who have received blows on the head are driven to a state of frenzy or mania by slight causes, which

---

[1] See an article by Dr. Wilks, Journal of Mental Science, April, 1874.

would produce little or no effect on an uninjured and healthy brain, such
as a very small amount of drink, or trivial matters exciting anger or grief.
From such transient attacks patients recover, and return to their normal
state of equipoise, to be thrown off their balance again by some other disturbing event. But when the change in the mind is insidious and gradual,
when acute symptoms are absent, and either quiet and concealed delusions, or a mere perversion and alteration of the whole man are alone to
be noticed, our prognosis must be extremely unfavorable, if we hear a
history of a blow or fall on the head.

There is yet another pathological condition brought about by a cause
external to the individual. This is the insanity developed in people,
chiefly men, who have been long exposed to the heat of the tropics,
whether they have had an actual sunstroke or not. Even here in excep-
tionally hot summers, such as that of 1868, not a few cases are directly attributable to this cause; and in my own experience I have met with a large
number of patients who, either in India, or in returning thence, have
shown symptoms of insanity. Sir Joseph Fayrer tells us[1] that the condition of sunstroke is intense pyrexia due to vaso-motor paralysis and to the
nerve centres being overstimulated and then exhausted by the action of
heat on the body generally. Failure of respiration, of circulation, and of
innervation generally result, and asphyxia follows. Recovery is frequent,
but in many cases is incomplete from structural changes having taken
place in the nerve centres. Of course, when we say that heat causes insanity, we are speaking of patients who have been subjected to a degree
of heat to which they are unaccustomed, and which by their race and constitution they are little fitted to endure. I am not now comparing the
natives of tropical countries with those of colder regions. The former
lead for the most part simple lives, and are temperate as regards the use
of alcoholic drink. The dwellers in cold climates, as Sweden and Nor-
way, are notoriously intemperate, and we have seen already how large a
part this habit plays in the causation of insanity. Here, as in so many
other conditions, we shall find that the man who is by inherited nature
of an unstable nerve organization will succumb to the influences of climate, particularly if, in addition to these two causes, he combines the
effect of intemperate habits, or of exhaustion produced by mental anxiety
or bodily illness.

Thus have I sketched certain varieties of insanity differing not in symptoms, but in what has been called their "natural history." Gathered
together, there does not appear to be a large number of these groups. There
are the patients whose disorder is caused by nervous weakness, and the
exhaustion of bodily ill-health, or such causes as pregnancy or lactation.
There is a group whose highly "irritable" and unstable nerve organiza-

[1] Practitioner, March, 1876.

tion is easily upset by such events as parturition, or puberty, or by moral causes such as fright or worry. There is a large number whose insanity is connected in some way or other with alcohol. Febrile diseases give rise to mental disorder in some, gout and rheumatism are connected with it in others, while brain degeneration and mental failure have their origin in such causes as syphilis, epilepsy, blows, tumors, chronic inflammation, and old age. In the last group the symptoms have much in common. There is in all a gradually advancing dementia: mind decays little by little, the memory gets worse and worse till nothing is left, not even the last recollections of the days of youth. But setting aside this degeneration and obliteration of mind, we remark that the other groups are not characterized by constant and uniform mental symptoms. In all we may find depression, in all excitement. Even in the varieties most constantly characterized by depression, as in climacteric insanity, we occasionally meet with excitement, and may see suicidal melancholia in a girl of the age of puberty. Some physicians have thought that, with certain of these pathological conditions, there is to be seen a peculiar assemblage of mental symptoms characterizing that variety of insanity. Dr. Maudsley has described the special symptoms of masturbating insanity,[1] but his description differs entirely from that of Drs. Bell, Schroeder Van der Kolk, and Skae.[2] Dr. Clouston thinks the symptoms of phthisical insanity are marked and peculiar, just as are those of the great majority of the cases of general paralysis or climacteric insanity; but the deepest melancholia may be seen in general paralytics, the greatest exaltation and most exalted delusions may be found in patients who are not paralytic, and hilarious mania may exist in a person at the climacteric period of life. Therefore, while the whole pathological state and condition of the individual may lead us to call the insanity climacteric, or puerperal, or alcoholic, yet his disorder is for all that melancholia or mania, or it may be first the one and then the other. That is, the mental symptoms at any moment are but the expression of the disordered state of his brain and whole nervous system at that moment, and so he may be one day maniacal and the next melancholic, or one day maniacal or melancholic and the next comparatively sane, with that curious daily alternation not unfrequently witnessed. As persons differ, so will their disorders differ, though as they can only differ within certain limits, there will be a strong similarity in the disorders, depending not on a similarity in the causes so much as on a similarity in the individuals themselves. I have often thought that insanity might be classified simply according to age. In the insanity of the young, at the age of puberty, and under twenty years, there are strongly marked features, because the majority of young people agree in attain-

---

[1] Journal of Mental Science, July, 1868.
[2] Bucknill and Tuke's Psycholog. Med., 3rd edition, p. 343.

5

ments and development. Their intellect is not fully developed, so intellectual delusions are absent. They are full of force and bodily energy, and their disorder is noisy and restless mania. But here and there we come across one who is not vigorous, but weak, whose bodily strength has been in some way exhausted, and then we meet with as decided melancholia as though the climacteric had been reached. The same individual becoming insane at various times may indicate by the form his insanity assumes the condition of his general nervous system, and the remedies and treatment appropriate at one time would be utterly inappropriate at another. Climacteric insanity comes on in those who are beginning to decline in years and strength. Hence the symptoms are alike, and are chiefly those of melancholia. But if we look at the enormous number of cases of melancholia which occur in patients not in their climacteric, but of every age from fifteen years and upwards; if we find symptoms similar to those described as phthisical insanity, or masturbating insanity, in patients who are not phthisical and do not masturbate, it is obvious that a like pathological condition exists in these various patients, though it may arise from different causes or assemblages of causes, just as we see the same symptoms arising from idiopathic or traumatic tetanus, and from strychnia poison. I believe that the mental symptoms observable in a patient at a particular time are most valuable indications and not to be neglected; that melancholia or mania implies a state of the nervous centres which must be taken into account in our prognosis and treatment; but that this insane state may be brought about in a variety of ways and by many different pathological causes. Can we, from our survey of these varieties of insanity, form any opinion as to the actual pathological state or disorder of the brain which exists in a recently insane person? We know that such disorder occurs in the young and in the old, in the feeble and in the strong, in those who are broken down by care and anxiety, and in others who have never known an hour's trouble or sorrow; that in some it vanishes in a few days or weeks, while in others, equally promising cases, it remains persistently to the end of a long life. How different is the insanity, and how different is the bodily appearance, development, age, and strength of the patients! And yet it is probable that the disordered condition is at first nearly the same in all.

The life and function of the highest cerebral centres, which are in other words the cells of the superadded convolutions, are disordered by interruption in their normal nutrition. As they are the most specialized and complex organs of all, so it must needs be that they are the most sensitive and most dependent on a proper and duly regulated supply of healthy blood. If there is an absolute defect, an anæmia of the whole bodily system, if there is an impoverished blood or a blood poisoned by deleterious ingredients, the effect must be visible in the functions of the brain. If the vessels are disordered, and the supply impeded thereby, there must be a proportionate result. And if the waste products are not duly removed, harm must follow, and the waste of such an active organ cannot be inconsiderable.

If we consider the brain as we would any other organ of the body, and apply to it the ordinary laws of disease, we find a close analogy between the phenomena we have been reviewing and those of disease generally. If we turn to Mr. Simon's article on Inflammation, in Holmes' System of Surgery, we read "that inflammation may have its starting-point from any *undue* production of textural death, because in every normal textural change *some* textural death is latent, some supersession of the organic material which acts, some going away of what is effete, some place-making for what is new. From any *undue* production of this textural death, inflammation, it seems, may originate, equally whether· the texture have received mechanically or chemically some direct or sudden deathblow of an instantly disorganizing kind, or have been more gradually brought to death by powers distinctly stimulant or depressant. Every determining cause of inflammation acts primarily after one of these two patterns. Like heat, it operates as a textural stimulant, or like cold, it acts as a textural depressant." If we go back to our ætiological varieties we shall find that this applies to insanity no less than to inflammation generally. The textural life of the brain is interfered with on the one side by depressing agencies, on the other by stimulating. By the former the functional activity of the parts is checked, by the latter there must be an *undue* production of textural death, and unless this be met by a proportionate textural renewal, there must be not only a temporary but a permanent loss of function. And when I speak of the morbid anatomy, we shall find that the attempts at textural renewal are to be seen in elements which before death had been engaged in efforts of overgrowth—to use Mr. Simon's words—to repair the textural death. What he says also as to the predisposing causes of inflammation, is closely applicable to our subject. "The influences which may be most distinctly recognized as predisponent to inflammation are chronic inanition, exhaustive disease, and old age; local disease of arteries, local obstruction of veins, and local defects of innervation; the fact of a part having been previously inflamed; an overfed state of body; the over-use of strong drinks; the ingestion by breathing or otherwise of decomposing organic matters; heat of climate, and perhaps other climatic conditions; specific incidents of bodily development, either morbidly occurring in particular persons and families, or normal to the human body at particular periods of life." And he goes on to say that molecular weakness of texture, *i.e.*, weakness in the sense, of readiness to die under the action of stimulants and depressants, is probably the prime local condition which predisposes to inflammation.

In the investigation of these phenomena of textural change, the physiologist, the anatomist, and the chemist must combine. It is certain that chemical changes and chemical mobility must correspond to the other changes which the brain undergoes. But as in inflammation, whether caused by depression or stimulation, there is a local hyperæmia leading to textural change, so there is in insanity a hyperæmia which implies a malnutrition of the brain cells and an impairment of that function which we call mind.

If there is one fact beyond all others certain in the prognosis of insanity, it is that the disorder, if curable, must be treated early; if it exists for a long period, whether under treatment or not, the chances of recovery are small. The reason of this is, that the *undue* production of textural death brought about by the stimulation and hyperæmia of months and years of over-activity of the brain-centres, implies a renewal of textural elements of a lower character, an over-growth of less complex and less highly organized material. In one form of insanity, and in one only, so far as I am aware, recovery may take place after years of mental aberration. This form is melancholia. Here the pathological condition is frequently one where hyperæmia, if it has existed at all, has lasted but for a short time. There has been a checking of the function of the parts rather than a stimulation with undue metamorphosis, and when the functional activity is again raised to its normal level, the patient is sane as before.

LECTURE V.

Morbid Appearances—In Recent Insanity—Meninges—Brain—Vessels—In Chronic Insanity—Vessels—Nerve-Cells—Nerve-Fibres—Neuroglia—Appearances in Brains of the Sane—The Insane Ear—Classification—Various Systems—Points to be Observed.

THE uncertainty of our speculations upon the pathological conditions of insanity is but little diminished by the changes noticeable after death in the brains of the insane. The mind of man is evident during life; after death we see only a mass of brain-substance, a portion of which taken from a philosopher may closely resemble that taken from a madman, an idiot, or even an animal. The signs of insanity do not stare us in the face when with the naked eye we examine the brain of one who has died of the disease. Still less apparent are they in the brain of many persons who live insane to threescore years and ten, and then die of other disorders. That which I have to say concerning the post-mortem appearances and the morbid products observed in such patients is said with the utmost diffidence, because it amounts in truth to very little, and rests on the researches and labors of others, not my own. I claim no credit for the discovery of the various lesions described. Concerning the latter, observers differ; in fact, the method of observation of each one differs from that of others, and according to the method so is the result observed. Yet not a little has been accomplished within the last few years, and it is to be hoped that each generation of observers will, by the aid of the labors of the last, and by further investigations of its own, give more and more accuracy to the phenomena they describe. The rough-and-ready post-mortem examinations of former days are of no avail at present. "We can no longer depend on the pound-weight, the foot-rule, or the naked eye as guides to a knowledge of the condition of the unhealthy brain, and unless the microscope is brought into play the autopsy must be considered imperfect." [1] Moreover, no observer is competent to pronounce as to what is normal and what abnormal unless he has studied to a considerable extent not only disordered but healthy brains. The goal of each microscopist's and pathologist's ambition is to discover the morbid products which correspond to the symptoms seen during life, to be able to say, that after acute delirium we shall find this, after melancholia that lesion, that in the brain of a general paralytic we shall undoubtedly see one kind of morbid product, and in that of a chronic demented person there will be another

---

[1] Bucknill and Tuke's Psycholog. Medicine, 3rd ed., p. 613.

equally special and peculiar. From all this we are as yet far removed.
But it is to be hoped that as laborers increase our stores will be garnered
in greater abundance. The harvest truly is ample. In the year 1882
there were no less than 4,785 deaths in the asylums of England and Wales
alone, but the inquirers to whose work and writings one may turn for
help are few indeed.

There are various modes of dying witnessed in an asylum. There is
the death which occurs perhaps in a week after a violent outbrake of acute
delirium. There is that which terminates a less acute attack of mania or
melancholia after some weeks or even months. There is the lingering
death from brain atrophy, with symptoms of gradually advancing fatuity
and paralysis, not to be confounded with the general progressive paralysis
of the insane. In all these death is due to brain disease as much as in cases
of epilepsy, tumor, or abscess. And there are besides all these the deaths
which are caused by other diseases supervening on recent or chronic insanity, where morbid products are seen in the brain though disorder of the
latter has not terminated life. What is there which we may expect to
find in these various classes of patients?

Now, scanty and imperfect as are the observations of morbid appearances found in acute and chronic cases of insanity, the tale they tell is
consecutive and certain, and is illustrated by the changes and disease of
other organs. I will mention first that which is observed, whether by the
naked eye or by the aid of the microscope, in patients dying of acute and
recent insanity, and then on reviewing the phenomena found in the brains
of patients who have been insane for years, I think it will be found that
the pathology of the one class tallies with that of the other.

A man becomes maniacal ; his mania passes into violent delirium, and
in the delirium he dies perhaps in a week; it is the most rapid mode of
death in a state of insanity. There may have been no bodily complication,
no assignable cause; the case is what is called genuine "idiopathic"
mania. We open the head, and expect to find sufficient to account for
death. All the organs of the body may be quite healthy, and yet that
which is seen in the brain does not seem enough to kill a strong man so
speedily. But in truth the appearances point to the great storm that has
raged there during the last week or fortnight of life—a storm that has
brought about death by its violence, though the traces may by some be
summed up as great vascularity or congestion of the brain and its membranes.

To take first the naked-eye appearances, we find signs of violent disturbances manifested in the meninges, the vehicles of the blood-supply of
the brain itself. "Sinuses and veins of pia mater full of blood—considerable serous effusions in subarachnoid space." This was in a female
who died of mania in fourteen days. "Pia mater much congested,
arachnoid slightly opalescent." This also was in a female; and where

you find the arachnoid opaque or the pia mater thickened, the brain substance underneath will generally be diseased. Opacity of the arachnoid is common: you will find it often in patients not insane; but in the cases of which I am speaking it denotes excessive action tending to meningitis. So violent may it have been, that not unfrequently we find effusion of blood from the rupture of small vessels between the membranes. And much serous effusion is commonly found, which probably precedes death only a short time, and accompanies the coma in which so many of these patients sink. In others who have had previous attacks, or have been excitable, semi-insane, and constantly liable to disturbance of the cerebral circulation, we find great thickening and eburnation of the cranial bones, and extensive adhesion of the dura mater. These appearances point to a low inflammatory and degenerative change which has been going on for some time, and you will frequently see them in patients not insane. When we remove the pia mater, and this in a recent case can generally be accomplished without difficulty, we find the same traces of violent action. The brain is not uniform and healthy in tint. In places it is discolored from pink to purple, and is often softened. Its structure, cells, tubes, and connecting tissue are obviously altered and damaged by the hyperæmia which has existed. There has been great vascularity: on slicing it we see many bleeding puncta, and the vessels have manifestly been dilated.

As might be expected, there are to be found in the brains of the insane extravasations of all kinds and degrees, especially in those who have died after a brief period of violent excitement. Blood-cysts are found in the cavity of the arachnoid, and stains and spots of extravasated blood are seen on the surface of the convolutions and in the substance of the brain; capillary apoplexies, extravasated blood-corpuscles, and yellowish or reddened brain matter, are appearances well known to those who have examined such patients after death.

The microscope confirms the condition which the parts present to the naked eye, and shows that often the capillary vessels are distended with blood, and blocked by it. We also may see minute apoplexies in the cerebral substance, which bear witness to the increased vascular action, and the obstruction thereto.

In a thesis read before the University of Oxford in 1867, I propounded the theory that the cause of delirium and death in acute and rapid cases of insanity, delirium tremens, and the like, is stasis of the capillary circulation, the result of pressure or inflammatory changes in the blood. In a paper already referred to (page 38), published in the British Medical Journal" of 23rd January, 1869, Dr. Charlton Bastian narrates certain post-mortem appearances found by him in the brain of an intemperate man, who died of erysipelas of the scalp, following a fall on the head, with violent delirium. These, I think, strongly confirm the conjectures I put forth in 1867—conjectures based chiefly on Mr. Lister's Experiments and

Observations on Inflammation, published in the Philosophical Transactions, 1859. "To the naked eye all parts of the brain showed a decidedly abnormal amount of vascularity, and this was particularly evident on
the venous side of the circulation. The veins of the surface of the lateral
ventricles, and those of the choroid plexuses, were all notably distended
with dark blood, though there was no obstruction in the venæ magnæ
Galeni or in the straight sinus. The lateral ventricles, however, contained
only a small quantity of pale serum. The 'red points' were very abundant wherever sections were made through the white substance of the
hemispheres. On microsopical examination of some minute vessels and
capillaries taken from the gray matter of the convolutions, every speci·
men looked at showed minute embolic masses in various parts of the
course of the small arteries and capillaries, of a most unmistakable nature,
though apparently of recent origin. Distinct masses, of irregular shape
and size, could be seen, made up of an agglomeration of white blood-corpuscles. In some cases the masses were small, and formed by the union of
three or four white corpuscles only; whilst in others large irregularly-
shaped aggregations could be seen within some of the bigger vessels,
which may have been made up by the mutual adhesion of two or three
hundred of such corpuscles. The largest mass actually measured was $\frac{1}{100}$
in. long by $\frac{1}{250}$ in. broad. In other parts large rounded bodies were seen,
whose nature was not at first sight obvious, though after a little careful
examination I became convinced that these had, in all probability, been
formed by the complete fusion of corpuscles into a single mass, which had
afterward undergone more or less of a granular degeneration."

In commenting on this case Dr. Bastian mentions others: one, a case
of rheumatic fever, with delirium; another, of double pneumonia, progressing favorably, when maniacal delirium set in, lasted about thirty-six
hours, and then gradually subsided. The latter was, in fact, an instance
of mania arising in the course of an acute disease, akin to that already
mentioned, which Dr. Weber has termed the "delirium of collapse." Dr.
Bastian was strongly impressed at the time with the probability that the
delirium, in the latter case at least, was due to some accidental plugging
of minute vessels of the brain by means of aggregated white blood-corpuscles, and to the consequent total disturbance in the incidence of blood-
pressure, and in the condition of nutritive supply in the convolutional
gray matter of the brain. "It was therefore with a feeling of considerable
interest that I proceeded to examine the brain in this case of erysipelas of
the scalp, which has been associated during life with delirium and stupor."

Dr. Clouston has recorded[1] another product, the result of stasis. In
the veins of the pia mater of a female who died after a short and acute attack of mania, he noticed, "small white pearly-looking bodies, that looked

---

[1] Journ. Mental Science, xx. 595.

at first like limited white thickening of the venous coats, but were found
to be masses of organized fibrinous material. In many places these were
attached to thin strings of the usual post-mortem clot, and the difference
between the two structures was very great. A microscopic examination
showed this difference still better. Instead of the ordinary white blood-
corpuscles caught up in the meshes of innumerable fine fibres of white
post-mortem blood-clot, those masses consisted of bodies like the white
blood-corpuscles, but much larger, with distinct nuclei and nucleoli, and
instead of the fine linear fibres, there were fusiform cells cohering strongly,
among which those bodies lay in regular parallel rows." The same
appearances in a less degree were met with in a general paralytic who died
comatose in a congestive attack. Dr. Clouston asks if it is possible that
in such cases the vaso-motor paralysis and blood stasis that form so essential a feature in the typhoid condition of acute insanity and the congestive
attacks of general paralysis, had gone on to a still further stage, when the
white blood-corpuscles began to adhere to the inside of the walls of the vessels, gradually accumulating, and becoming organized into the masses
described.

This extreme condition of stasis, it is to be remarked, is seen after
death, and, according to my view, it is the immediate cause of the effusion
and coma which lead to death; but it is probable that a less degree may
exist in patients who do not die, in whom it may subside after a long or
short time, leaving more or less traces according to its presistence. The
condition at first is one of great hyperæmic activity; the condition of
stasis comes later. The delirium so often witnessed is due, in my opinion, to the hyperæmia, not to the stasis. For we frequently see the most
violent delirium arise and pass away again in a very short space of time,
even in a few hours. It is hardly to be imagined that a condition of stasis
would subside so rapidly. A patient may rave unceasingly for some hours,
may then fall asleep either from exhaustion, from the influence of some
drug, or from the derivative effect of a wet pack; he sleeps naturally, not
in a state of coma, and wakes sane. This hyperæmia may be ascribed to
instability of nerve-cells, whereby the controlling influence of the vaso-
motor centres is overcome, or it may be caused by irritation propagated
from a distant organ. It is, however, the cause of the delirium, and may
subside without stasis, or go on to a greater or less degree of the latter.
If the obstruction is great, death ensues; if this does not take place, the
products of the stasis may be gradually removed, or may become more or
less organized and permanent.

Dr. J. Batty Tuke has recorded the fact [1] that it is on the superior surface of the convolutions that we find the evidence of hyperæmia and its
consequences. " On removing the calvarium and dura mater we almost

---

[1] Morisonian Lectures, p. 55.

invariably find the arachnoid milky, opalescent, thickened in a greater or
less degree, and the pia mater injected most markedly over the area of
convolution exposed; for when we remove the brain from the skull, the
portions below the line made by the saw are comparatively healthy to all
outward appearance.   We find that the focus of disease is at the extreme
vertex, at the point where the fissures of Rolando meet, extending backward and forward along the line of the longitudinal fissure, and that the
manifestations of morbidity fade away laterally, and are gradually lost above
the level of the horizontal limb of the fissure of Sylvius, and that, except
under specific conditions, the base of the brain is to all appearance as
healthy in the chronic maniac as in the most mentally stable subject. And
what is true of the naked eye appearances is true also of the microscopic,
for, with the exception of the cerebellum, microscopic lesions are more
pronounced in parts subjacent to thickened membranes than elsewhere."
Dr. Tuke was formerly of opinion that the secondary lesions are caused
by the obstruction of the cerebral lymphatic vessels, owing to the artery
occupying in its hyperæmic condition the whole of the cylinder in which
both the arteries and lymphatics are contained.   Thus the effete and
superfluous waste products are not carried off, "the brain substance and
membranes become œdematous, the pia mater is displaced by the serum,
which, oozing through the tissues, raises it from the convolutions within
and without the sac of the arachnoid, fluid becomes arrested, and the
ventricles of the organ become dropsical."   But he now admits that the
evidence concerning perivascular lymph spaces is imperfect.[1]

When we examine the brains of patients who have died after a long
period of insanity, and seek for traces of their malady, our labors are better
rewarded.   We find degeneration of all the structures which join in the
working of the brain, and an increase of the lowest tissues, together with
a growth of adventitious and abnormal products, which supplant the
healthy structure, and subvert its function. We find, in fact, the same
evidences which testify to degeneration and decay of the other organs of
the body; and, so far as we can observe, these changes appear to be the
ultimate end and result of the acute disease which was first manifested in
symptoms which exist in the early stages of almost every variety of insanity.

I have said that the appearances found in those who die of acute
insanity point to a violent disturbance of the cerebral circulation.   The
changes we meet with in the chronic insane point to the results of this
disturbance, to damage of the vessels, to extravasation of their contents,
to a consequent interference with their life and function, and nutrition of
the tissues they supply.   Degeneration and atrophy have wasted the cells
and tubes, while more lowly organized structures—bone, connective tissue,

---

[1] Bucknill and Tuke's Psycholog. Medicine, 4th ed., 1879.

colloid and amyloid corpuscles—flourish and abound, as they ever do where the normal tissues and the normal nutrition are at fault.

What, then, are the changes observed in the blood-vessels? Various pathologists have directed their attention to these, and have recorded not a few deviations from the healthy state.

Dr. Batty Tuke, in his Lectures enumerates five morbid conditions of the vessels in the order of their incidence. 1. Simple dilatation. 2. Exudate deposits. 3. Opacity and thickening of their hyaline membrane. 4. Dilatation of the retaining canal. 5. Hypertrophy of the muscular coats. With the observations of Dr. Batty Tuke I shall incorporate those of Dr. Herbert Major, as given by him in the second and third volumes of the West-Riding Asylum Reports.

Dilatation may be observed in many patients, even where there is no thickening of the walls. It may be found in both recent and chronic cases, and is a necessary sequence of prolonged hyperæmic action. Dr. Major remarks that it is the condition most commonly present in general paralysis, and that it also characterizes the vessels seen in chronic brain-wasting.

Morbid deposits are seen adhering to the walls of the vessels. These are hæmatoidin, which is seen in the form of large amorphous masses of a dirty yellow material. This substance, Dr. Tuke says, he has generally found in small quantities on the vessels of most sane subjects, but in a manifestly less degree than in the insane; that in the non-lunatic subject it appears to depend to some extent on age, and more especially on the nature of the disease which has caused death. In fever cases, in which the insanity of delirium and coma had supervened, it is pretty well marked. Hæmatoidin may also be found in the form of crystals, especially at the bifurcations of vessels. Another morbid deposit, noticed both by Dr. B. Tuke and Dr. Major, is a very fine molecular material found in the smallest capillaries, homogeneous in structure, sometimes of a slightly pale color; more frequently colorless, in many ways suggesting a fatty nature, though the latter is not borne out by chemical tests. Of this Dr. Major says that it is highly refractive, and unaffected by carmine. Syphilomatous deposits are found around the walls of the cerebral arterioles, which, being converted into fibrous tissue, cause the transverse sections to appear greatly hypertrophied, and this may go on to almost complete occlusion of the vessels. Dr. Clouston gives [1] a section of the brain of a man who had labored under syphilitic insanity, with slow arteritis affecting the vessels supplying the anterior and part of the middle lobes of one hemisphere. Absorption of nearly all the white matter of the centre of these lobes had taken place, the gray matter of the convolutions being left intact, and forming a wall round the fluid mass. The greater vascularity and vitality

---

[1] Clinical Lectures on Mental Diseases, Plate V., p. 42.

of the gray matter as compared with the white are thus illustrated, and also the different sources of the blood supply of each.

The hyaline membrane, which is described as a prolongation inward of the pia mater, forming the wall of the lymphatic spaces which exist around the vessels, becomes opaque and fibroid instead of being transparent and hyaline. It may serve for the deposit of fatty granules and hæmatoidin, or may itself be puckered and thickened, and in this state may be traced with unusual distinctness.

The canals in which the vessels are contained, the so-called perivascular canals, are sometimes found dilated. "In chronic cases," says Dr. B. Tuke, "more especially in epileptics and general paralytics, the transversely cut vessel is seen surrounded by a clearing of unoccupied space, with radiating trabeculæ of connective tissue extending between the hyaline membrane and the cerebral substance. In extreme cases the cylinder has been found from four to six times the calibre of the contained vessel." These clear spaces are caused by the dilatation of the congested vessel as it produces the expansion of the surrounding parts. Dr. Major, speaking of a case of chronic brain-wasting, says that perivascular canals of considerable size are among the conditions presented, but that it is much to be regretted that by none of the means at our disposal can this point be studied while the brain is in a fresh state, before hardening re-agents have introduced a source of fallacy.

A thickening of the coats of the arteries was described by Dr. Sankey some years ago.[1] It may be due to hypertrophy of the muscular coat, the result of efforts to overcome obstructions in the ultimate capillaries. Dr. B. Tuke says that the circular fibres are those affected, that it is only present in old-standing cases, and is best marked in those who have been the subjects of disease implying hyperæmia, e.g., general paralysis and epilepsy. Dr. Major, in his examination of four patients who died of general paralysis, did not find thickening of the walls a marked feature. Another change especially noticed by Dr. Major in the brain of one who died of chronic atrophy is enormous proliferation of nuclei. "In health these bodies are few in number, oval in form, and not deeply stained by carmine; but here, in addition to these, the whole vessel is seen to be crowded with small nuclear bodies, closely but irregularly scattered over the vessel. They are exceedingly minute, round in form, and deeply stained—quite different, therefore, from the larger oblong nuclei which normally exist. It is in the deep rather than in the superficial part of the cortical layer that the disorganization of the vessels is most distinct. I have frequently observed a small vessel in which at the commencement of its passage downward I could detect little or no alteration from the normal condition, soon alter its character, and assume more and more conditions undoubtedly morbid."[2]

---

[1] Journal of Mental Science, January, 1869.
[2] West-Riding Asylum Reports, iii. p. 99.

A looped, tortuous, and varicose condition of the vessels has been noticed by many observers, especially in cases of general paralysis. It may be found where there is no hypertrophy, and indicates that the vessel has undergone great congestion and overstraining, and has remained distorted, twisted, and often dilated, after the excessive action has subsided.

The nerve-cells may be met with in a condition of *atrophy*. Such a state was found by Dr. Major in cases of general paralysis. "The body of the cell is usually shrunk, the wall being closely applied to the nucleus, which is of large size, more or less round, and but slightly stained. In the more advanced degenerative stages no cell wall is observed, the nucleus alone, or surrounded in part by molecular *débris*, being all that is left. There exists, also, another condition which has been described by some as 'inflated.' The size of the cell is not much altered, but it exhibits an irregular inflated appearance, this not being due to excess of pigment or fatty degeneration. It would seem probable that this is the earlier stage, which is succeeded by the shrinking and atrophy before described." In another case of general paralysis Dr. Major observed quite an opposite condition, which we may call *hypertrophy*. "It consists in the presence of nerve-cells of immense size, situated about midway in the depth of the cortical layer. Standing out like giants among the other cells, they present an appearance which cannot fail instantly to attract attention. In shape they vary considerably, the majority approaching more or less the pyramidal form, but often being very irregular in their contour. Their branches are both large and numerous. I have counted as many as eight or ten proceeding from a single cell. The nucleus is large, but not proportional to the size of the containing body, the density of which masks the contents. In number, these large cells, when compared with the others, are relatively few—usually not more than a dozen or so being seen in a specimen, but sometimes fewer, and occasionally none at all. I find them most numerous in the parietal region, then in the occipital, and rarest in the frontal lobes. As to their pathological significance, I am unable at present to form an opinion; the point is one for further inquiry."[1] Drs. Batty Tuke and Rutherford also pointed out, in a paper read before the British Association in 1870, that in cases of senile insanity, where the cells of the outer layers had undergone atrophy, the cells of the two internal layers appeared to be much increased in size.

Degeneration of various kinds are found in the nerve-cells. There may be a want of distinctness in their outline and branches; their prolongations are torn and few in number, and the cell rounded and deformed. The nucleus partakes of the general change. This body, as seen in the large pyramidal healthy cell, follows more or less closely the form of the cell in which it lies; here, on the contrary, with very few

---

[1] West-Riding Asylum Reports, iii. p. 109.

exceptions, it is round or oval. But though altered in form, it is invariably of large size, sometimes completely filling the enclosing cell, or even causing a projection from it (Major). Fatty and pigmentary degeneration is not uncommon in general paralysis, and also in senile atrophy. In a case of the latter Dr. Major says: " The degeneration is most observable in the large pyramidal cells, in which, in its earlier stages, it gives rise to a slight bulging of the cell, the nucleus, however, preserving its normal position and characters. 'In more advanced stages the bulging and deformity are found to be greater, the branches are no longer visible, and the nucleus is atrophied and pushed out of position. In the last stages the whole cell is reduced to a simple mass of granules, no branches, cell-wall, or nucleus remaining.'' [1]

In nerve-fibres, says Dr. B. Tuke, differences will be found as to the power they possess of resisting pressure; in some cases they retain their normal condition under the covering glass, in others they become readily ampullated. [2] Drs. Tuke and Rutherford found in a case of chronic dementia, complicated with chorea, that the nerve-fibres of the anterior and posterior roots of the spinal nerve had undergone a pigmentary degeneration similar to that noticed in the nerve-cells. They may also present fusiform or oval swellings, presenting the appearance of the so-called amyloid bodies. [3]

The morbid changes, however, are not confined to the blood-vessels, the nerve-cells, or nerve-fibres. The neuroglia or connective tissue, which binds together all the foregoing, is the seat of various important lesions. "When the earliest symptoms present a gradually increasing apathy, torpor, and stupidity, quickly followed by general bodily symptoms, loss of expression, heaviness, and immobility of the facial muscles, a slobbering mouth, speech thick and slow, impairment of reflex action, considerable anæsthesia, hæmatoma auris, and the passage of fæces and urine apparently involuntarily,—we may look to the neuroglia as the seat of the primary lesion. It is a condition of true primary dementia, which may perhaps be best described as in every way lacking acute symptoms, but accompanied by rapid decadence of the general system. It is closely allied to what is often spoken of vaguely as organic brain disease. In all such cases we find a sclerosis, and generally a colloid degeneration of the neuroglia.'' [4]

There are at least two forms of sclerosis. There is the *disseminated sclerosis* or *gray degeneration* thus described by Drs. Tuke and Rutherford: [5]—" We have observed this lesion only in the white matter of the brains examined by us. As to the nature of the morbid change, our

[1] *Op. cit.*, p. 106.
[2] Bucknill and Tuke's Psychol. Med., 3rd Ed., p. 631.
[3] Dr. Batty Tuke. Quain's Dict. of Medicine, p. 720.
[4] B. Tuke's Morisonian Lects., p. 64.
[5] Ed. Med. Journal, October 1869.

observations lead us to agree with Rokitansky in regarding it as primarily a modification of the connective tissue. In the spinal cord, medulla oblongata, and pons, it appears to us that the connective tissue or neuroglia is a nucleated, transparent, homogeneous, non-fibrillated matrix, representing the fibrillated connective tissue of the spinal cord. Owing to the extreme fineness of the nerve-tubes of the white matter of the brain, as compared with those of the spinal cord, an inquiry into the diseased conditions of the white matter of the former is much more difficult than in the case of the latter; but a careful inspection of numerous finely prepared sections, by means of a magnifying power of 800 diameters linear (Hartnack's immersion lens No. 10, eye-piece No. 3), has resulted in the demonstration that fibrillation and increase of the neurogliar matrix, together with proliferation of its nuclei, are the essential changes in gray degeneration, as Rokitansky has already pointed out. Sometimes proliferation of the nuclei precedes fibrillation of the matrix, at other times the converse holds good. Sometimes there is a marked proliferation, in the nuclei of the capillary walls in the diseased tracts, but we have not been able to confirm Rindfleisch's observation that the diseased process invariably starts from these. Indeed, our specimens show that the morbid change just as often begins at a distance from, as in the immediate neighborhood of the vessels. Regarding the fate of the nerve-tubes in the diseased tracts in the white matter of the brain, our observations, owing to the fineness of these elements, scarcely enable us to speak with confidence. They appear, however, to undergo atrophy, and this need scarcely be doubted, seeing that they certainly do so under similar conditions in the spinal cord."

There is also another form first described by Drs. Tuke and Rutherford in 1868, and called by them *miliary* sclerosis. The following account of it is given by Dr. B. Tuke, in his Morisonian Lectures (p. 66): "Miliary sclerosis is a disease of the nuclei of the neuroglia, and its progress is marked by three stages; in the first, a nucleus becomes enlarged and throws out a homogeneous plasm of a milky color, and apparently of a highly viscid consistence, for the long axis of the spot is almost always in the direction of the fibres, which are displaced by its presence instead of being involved in it; thus indicating that its density is considerably greater than that of the cerebral matrix. In the centre of these semi-opaque spots a cell-like body is generally discoverable, possessing a nucleus; this is the original dilated nucleus of neuroglia. In the largest patches more than one cell can be seen; whether these arise from division of the first nucleus involved, or from the original implication of more than one nucleus, has not been determined; but from the fact that multiple cells are seen only in the largest spots, it is most likely that the latter hypothesis is correct. Occasionally several neighboring nuclei become diseased simultaneously, coalesce, and form a multilocular patch of considerable extent;

the largest spot which has yet been figured is the one-fortieth of an inch in its longest diameter. During the second stage of development the morbid plasm becomes distinctly molecular in character and permeated by fibrils. It is probable that at this period a further displacement of the contiguous tissues take place, as a degree of induration of the compressed fibres and blood-vessels which curve round the diseased tract is indicated by the increased amount of coloring material which they then absorb. At this stage the morbid tracts present the following appearances. As a rule the spots are unilocular, occasionally bilocular, and in rare instances multilocular; but whatever their condition in this respect is, they possess the same internal characteristics. A thin section prepared in chromic acid viewed by the naked eye shows a number of opaque spots irregularly distributed over the surface of the white matter; they are best seen in a tinted section, as they are not colorable by carmine. When magnified by a low power they have a somewhat luminous pearly lustre, and when magnified 250 and 800 diameters linear they are seen to consist of molecular material, with a stroma of exceedingly delicate colorless fibrils. They possess a well-defined outline, and the neighboring nerve-fibres and blood-vessels are pushed aside and curve round them. In well-advanced cases the plasm seems denser at the circumference of the spots than at their centre, and a degree of absorption of the contiguous nerve-fibres is evident; this solution of continuity is only noticeable at the point where lateral expansion is greatest. The spots are generally colorless, but in some instances they are of a yellowish green tint, which may be attributable to chromic acid. They vary much in size; multilocular patches are $\frac{1}{40}$ of an inch to $\frac{1}{100}$ of an inch in diameter, the unilocular vary from $\frac{1}{200}$ to $\frac{1}{800}$ of an inch. As many as eleven locules have been noticed in one patch, separated one from the other by fine trabeculæ of tissue. The nerve tissue of a section containing spots of miliary sclerosis in the second stage, when removed from spirit and allowed to dry, shrinks from the diseased patches and leaves them elevated and distinctly separated from it; so much so, that they can be picked out with the point of a knife. It is still doubtful whether this lesion can be detected in recent specimens. In two cases in which, by the chromic acid process, miliary sclerosis in the second stage was demonstrated, the recent white matter, when squeezed out under a covering-glass, exhibited spaces containing a clear maerial in which some rounded nucleated cells were visible. It is, however, only by prepared sections that its presence can be definitely ascertained.

" In the third stage of miliary sclerosis the molecular matter becomes more opaque and contracts on itself, the boundaries become puckered and irregular in outline, and the material often falls out of the section, leaving ragged holes. These holes cannot be mistaken for empty perivascular canals, which, whatever their size, are smoothly rounded or oval.

When in this condition the morbid products of miliary sclerosis are distinctly gritty, effervesce immediately on being subjected to the action of nitric acid, which produces no such appearances as are evolved by its application in the second stage." [1]

These holes in the brain have been observed by various microscopists, and different opinions have been held concerning them. Dr. Lockhart Clarke describes them as seen by him in the white substance of the convolutions and optic thalamus of a general paralytic patient. [2] There were numerous cavities of a round, oval, fusiform, crescentic or somewhat cylindrical shape.

"For the most part they were empty, had perfectly smooth walls, without any lining membrane, and seemed as if they had been sharply cut out of the tissue. A few, however, were found to contain what appeared to be the remains or *débris* of blood-vessels, mixed with a few granules of hæmatoidin. One or two were found to communicate with the surface of the convolutions through the natural fissures between them and to contain a perfect blood-vessel, with its branches. On removing the blood-vessel, the wall of the broad but shallow cavity was seen to be perforated by a multitude of minute orifices, through which the finer branches of the vessel had passed. These latter circumstances, together with a comparison of the shape and course of some of the natural fissures transmitting blood-vessels from the surface, render it almost certain that at least the greater number of these cavities were perivascular spaces or canals, which originally contained blood-vessels, surrounded by their peculiar sheaths, and which subsequently became empty by the destruction and absorption of those vessels."

Doubtless you are aware that Dr. Dickinson has discovered similar holes in the brain and spinal cord of diabetic patients. The preparations in his possession appear to throw great light on the mode in which some of these holes are produced, and illustrate not a little the pathology of insanity.

In more than one a blood-vessel is seen in a state of great congestion, so great that the blood-corpuscles have apparently transuded through the walls of the vessel, and invaded the tissue in which it is embedded. If the patient does not die in this stage, the acute congestive stage of insanity, we may suppose that the exuded blood becomes absorbed, and then a hole is left behind.

From the observations of these various observers one may gather that the holes are formed in more than one way; that some are the result of

---

[1] Some doubt has been raised as to the nature of these so-called miliary spots, *vid. Diffused Cerebral Sclerosis*, by Dr. T. W. M'Dowall, Journal of Mental Science, xxv. 490. Dr. Plaxton thinks this appearance is due to post-mortem change or to the mode of preparing specimens for the microscope. Journ. Ment. Sc., xxix. 27.   [2] Journal of Mental Science, xv. 499.

the miliary sclerosis described by Dr. Batty Tuke, while others are caused by the dilatation of the vessels and their subsequent absorption or destruction.

There is yet another morbid condition of the nuclei of the neuroglia called by Dr. B. Tuke " colloid " degeneration. It is, he says, one of the most important degenerations of the neuroglia, not only on account of its degree, but of the frequency of its presence. This and miliary sclerosis are the lesions spoken of as ever present in true primary dementia. He thus describes it;[1] " In its earliest stages this abnormal condition shows itself in circumscribed semi-translucent spots, varying in size from the $\frac{1}{4000}$ to the $\frac{1}{2000}$ of an inch in diameter; they have well-defined irregular edges, and their contents are molecular in appearance. In fresh specimens, however, this molecular appearance is not observable, and colloid bodies appear as round or oval in form, having a distinct wall containing a clear, homogeneous, transparent, colorless plasm, and occasionally showing a small nucleus but no nucleolus. Colloid bodies are not colorable by carmine. They appear first in the white matter immediately contiguous to the cortical substance, but as the disease advances they become diffused outward and inward. In extreme cases the appearance of sections containing them may best be compared to a slice of sago-pudding, for they exist in such large numbers as almost completely to fill the field of the microscope, separated slightly from each other by a fine granular material. Although readily recognizable when set up in Canada balsam or turpentine, the characteristics of colloid degeneration are best brought out by glycerine. I feel stongly inclined to regard this as a form of degeneration of the nuclei of the neuroglia; it is first seen and best marked in the white matter, but in certain specimens in which it occurs in the gray matter, cells have been seen which are undergoing or have undergone changes in many respects resembling those noticed in the nuclei of the neuroglia. It is not associated with proliferation of nuclei. Careful study of a large number of specimens leads to the conclusion that the nuclei are the original seats of the disease, for in all cases in which colloid degeneration shows itself, they are to be seen more or less departing from normality; in fact, it may be safely stated that they are always unhealthy, and appear to merge gradually into the colloid condition."

Having said thus much upon the morbid appearances revealed in the insane brain, I would draw your attention for a moment to peculiarities observed by competent microscopists in the brain of the sane. Dr. Major, whom I have already frequently quoted, examined the brain of a man aged 47, who died from the effects of a compound fracture of the leg, and was a perfectly healthy and intelligent man. The brain presented no features of abnormality; there was no wasting of the convolutions, no thickening of membranes, and no appearance of vascular disease. Never-

---

[1] Morisonian Lectures, p. 68.

theless, on examination, it was found that in many instances the pyramidal nerve-cells of the cortex were morbidly affected. The abnormality consisted in the accumulation of yellow granules in the interior of the affected cells, in no case going to the extent of producing destruction of the corpuscle, but at most a slight bulging and alteration in its form. The nuclei and nucleoli appeared to be unaffected. The condition above described could only be seen among the large pyramidal cells of the deeper layers of the gray matter, and not in the smaller superficial bodies. At the posterior extremity of the occipital lobes, the nerve-cell, with rare exceptions, presented no evidence of any morbid process. These appearances indicate, according to Dr. Major, that *some* of the nerve-corpuscles of the cortex may be morbidly affected without any appreciable mental impairment.[1] Dr. Obersteiner, in a paper read before the Medical Society of Vienna,[2] says that the vessels of the brain in healthy persons are rarely normal; that in all brains he has found fatty granules more or less, especially in the adventitia of the smaller vessels. These he regards as the remains of fatty granular cells constantly to be found in the brain, especially on the vessels in early life. He also found in the lymph-sheaths of the cerebral vessels yellow and brown pigment. Fatty and calcareous changes were not unfrequently seen in the muscular coats of the vessels, and often the middle coat was in a state of degeneration.

There is one morbid appearance which may be mentioned here. It is called the " insane ear," or *hæmatoma auris*, and is a tumor or swelling caused by an effusion of blood under the perichondrium, commencing generally in the helix. It may come on gradually, or even quite suddenly in a night, may attain a large size, and appear at the point of bursting. I have seen a slight oozing, but it seldom bursts. Slowly it shrinks and becomes absorbed, leaving the ear withered and shrivelled. To prevent this shriveling, Dr. Hearder recommends[3] that the inner surface of the pinna be painted with a blistering fluid. In six consecutive cases so treated, he tells us that the diseased action was arrested by this method, and that, although the results of the morbid agency were traceable in permanent thickening of the parts, yet that there was none of the shriveling and obtrusive distortion which ensues when the affection has run its course uncontrolled. It occurs more frequently in the left than in the right ear, and is generally seen in chronic patients, generally paralytics, and those suffering from chronic mania and dementia. It is about four times as frequent in men as in women. Controversy exists as to its origin, some writers asserting that it is the result of injury, others denying it. In my own experience I have met with " insane ears " which were undoubtedly the result of violence. In one case the patient fell out of bed, striking

---

the ear. In another, it followed the holding of the head, necessitated by refusal of food and medicine, and consequent forcible feeding. In one of the cases mentioned by Dr. Hearder, it followed a fall. Dr. Farquharson, lately the medical attendant of Rugby School, says [1] that he saw several cases of this affection in boys, the result of violence received at football. The characteristic swelling was followed by the well-known shriveled and puckered appearance. I am far from thinking, however, that it is always, or even generally, due to violence. Probably it is the result of great vascular excitement, and of a weakness of the vessels of this part. The greater frequency of its occurrence in the left ear is supposed, by Dr. Lennox Browne, [2] to be due to the nearer position of the left common carotid to the heart, and the more direct and less impeded arterial supply to the left than the right side of the head. There is another question, Is it peculiar to the insane, or is it found in the sane? Dr. Lennox Browne, though he inclines to the former opinion, mentions that he has seen it in a sane though excitable man,—a city missionary. And, besides the instance adduced by Dr. Farquharson, various other observers, as Fischer and Gruber, concur in the view that it is not peculiar to the insane, though far more commonly seen in them.

CLASSIFICATIONS OF INSANITY.

Great differences of opinion have existed for many years as to the classification of insanity and insane patients, and various systems and principles have been laid down. The old writers divided the disorder into Mania and Melancholia: Arnold (1782) into Ideal and Notional insanity; Pinel makes four divisions, Mania, Melancholia, Dementia, and Idiocy; Esquirol adds another, Monomania. Guislain's classification runs thus: 1. *Phrenalgia,* or Melancholy; 2. *Phrenoplexia,* or Ecstasy; 3. *Hyperphrenia,* or Mania; 4. *Paraphrenia,* or Folly; *Ideophrenia,* or delirium; 6. *Aphrenia,* or dementia. Dr. Conolly speaks of Mania, Melancholia, and Dementia; Dr. Prichard of Moral and Intellectual Insanity; Dr. Noble's division gives Emotional, Notional, and Intelligential disorder. Then, again, Mr. Morel rejects all these divisions as artificial, and based merely upon the symptoms of the disease, or on supposed divisions of the human mind, and he proposes to divide it strictly according to its ætiology into six groups: 1. Cases of Hereditary Transmission; 2. Toxic Insanity; 3. Those resulting from the transformation of other neuroses; 4. Idiopathic Insanity; 5. Sympathetic Insanity; 6. Dementia. I will not stop to comment on the imperfections of this classification; suffice it to say, that it is not even what it professes to be—ætiological.

One of the latest writers, so far as I am aware, who divide insanity

---

according to the mental symptoms is Dr. Maudsley,[1] whose classification is as follows:—

Affective Insanity or Insanity without Delusion.	{ a. Instinctive. b. Moral.

Ideational Insanity.	Melancholia.	{ Acute. Chronic.
	Mania.	{ Acute. Chronic.
	Monomania.	
	Dementia.	{ Acute. Chronic.

Amentia.	{ Imbecility. Idiocy.	{ Moral and Intellectual.

At the International Congress of Alienists, held in Paris in 1867, there was adopted a system of statistics prepared by a committee appointed for that purpose, and the following classification was put forward as denoting the typical forms of the disease:

I. *Simple Insanity*, embracing the different varieties of mania, melancholia, and monomania, circular insanity and mixed insanity, delusion of persecution, moral insanity, and the dementia following these different forms of insanity.

II. *Epileptic Insanity*, or insanity with epilepsy, whether the convulsive affection has preceded the insanity, and has seemed to have been the cause; or whether, on the contrary, it has appeared, during the course of the mental disease, only as a symptom or a complication.

III. *Paralytic Insanity.*—The commission regards the disease called general paralysis of the insane as a distinct morbid entity, and not at all as a complication, or termination of insanity. It proposes, then, to comprehend under the name of paralytic insane all the insane who show in any degree whatever the characteristic symptoms of this disease.

IV. *Senile Dementia*, which we should define as the slow and progressive enfeeblement of the intellectual and moral faculties, consequent upon old age.

V. *Organic Dementia*, a term by which the commission means to designate a disease which is neither the dementia consequent upon insanity or epilepsy, nor paralytic dementia, nor senile dementia, but that which is consequent upon organic lesion of the brain, nearly always local, and which presents, as an almost constant symptom, hemiplegic occurrences more or less prolonged.

---

[1] Pathology of Mind, p. 132.

VI. *Idiocy*, characterized by the absence or arrest of development of the intellectual and moral faculties. Imbecility and weakness of mind constitute, hereof, two degrees or varieties.

VII. *Cretinism*, characterized by a lesion of the intellectual faculties, more or less analogous to that observed in idiocy, but with which is uniformly associated a characteristic vicious conformation of the body, an arrest of the development of the entirety of the organism. Outside of these typical forms there are such as—

1. Delirium tremens.
2. Delirium of acute diseases: traumatic delirium.
3. Simple epilepsy.

A committee of the Medico-Psychological Association appointed, in 1869, for the purpose of taking into consideration certain questions relating to the uniform recording of cases of insanity, and the medical treatment of insanity, recommends that cases should be classified according to two methods—1. That depending on the bodily causes and natural history of the disease, as proposed by Dr. Skae; 2. That proposed by the International Congress of Alienists, as given above.

The Table on the opposite page is put forth by this committee as a specimen of the way in which cases may be arranged.

Dr. Daniel H. Tuke thus arranges the forms of insanity:—

DISORDERS OF THE MIND INVOLVING,

Intellectual.

CLASS I.—*The Intellect or the Ideas.*

Order 1. Development incomplete. { Idiocy. / Imbecility

Order 2. Invasion of disease after development. { Dementia. / Delusional Insanity. / Monomania. / Mania.

Emotional and Volitional.

CLASS II.—*The Feelings and Moral Sentiments.*

Order 1. Development incomplete. } Moral Imbecility.

Order 2. Invasion of disease after development. { Moral Insanity. / Melancholia. / Exaltation.

CLASS III.—*The Propensities, Instincts, or Desires.*

Order 1. General.   Mania.

Order 2. Partial. { Homicidal Mania. / Suicidal. / Erotomania. / Dipsomania, etc.

PREDOMINANT FEATURES.

	a Acute Delirium and Incoherence.	b Simple Excitement.	c Simple Depression.	d Stupor.	e Hypochondria.	f Strong Suicidal Impulses.	g Remittency or Intermittency.	h Chorea.	i Hallucinations.	k Enfeeblement.

CURABLE.

1. Insanity of Pregnancy, ... Mania,
2. " " . . . . Melancholia,
3. Insanity of Childbirth, ... Mania,
4. " " . . . . Melancholia,
5. Insanity of Lactation, ... Mania,
6. " " . . . . Melancholia,
7. Climacteric Insanity, ... Melancholia,
8. " . . . . Mania,
9. Insanity from Uterine Disorder, . . Melancholia,
10. " " . . Mania,
11. Insanity from Tuberculosis, . . . Melancholia,
12. " " . . . Mania,
13. Insanity from Masturbation, . . . Melancholia,
14. " " . . . Mania,
15. Insanity from Alcoholism, . . . Melancholia,
16. " " . . . Delirium Tremens,
17. Delirium Tremens . . . . Mania.
18. Post-Febrile Insanity, ... Melancholia,
19. " " . ? . Mania,
20. Hysterical Insanity, ...

INCURABLE.

1. General Paralysis, ... Paralytic Insanity
2. Epileptic Insanity, ... Epileptic,
3. Senile Insanity, ... Senile Dementia.
4. Paralytic Insanity, ... Organic Dementia,

88 INSANITY AND ITS TREATMENT.

Dr. Bucknill has given an elaborate classification, which is too long for reproduction, but will be found in an appendix to the last edition of Drs. Bucknill and Tuke's, "Psychological Medicine." He gives three classes of *Psychical Phenomena,* melancholia, mania, and dementia, subdivided into twelve sub-classes. He also names seven orders of *Pathogenetic Relations,* simple insanity, allied, sequential, concurrent, egressing, metastatic, and climacteric, which are subdivided into genera of *Pathogenetic Relations.*

I have already alluded to the classification of the late Dr. Skae, which, as finally modified by him, stood as follows:—

Idiocy, ⎫ Intellectual.	Rheumatic Insanity.
Imbecility, ⎬ Moral.	Podagrous "
Insanity with Epilepsy.	Syphilitic "
" of Pubescence.	Delirium Tremens.
" of Masturbation.	Dipsomania.
Hysterical Insanity.	Insanity of Alcoholism.
Amenorrhœal Insanity.	Malarious Insanity.
Post-connubial Insanity.	Pellagrous "
Puerperal Insanity.	Post-febrile "
Insanity of Lactation.	Insanity of Oxaluria.
" of Pregnancy.	Anæmic Insanity.
Climacteric Insanity.	Choreic "
Ovarian Insanity.	General Paralysis with Insanity.
Hypochondriacal Insanity.	Insanity from Brain Disease.
Senile "	Hereditary Insanity of Adoles-
Phthisical "	cence.
Metastatic "	Idiopathic ⎰ Sthenic.
Traumatic "	Insanity, ⎱ Asthenic.

Dr. Batty Tuke, in his Morisonian Lectures, divides the conditions productive of insanity into three:—I. *Idioencephalic* disease, or disease affecting primarily the cerebral tissues, and this he subdivides into four heads:—1. Traumatic; 2. Adventitious products; 3. Over-excitation of the brain; 4. Defective organization; II. *Evolutional* conditions of the body concurrently implicating the brain, subdivided into—1. Pubescence; 2. Pregnancy; 3. The puerperal state; 4. The climacteric period; 5. The senile period. III. *Morbid* conditions of the body concurrently implicating the brain, subdivided into 1. The diathetic; 2. The toxic; 3. The metastatic.

M. Auguste Voisin divides insanity into six classes.[1]

I. Acquired insanity (*folie acquise*), which comes on in an individual hitherto sane.

[1] Leçons cliniques sur les maladies mentales professées à la Salpetrière, 1876.

Of this there are four varieties :—

(a) *Folie Primitive*, or Idiopathic Insanity, which is divided into

> That which is unaccompanied by an appreciable morbid lesion.
> That depending on morbid lesions, viz.,
> 1. Active congestion. Hyperæmia.
> 2. Passive congestion.
> 3. Simple anæmia.
> 4. Secondary anæmia.
> 5. Atheroma.
> 6. Tumors.

(b) Insanity following a nervous affection, as epilepsy or hysteria.

(ɔ) Sensorial Insanity.

(d) Sympathetic Insanity.

II. Congenital Insanity (*folie native*), where the disorder has appeared at a very early age.

III. Insanity from drink or poison.

IV. Cretinism, idiocy, and imbecility.

V. General Paralysis.

VI. Senile Dementia.

Dr. Clouston[1] uses a symptomatological classification, of which the following are the heads:—

1 States of Mental Depression (*Malancholia, Psychalgia*).

2. States of Mental Exaltation (*Mania, Psychlampsia*).

3. States of Regularly Alternating Mental Conditions (*Folie Circulaire, Psychorythm, Folie à Double Forme, Circular Insanity, Periodic Mania, Recurrent Mania, Katatonia*).

4. States of Fixed and Limited Delusion (*Monomania, Monopsychosis*).

5. States of Mental Enfeeblement (*Dementia and Amentia, Psychoparesis, Congenital Imbecility, Idiocy*).

6. States of Mental Stupor (*Stupor, Psychocoma*).

7. States of Defective Inhibition (*Psychokinesia, Hyperkinesia, Impulsive Insanity, Volitional Insanity, Uncontrollable Impulse, Insanity without Delusion*).

8. The Insane Diathesis (*Psycho-Neurosis, Neurosis Insana, Neurosis Spasmodica*).

He also gives a modification of Dr. Skae's clinical classification.

Examining these various schemes of classification, we find them to be based on one or other of three principles. Either they are framed according to the mental pecularities of the patient, his exaltation, his depression, his imbecility; or they point to a disorder of one or other of the portions into which the human mind is by some authors divided; or, the mental symptoms being put entirely aside, the malady is classified accord-

---

[1] Clinical Lectures on Mental Diseases, p. 19.

ing to its pathological cause and its relations to the bodily organism. Objections are easily raised to any one of these plans. A patient, it is said, may be melancholic one week and maniacal the next; therefore melancholia and mania are not scientific divisions. Most true is it that a patient may have no delusions one week, but may have so far advanced as to have plenty in the week following; therefore it may be said that affective and ideational insanity are not true divisions. Then, if we take causes as our basis of classification, we may find two patients whose insanity springs from the same cause, yet they are in every shape and way the opposite one of another, requiring different treatment, differing as regards diagnosis and prognosis, the one hopelessly incurable, the other bidding fair to recover. Can we adopt such a basis as this? If not, what are we to look for to guide us in our attempt?

It appears to me, that if we classify not the so-called forms of insanity, but insane patients, we shall be reminded practically of certain points which otherwise we might overlook. We wish, of course, to ascertain for the purposes of our classification as many as possible of the conditions of the patients before us. If the conditions of any two were precisely alike, the insanity would be identical; but as no two people are alike, no two people's insanity is alike. If we have before us a dozen patients whom we are to classify, and we find that four of these are in an extreme state of depression, four are delirious and in a state of furious mania, while the remaining are gay and exalted, presenting the well-known symptoms of general paralysis, it is plain that there must be allied conditions existing in the members of each one of these groups which bring about the peculiar features of it, and which do not exist in the other groups. What these conditions are we may not be able scientifically to determine, but we may be sure that those which give rise to melancholy in a man of fifty are not the same as those which exist in a young man suffering from acute mania at twenty-five. Yet we may, according to some, group them together, and give to each the name of idiopathic insanity. I maintain that mania or melancholia denotes a group of conditions, most of which are unknown to us, though some may be ascertainable; that in our scientific classification the sum-total of these conditions, which is presented to us by the whole of the symptoms evinced by the patient, is not to be laid aside in favor of some one condition or cause, whether proximate or remote. As physicians engaged in the cure and treatment of insanity will never be able practically to lay aside the classification of mania and melancholia, and will always be compelled to treat melancholy as one thing and mania as another, so I believe that as pathologists they will comprehend under these general names a multitude of conditions which must be assumed, but cannot at present be demonstrated, but which year by year will be more and more differentiated and specialized, not by fixing our attention upon one, and one only, in each case, but by looking on every case as the result of an infinite number. One objection to the divisions of mania and melancholia is, that many patients cannot be ranged under either of these heads. They either hold a position midway between the two, so that they may be called by one maniacal, by another melancholic, or they cannot be said to be at all maniacal or at all melancholic, their insanity being denoted either by a total loss of mind, such as we call dementia, or by a mere assemblage of delusions, or even by one delusion without emotional display of any kind. For this state the term monomania has been invented, but it is applied also to other varieties of unsoundness of mind. Now, with regard to the above objection, I would say that such patients are for the most part chronic and incurable, presenting to us the results of former attacks and pathological states. As the damaged valves of the heart point to a long past condition of endocardiac inflammation, so the fixed delusions or hallucinations of an incurable monomaniac point not to a present but to a past pathological state. Were we always called upon to examine and classify patients in the very earliest stage of their insanity, we might possibly classify them according to the pathological origin of their disorder; but after years of alienation we necessarily lose sight of the origin and original condition, and our attention is directed to the mental symptoms. For the purpose of treatment and safe custody we cannot ignore these; we are obliged practically to classify our patients as melancholic or maniacal, paralytic or demented. To take the illustration already used, we recognize as most important the distinction between mitral and aortic disease, yet, pathologically, we might say that each may spring from rheumatic endocarditis, and therefore the division is an accident, and not scientific. But it is one thing to lay down the pathological varieties of insanity, as I have attempted in my third and fourth lectures; it is another to classify insane persons who may be set before us in all stages and periods of the disorder. If we can examine the individual at the outset of the disorder, and thoroughly ascertain his history, we may be able to lay down with considerable accuracy his pathological condition. He is not yet in a state which warrants the name of mania or of melancholia, still less of dementia; but his state is clearly aberration of mind and disturbance of brain function, depending on some conditions or causes such as I have been describing to you. But as the disorder becomes more marked and systematized, it will be found to assume one or other of certain forms, and to be accompanied by certain symptoms which have gained for it the name of mania, or melancholia, or acute dementia; and as the treatment and prognosis must vary according to the symptoms, I shall pass in review some specimens of these different patients, that I may be enabled to give you some practical advice as to what you are to do when called upon to treat them. To classify insanity perfectly, we ought to be able to connect the symptoms of exaltation or depression with the pathological history of the individual, but this at present we cannot do. We are obliged to make two classifications, to lay down abstractedly a pathological classification of insanity, such as I have endeavored to give, and on the other hand to describe according to the most prominent and important symptoms the various patients we have to protect and cure. Classifying not the disorder, but patients, I would reverse the order suggested by the committee of the Medico-Psychological Association, and note in the first place the mental symptoms observable at the time of inspection, and afterwards assign to these their pathological significance, if the history of symptoms enable us to do so. As in all diseases, the immediate symptoms must direct the immediate treatment, though the pathology will also have an importance which it is hard to over-estimate.

LECTURE VI.

Causes of Insanity—Predisposing Causes or Tendencies—Hereditary Predisposition—Prognosis—Statistics—Age—Sex—Condition of Life—Is Insanity on the Increase?—Exciting Causes—1. Moral—How to be Avoided—2. Physical—Prevention of the Recurrence of Insanity.

IN my lecture on the Pathology of Insanity I pointed out that each case at its commencement ought to be examined as a whole and all the various conditions considered, which, by preceding it, become the cause. These conditions are, some of them, extremely complex, some comparatively simple. They often require themselves to be resolved into simpler causes, and for this reason it is necessary that we should investigate them at greater length than has been already done; and first I must set before you certain states which are often called predisposing causes of insanity, such as sex, age, degree of civilization, inherited taint, and the like. It is clear that these can only be called causes in the sense of their being concurrent conditions of the individual who for the time is insane. A man in one or other of these states has greater tendency to become insane, if other circumstances also tend to produce insanity in him. The latter may be the result of a number of tendencies which may exist separately in others without producing anything of the kind, but which, concurring in him, are the cause of it; or these tendencies may remain for years unproductive of evil, till some external circumstance completes the series, and overthrows the stability of the mind. Speaking generally, we may examine the causes of insanity under the heads of *tendencies,* or, to use a commoner term, *predisposing causes* and *events,* more or less accidental to the individual, such as are generally called *moral and physical* causes. It is not my intention to examine these various causes with the idea of connecting them with the pathological state of the brain in insanity. I have already considered this, and have spoken of some from this point of view: I mention them now in order to make some suggestions of a practical nature.

The first *tendency* which demands your attention is hereditary transmission, for it is of all the most potent and ought always to be kept in view by those aware of its existence, whether medical men, parents or guardians. Here is a cause of insanity which cannot be got rid of, a part and parcel of the individual's constitution and being; consequently his surroundings must be adapted to it; and so far as we can we should avert such events as are likely to upset his mental balance. Much may be done in this way if friends will only look the threatening evil in the face and not try to hide it away in the vain hope that no one will ever know it.

The first remark to be made is that children may inherit insanity from parents who are not insane; and this we can explain in two ways: *first*, although the parents may not have been insane insanity may have existed in their parents and reappeared in the grandchildren, skipping a generation; *secondly*, though the parents may not have been insane they may have been the subjects of neuroses which in their progeny become insanity; they may have been chronic drunkards, epileptics, hypochondriacs, weak-minded, or have indicated their nervous condition by chorea, stammering, and the like. The reverse of this is also true: insane parents, either or both, may have of a number of children some insane, others idiots, others epileptics, deaf mutes, or nervous, and some perfectly sane and sound. Two laws of nature are concerned in the production of these phenomena. One is, that peculiarities and abnormalities are apt to recur in descendants for many generations; the other, that there is always a tendency to return to the type of health in beings which have sufficient vitality to perpetuate their existence, and carry on their race for successive generations. We could not breed an insane family, of which all the members should be insane for generations. We should have sterility and extinction, or a return to a healthy type. Were this not so, the numbers of our lunatics would be ten-fold what they are. We may see one child in a family insane and the others sane. The one has inherited more of the ancestral defect than the others, but in the descendants of the latter the family taint may again reveal itself. All these points you will have to consider if you are consulted as to the marriage of persons of whose families some member is insane. This question will come before you in various ways. You will be asked whether you can sanction the marriage of a man or a woman who has once been insane; also, whether you advise that of a person whose father, mother, brother, sister, or more distant relative, is or has been insane. You will be consulted on both sides, on behalf of the individual, and also by him or her who is about to enter into matrimonial relations with the tainted family. Here professional confidence and ethics are involved, and cases arise of no little difficulty. It is not easy to lay down any rules for your guidance. This much I may say, that if any one has already had an attack of insanity, there is always good reason for thinking that he or she will have another sooner or later; consequently, whoever is about to marry such a man or woman ought, beyond all question, to be informed of the preceding, and the chances of subsequent attacks. You will find, however, that such matters are kept profound secrets; and not once or twice, but over and over again, complaints have been made to me that they not only were not made known to the intended, but that the most flagrant falsehoods were told when the direct question was asked. There is also another consideration—the prospects of the children. Where a woman has been insane, her insanity is so likely to recur during pregnancy, or after parturition, or even during the first excitement of nuptial intercourse, that I never could bring myself to consent to the marriage of such a one; in addition, her children would run a great risk of being insane, nervous, epileptic, or idiots. A man is not exposed to so many causes of insanity as a woman, but his children are also liable to be affected with the inherited taint. Of course, if the woman is past child-bearing, there is far less objection to her contracting marriage. Then it is for the husband only to say whether he chooses to encounter the risk of marrying one who has already shown symptoms of the disorder.

If we have to consider the marriage of persons who have never themselves shown any signs of derangement, but whose parents or brothers, or sisters, have been, or are, insane, we have a much more complicated problem to solve. Much must here depend on the number of individuals in a family who have been, or are, insane. I have known a family in which, out of nine sons and daughters, six have shown unmistakable signs of mental disturbance; marriage with any one of these children should certainly be avoided. But we may see other families in which perhaps one member is insane and the rest perfectly sound. In this case we can only argue in view of the particular individual. I know brothers of insane men and women who are, in my opinion, as little likely to become insane as any of my acquaintance, but the transmitted taint may crop out in some of their children, even if the majority escape. I would not go the length of forbidding every one to marry who had an insane relative, for the number thus barred would be immense; but it must be considered that there is a certain element of risk, and this at any rate should be clearly set before the person who is thinking of entering into the union.

If we are asked whether the risk is greater to an individual when either of his parents has been insane than when a brother or sister has, what reply are we to make? Much must depend on the family history considered as a whole, and no answer can be given absolutely. We must look at the brothers and sisters of the parent and ascertain their mental health. It may be found that so many of them have been affected that the family must be held to be very strongly tainted with insanity, and therefore there will be a strong probability that the parent will transmit the disorder to the children. On the other hand, a child may be insane, and yet in the preceding generation the signs of neurotic disease may be wanting, and out of a number of children only one may be affected. Not unfrequently we may see one child insane when both parents have been healthy, together with the remaining brothers and sisters. I have known a father insane, and yet a large family of sons and daughters had all reached adult life without any trace of disorder; and in another family, five out of eight children became insane, though both parents reached old age without being affected. Here the chances of the remaining children would be decidedly bad, much worse than those of the former family, not-withstanding the insanity of one of the parents. The degree in which each generation shows the prevailing taint must be our guide in estimating the chance of an individual, and in the case of an only child the difficulty is much increased. The risk is greater for girls than for men, for obvious reasons; also, it is greater for a brother who has an insane brother, than for one who has an insane sister; and conversely, an insane sister increases the risk of a girl more than an insane brother. It is, I believe, a fact borne out by statistics, that daughters inherit insanity from an insane mother in greater proportion than sons, and in the same ratio sons inherit it from an insane father. The risk, therefore, to a daughter or son would be determined to some extent by the sex of an insane parent. I said that peculiarities are apt to return in descendants through many generations. The chance, however, of their appearing in any one individual is lessened in proportion to his distance from the diseased ancestor. Insanity is most frequently derived from parents, but it may come from grandparents, great-grandparents, or progenitors even more remote. When it comes from grandparents, it is called *atavic*, and is said to have skipped a generation. This is nothing but an instance of a defect reappearing at a longer interval than the first generation, and is only a variety of the general law. For we often find that although some one or more of the children may escape, yet others will show signs of insanity or other neuroses. If the descendants of these healthy children are affected, we may call it atavism, but the disorder cannot be truly said to have missed a generation if uncles and aunts were affected. I do not believe that we ever find a number of children of insane parents all entirely exempt from insanity or allied disorders, which afterward are manifested in the grandchildren, the intermediate genera-tion being completely unscathed. The fact is, that we have very imper-fect data on which to base our laws of heredity. The paupers who come to our county asylums know little of the history of their family beyond their nearest of kin; the well-to-do portion of the community deny the existence of hereditary insanity in their families in a way which must be heard to be believed; and as nothing is called insanity but that which necessitates incarceration in an asylum, and as other neuroses are disre-garded altogether, it is clear that our information is valueless. You may, possibly, some day be able to follow the fortunes of a family in which in-sanity is manifested for two or three generations. A carefully-recorded history of such a one would be of the highest value.

I have said that among the descendants of insane patients there is the tendency to revert to the healthy type, this being effected by the intro-duction of vigorous germs from other stocks, while a certain portion, not healthy enough for this, descends by gradual stages to sterility, idiocy, and extinction. Hence two things; first, we are not likely to have to look back very many generations to find the source of the inherited disease;

secondly, it follows that people must be continually contracting that which becomes in their descendants hereditary insanity—for hereditary insanity is not an entity to be acquired and transmitted *per se*. It must itself have had a cause and a beginning, and this we shall have to seek among the pathological causes and conditions of nervous disorders.

If we have the opportunity of knowing and observing all the children in a family tainted with insanity, it will not be difficult to point out, even at an early age, those individuals in which it is most likely to be developed. Children may show signs of a nervous temperament almost from birth. Convulsive attacks, night horrors, a tendency to spasmodic ailments, chorea, or epilepsy, mark out those who inherit, beyond others, the hereditary weakness. These we should specially guard: their future and external surroundings must be regulated so as to preserve them from too great responsibilities, too continuous labor, too sudden changes of fortune. We must, so far as is possible, create their circumstances, and not leave them to be the creatures of circumstance entirely beyond their control. Although we cannot guarantee any one against becoming insane, we can point to those most threatened, and it may be necessary for such, if they be women, to pass through life without encountering the perils of child-bearing.

But, beside this, there is another practical point in connection with the question of hereditary insanity. When we are called to an insane patient, and are told that there is insanity in the family, how will the information affect our prognosis? Many persons think that a patient has no chance of recovery if the disorder is inherited. Is this so? I have found in my own experience that those affected by inherited insanity recover, at any rate in the first attack, quite as often as others. Being unstable by nature and constitution, they are thrown off their balance by something that is often trifling and removable, or there may be no assignable cause. So by dint of seclusion and quiet they regain their former equilibrium, probably to again break down at some future time. Such people, as you may conceive, recover more surely than those who have brought on insanity by years of alcoholism, syphilis, or sexual excess. And possibly for the same reasons the death-rate is less in them, at any rate in the first attack. There is unquestionably a tendency to subsequent recurrence of the malady, but this exists in all who have been insane, whatever may have been the cause.

Statistics on the subject are valueless. One author attributes 10 per cent. of cases to this cause, another no less than 90. This arises from lack of information on the part of some friends, and the willful concealment of others, and also because some statisticians seek for insanity only, taking no account of other neuroses, such as epilepsy or paralysis. If we make the attempt, we shall soon find how difficult it is to get an accurate account of the health of the father and mother, and grandfather and grandmother of any one patient.

7

As insanity may be engendered of nervous disorders of other kinds in the parents, or as the parent's insanity may appear as the child's chorea, so may insanity of one kind in the parent reappear as insanity of another kind in the children; in fact, the latter may represent every variety of the disorder. As I have already said, the particular character of it, the mania or the melancholia, depends on the constitutional strength or weakness of the individual at the time of the outbreak, and the same person may be at the one time maniacal, at another melancholic. It is true that we frequently see the same form in successive generations, e.g., suicidal, melancholic or hereditary drunkenness, but this can only be looked upon as a coincidence, if we consider the vast number of cases where the form is different, and where various children are variously affected.

The tendency to become insane is greater or less according to the age of the individual, and the character of the insanity is also determined to a considerable extent by it. We find insanity, it is true, at all ages, but in the first decade of life it is rare. In the second it is more frequent; the mind is developing, the child is growing into the man or the woman, is acquiring knowledge, "looking before and after." But the next decade produces even more cases, and the period between twenty-five and forty years is that in which the number attains its maximum; this we recognize as the time of highest development, the prime of vigor, the height and climax of all hope and forward-looking, the time when strength is strained to the utmost in the battle of life. After this, in each successive decade, the number declines, just as in youth it rose.

According to the time of life, variations in the nature of the insanity are observable. In childhood, we find it displayed in violence of temper and act, in irregular and paroxysmal attacks, often of a convulsive character, alternating with cataleptoid states, recalling to our minds the choreic and convulsive condition of other children. The mental symptoms are not those of the fully-developed mind, as delusion, but perverted feelings, hatred of relatives, wanton and indecent behavior, cruelty and destructiveness, and hallucinations of the senses—such as we often witness in the dreams and nightmares of the young. After puberty, we may find more of the ordinary insanity of adult life; but this will be generally attended with violence and mania rather than by depression. Of fifty-seven boys and girls admitted during five years into St. Luke's Hospital, all of whom were below the age of twenty, only eight were melancholic. Between the ages of twenty and forty we meet with violent and acute mania and acute delirium. This is especially the period at which we should expect these forms of insanity. Later in life, in the time of waning strength and declining vigor, both bodily and mental, we find melancholia prevail. Fear and religious despondency constantly accompany the weakened nervous condition of the old, while later still we see the mental faculties giving way, and dementia and vacuity, rather than insanity, come over the sufferer as he sinks into second childhood.

These are, speaking generally, the forms we met with at each epoch of life, but exceptions to them are not unfrequent. We may find melancholia in youth and early manhood; we may meet with acute delirium in patients past their climacteric. And when we do, the prognosis is bad; in the latter case especially. I have found melancholia in the young far more difficult to eradicate than mania. It is generally prolonged, even if recovery takes place at last. Melancholia after parturition is often extremely obstinate, and yields with difficulty to remedies.

What has the sex of the patient to do with the chance of insanity showing itself? Do more men or women become insane? Here authors differ, and statistics mislead. The records of various countries would seem to show that in some more males become insane, in others more females. Into the figures and tables, and the many fallacies to be eliminated, I will not enter. In countries where the males predominate in the asylums, it is said that this is due to the fact that females are more easily managed at home, and therefore not sent so often to the asylum. On the other hand, the preponderance of females in our own asylums is explained by the lesser mortality among them, whereby they live a long time, swelling the lists of those in confinement. My own opinion is, that we have no accurate statistics on this point, for only those relating to new cases are of value. The comparative numbers of patients under confinement at a given time must necessarily be influenced by a variety of circumstances. In the Report of the year 1883, the Commissioners give, as the number of the insane in England and Wales, in and out of asylums, 4,127 private male patients, and 3,796 private females; while of paupers there were 30,355 males, and 38,487 females. I explain this discrepancy in this way: more males become insane than females, but die in much larger proportion, consequently the number of pauper females goes on increasing from the accumulation of chronic cases. In the well-to-do classes, however, so many females are kept at home that the male population of asylums predominates. The deaths during the year 1882, in asylums and hospitals, were 2,703 males and 2,082 females; and this is probably about the usual proportion in which the sexes die while insane.

Probably the difference in the number of the two sexes who become insane is not very material, but I think the males must be the larger body; otherwise, considering the mortality, they would fall below the number of the females further than they do. It might be thought that females are more likely to become insane, inasmuch as they are manifestly more prone to many nervous disorders—to chorea, hysteria, and allied neuroses. Yet, for all this, they are much less prone to serious brain disease. Men are the chief sufferers from apoplexy, hemiplegia, softening, and the like; and among the insane we find general paralysis attacking at least ten men for one woman, and among women of the higher classes it is quite unknown. It is this fatal form which swells the mortality of the male insane; and it has been said that, if we except the deaths from general paralysis, more women die than men.   But I turn to the statistics of St. Luke's Hospital, where formerly only curable cases were admitted, and where no paralytics were kept to die; out of 7,311 males and 10,778 females admitted, 808 males died, and only 573 females, *i.e.*, double the proportion of males.

Among the so-called predisposing causes of insanity are ranked the condition of life of the patient, and the degree of civilization.   Are civilized nations more prone to insanity than barbarous? are the poor more affected by it than the rich? the single than the married?   Here is a field for much speculation, profitable, however, to the philanthropist and the philosopher, rather than to the practical physician; for it is obvious that when we discuss the chances of a poor man or a rich man becoming insane, we take into consideration the whole surroundings of the individual and his ancestors, and include in our survey circumstances that are entirely beyond the reach of the physician.   It is equivalent to inquiring whether education or its absence tends toward insanity, or good food or scanty, or hard work or little work, or head-work or hand-work.   Though of little practical value, inasmuch as there must ever be poor and rich, yet the question is interesting, and when rightly considered and compared with carefully-scrutinized statistics, it is valuable in its bearings on the general history of our country.

And first with regard to civilization: of this we may say at once, almost from *a priori* consideration, that it tends to the production of insanity.   The life of civilized man is a highly specialized and complex life; the variations of its surroundings are endless, and it requires to be adjusted to these unceasingly.   And as is the life, so must be the brain: the brain of the savage is a simple structure compared with our own, as the whole organism of one of the lower animals is simple, compared with the complex human system.   Whether we consider the emotional or the intellectual traits of .the primitive man, we find that the prevailing characteristic of both is a childlike simplicity.   Emotionally the savage is impulsive and improvident, intellectually he is deficient in reflectiveness, in the power of generalizing, and capacity for abstract ideas, herein resembling the child of the civilized.   As we find insanity occurring but rarely in the children of our own country, so must it occur but rarely in savage races, more rarely indeed, for there we shall not meet with the transmitted taint to which is due almost all the insanity of early life that meets us at home.[1]   To say that our specialized and complex brain is more apt to be disordered than that of lower men, is no more than saying that a compound piece of mechanism is more likely to get out of order than a simple one.   Nevertheless, it is also true that much of the insanity of civilization

[1] *Vide* Herbert Spencer's "Principles of Sociology," chaps. vi. vii.

might be prevented. It grows out of the evils and vices of civilization, just as fevers and such like disorders are engendered by the crowding of populations. Even education, which, properly conducted, ought to bring strength to the mind and lessen the liability to insanity, may bring danger to individuals in many ways. The cramming for competitive examinations, which now goes on everywhere, is fraught with peril to many boys, and still more to girls, for the latter have to prepare for them at the very trying time of puberty, or shortly after. Although it must always be that the hard-working brain of civilized man is more prone to disorder than that of the childlike savage of the wilderness, yet it is to be hoped that the preventable sources of insanity may be by degrees diminished, just as sanitary knowledge and laws will reduce the mortality from fevers, scarlatina, smallpox, and the like.

If we compare civilized with uncivilized countries by means of the statistics of insanity, we shall labor in vain. We have enough to do rightly to examine the statistics of our own country. There are not half-a-dozen other countries in the world, civilized or uncivilized, whose statistics are comparable with our own. In the year 1844, there were in the asylums of England and Wales 11,272 lunatics, and in workhouses and private dwellings 9,339. In 1858, there were in asylums, 22,184; in workhouses, etc., 13,163. On 1st January, 1883, in asylums, 52,730; in workhouses, etc., 24,035. That is, the registered lunatics of England and Wales, who in 1844 numbered 20,611, in 1883 reach the total of 76,765. In other words, in 1859 the proportion of lunatics and idiots to the population was 18.67 to 10,000 persons, while on the 1st of January, 1882, it was 28.68.

Closer investigation, however, shows us that this apparent increase of insanity is not so formidable as at first sight it seems. Nay, many competent inquirers have held that there is no increase at all, but that the ever-growing number of those in confinement is due to more stringent application of the lunacy laws, to the sequestration of many patients who formerly would have been kept at home, to the prolongation of life of those who are cared for in the comfortable asylums of England, and to various Acts of Parliament which have led to an increase of the numbers of registered lunatics. For if we look at the numbers, not of those remaining under care and treatment, but of those admitted every year for the last fifteen years, we find that the admissions of private patients have actually decreased, that in 1861 they were 3,061, while in 1882 they were only 2,212. The paupers, however, who in 1861 were admitted to the number of 6,268, reached in 1882 to the total of 11,369. But the increase of the latter has not been regular. In certain years it has been large and abrupt, as in 1865 and 1875, owing to legislative enactments which induced parishes to send all harmless imbeciles and lunatics to asylums rather than keep them at home.[1]   My own opinion is that insanity is not increasing among the upper classes, but that it is to some extent, though probably not a large one, amongst the lower.   Let us see if we can discover any sources of the disease existing among them, to which the richer portion of the community is less exposed.   One thing we may at once observe: there is a degree of drunkenness among the lower classes of this country which is not to be found in the higher.   Those who read the accounts of the habits and customs of the richer classes at the close of the last and the beginning of the present century, must be aware that the gentlemen of our own day are, as regards temperance, entirely different from our grandfathers, with whom intemperance was the rule rather than the exception.   It cannot but be that amongst these classes the children of our own times must benefit largely in all that concerns their nervous condition by this change in their parents' life.   But it is to be feared that our lower orders are as yet but little reformed in the matter of drink. Yet much may be hoped from the exertions of blue-ribbon armies and other temperance societies, from the establishment of coffee-taverns, and the closing of gin-palaces.   Certainly at no period of the present century has the cause of temperance flourished as it does now; and those who have long deplored the spectacle of a drunken working-class may now hope that the tide at last has turned.   Hitherto, as I believe, the ever renewed insanity of our lower classes has sprung mainly from this cause. For as insanity has a tendency to die out like other diseases—to cause the extinction of a race, or itself to be overcome by the greater vigor of some of the stock—it is clear that the enormous insane population of our country must owe its insanity to ever-present causes,—it cannot all have been inherited from our great-grandfathers.   And if we could accurately ascertain the statistics of insanity in other countries, civilized, semi-civilized, or barbarous, I think it is probable that we should find insanity in proportion to the use of intoxicating liquors or substances.   Secondly, poverty itself bears a part in the causation of insanity.   The poorest counties in England contribute the largest numbers to their respective asylums.   According to the late Dr. Thurnam, Wiltshire stands in the unenviable position of having a greater proportion of insane paupers than any other county of England and Wales.   Those most nearly approaching it are Gloucester, Oxford, Berks, and Dorset—all agricultural counties. According to the poor-law returns, Wiltshire has a larger proportion of pauperism than any other county, viz., 1 in 12; the next in order being Dorset, Oxford, Gloucester, Berks, and Hereford.

The fact that in Ireland, with a decreasing population, we find an increase in the numbers of the insane, may aid us somewhat in discussing the causes of insanity.   On the 31st December, 1846, there were 6,180 reg-

[1] Dr. Maudsley, Journal of Mental Science, xxiii. 45.

istered insane persons and 6,217 unregistered, the latter return being furnished to Government by the constabulary. At this time the population
at large amounted to 8,175,124. In 1871, when the population had fallen
to 5,402,759, the registered insane were 10,767, and the unregistered
7,560. The increase was continuous and of regular growth, and many of
the causes which are supposed to have swelled the numbers of the Eng-
lish insane are inapplicable here.

That the poor must be more prone to become insane than the rich is
consistent with the pathology of the disorder. Insufficient food is an
acknowledged source of defect in the nerve-power, and with this there
must be the concomitant anxiety, the care for to-morrow, the spectacle of
the family stinted of the necessaries of life, and want of early medical
advice and treatment. Poverty of this kind must weigh heavily on the
mind. When we say that the poor are more disposed to insanity, we
virtually mean the poor in any class of life. There may be well-to-do
people in all ranks of life, who, although they would not be classed among
the rich, do not suffer from poverty. Amongst these well-to-do folks of
the artisan class the great causes of insanity are, first, the want of education, which leads to the second, drink. Whether either the one or the
other of these can be diminished by legislation is one of the questions of
the age. If we could lessen drunkenness we might close some of our
asylums; till we do this we shall have to enlarge them.

I have thus glanced at some of what are called the predisposing causes
of insanity, which are in truth the conditions of certain classes of the
community, and can only in a sense be considered causes. But there are
others, special to the individual, which are called *exciting causes*, and,
whether preventable or not, frequently bring about the particular attack
of insanity. Of many of these I have already spoken in discussing the
pathology of the disorder, but some few words still remain to be said.

In speaking of one section of these exciting causes—namely, the *moral*
or *psychical*—I divided them into those which operated rapidly, causing a
great shock or fright, and those which produced insanity by anxiety or
worry spread over many years. Of these, the former, I think, is more
frequently the active agent in the causation of insanity. A sudden calamity, loss of a dearly-loved relative or friend, reverse of fortune, politi-
cal catastrophe, or a shock or fright preceding from some awful spectacle
or violent quarrel, or near approach to death—these are the things which
unhinge the mind, throw it from its balance, and may render even a tolerably strong-minded person incapable of taking care of himself for a time.
The causes that operate, not suddenly, but slowly, may produce insanity,
or other brain disease, the result of over-fatigue and work. The brain
becomes worn out, and "softening" and such like conditions are the
result.

The pathological causation of insanity I have not here to consider.

When I speak of the causes, it is with the view of examining into such matters as may be prevented, or taken away from those who are insane or threatened with insanity. For it is certain that psychical causes preponderate in the production of insanity, because the persons who most frequently become insane are they who are strongly predisposed thereto, and are easily thrown off their balance by mental influences. Those who are perfectly free from all predisposition and taint do not become insane from losses, worry, or work. They suffer, and are strong; they buffet vigorously the waves of adverse fortune, and when the malevolent tide again ebbs, they are none the worse for the contest. But it behoves us to counsel and advise those, or the friends of those, who by hereditary taint or previous attacks of insanity are predisposed and liable to its invasion, to avoid to the utmost all that is likely to cause uncertainty, harass, or reverses in their daily life. There is much, of course, that no one can foresee, avoid, or escape. Health of self or family is a matter of uncertainty. Accidents may happen to wife or children, shocking sights may occur, but, nevertheless, in the choice of employment and method and rule of life, much may be done to avoid that which is to some a fertile cause of insanity. Supposing we have to deal with a young man, whose parents, or brothers, or sisters have been insane, or who at an early age has shown that he is himself not free from the family taint—and insanity in the young is, as I have said, almost invariably due to hereditary influences—we should advise that such a one should be put to some occupation or calling not attended by any great harass or responsibility, one of which the duties and work are of a routine character, affording a fair opportunity of holiday and recreation. He should not follow the profession of a lawyer or a doctor, for in them he will find hard and constant work, heavy responsibility, and the necessity of appearing in public; and his work and anxieties will follow him to his fireside and hours of sleep. In the church he will or may be assailed by religious doubts, by a sense of duties insufficiently discharged, and by all that tends to religious melancholy. In the army he will be exposed to the temptations of an idle life and the vicissitudes of climate. No post is so suited to these individuals as that of a government office. The hours are light, the responsibility not formidable, the holidays long, and if the emolument is not large it is at any rate certain, and certainty is above everything desirable. For the latter reason the property of such persons should not consist of doubtful or speculative investments, which may expose them to great anxiety or possible ruin, but they should be induced, even at a loss of income, to place their capital in solid and safe securities. Many anxieties and worries which are every day occurring might certainly be avoided by a little prudence and forethought.

It is no doubt of immense service to many such men that they should be happily married at an early age. The tendency to eccentricity and a solitary life is thereby counteracted, as well as other habits or vices. But everything will depend upon their finding wives who will be truly help-meets to them. Doctors can only speak in the most general terms on such a subject; but it is right that parents and guardians should put a stop so far as they can to the growing attachments of a couple in no way fitted for each other. The wife of such a man should be healthy, one who will not cause him constant anxiety, broken nights, or the expense attendant upon bad health, who will not breed sickly children to fade and die, or grow up a perpetual misery to their parents; and, above all, she should not be a relation.

If the object of our solicitude is a girl, what can we do to keep her from harm? The first question is, Shall she marry? Here I would say that no girl who has ever shown any symptoms of actual insanity ought ever to marry at all. Insanity is not less likely to return because the attack has been transitory. If it has come on at an early age without assignable cause, it is because it is inherited, and it is very likely to reappear during pregnancy or after childbirth.

But if there have been no symptoms in the girl herself, but she comes of parents, one or both of whom have been insane, or if she has brothers or sisters insane, what are we to say? We must be guided by what we know and see of her, by her *physique*, her history, especially the history of her infancy, the absence or presence of "fits," "nervousness," "hysteria," habitual sleeplessness, or irritability, especially at the catamenial time. If a girl who is predisposed to insanity does marry, it is important that she should not marry a poor man, whose life will be a constant struggle for existence; nor a sickly man, whom she must always nurse; nor one whom she must follow to tropic climates; nor one who is violent, irritable, or jealous. If she be obliged to do something for her livelihood, and does not marry, her work should, if possible, be free from heavy responsibility. It is obvious that we shall have very little to do with the arranging of all this. If we express our opinion, we often say our say, and say it in vain. And yet it is well to turn such subjects over in our minds, and to be ready with an opinion upon them. What I have said applies to the rich rather than the poor: it is about the rich and not the poor that we shall be consulted.

The physical causes of insanity are in many cases beyond prevention. We can only deal with them as they arise. They may be the sequel of other disorders or accidents, or may be due to the time of life.

Insanity, the result of alcohol, opium, or haschish, is no doubt preventable; but we have to deal with acquired habits in the majority of instances when brought into contact with persons who take these substances. When we have the opportunity of supervising the management of those predisposed to insanity, it is important to check at the outset that which may grow to be a habit hard to be abandoned. First and foremost they must be watched while children, lest they give themselves up to

masturbation, for they will carry it to excess, and will be the least likely
to abandon it, if it be once confirmed—a truth which applies equally to
girls and boys.   Early habits of drinking also are frequently contracted
by weak-minded people, both males and females; and we constantly find
that habitual drunkards, as well as dipsomaniacs, are the offspring of
insane or epileptic parents.   We must be careful lest we encourage a love
of drink in "nervous" people, especially women, by an incautious administration of stimulants for the cure of hysterical, hypochondriacal, or neuralgic symptoms.

That which I believe to be the most common cause of general paralysis
of the insane—viz., sexual excess—is a matter over which we have very
little control.   The disease does not commence in boyhood, rarely before
the age of twenty-five.   It occurs as often amongst married men as
amongst single, and frequently until it shows its presence there is nothing
to call for warning.   Such warnings are rarely heeded, but it may be in our
power to give them to many, both husbands and wives, for general paralysis is not the only ill that springs from unrestrained sexual intercourse.

In considering the causes of insanity, we shall have to bear in mind
not only the prevention of insanity in a patient who has never been
insane, but also the prevention of its recurrence in one who has had an
attack.   Here our difficulties will be in some respects greater, in others
less.   The same cause of worry or grief may not occur again, or may be
removed or counteracted; the same bodily illness, as fever or measles, may
not recur.   Attention to the health, to diet, and regimen may remove
some causes—syphilis may be eradicated, drink avoided.   But, on the
other hand, we have the fact to contend with, that insanity has really existed in the individual, and beyond all question has a tendency to recur
either upon a very slight cause, or without any assignable cause whatever.
This it is which makes it so hazardous a thing to marry a man or woman
who has ever shown any symptoms of the disorder.   When a patient leaves
an asylum, we are generally asked, What is to be done? what is to be
guarded against? and our answers are for the most part very general:
Avoid hard work, all that calls for great emotional exhaustion, avoid disappointment.   This may seem foolish counsel, but those who cannot bear
disappointment should not encounter it.   How many do we see aspiring
to that to which they can never attain, trying for appointments and the
prizes of life which are utterly out of their reach!   Quiet and contentment
are the qualities which favor health of mind, but it is not for us to give
these to the restless spirit of a patient saturated with insanity.

Men and women become insane because it is in their nature and constitution to develop insanity, and when we hear that this or that has caused
their insanity, it is often their restless and half-crazy brain that has made
mountains out of molehills, and given an objective existence to troubles
and vexations which exist in their minds subjectively, and have no outward reality whatever.

LECTURE VII.

The Symptoms of Insanity—The False Beliefs of the Insane—Definitions of
Terms—Delusions—Their Rise—Varieties—Hallucinations—Their Seat—Hallucinations of Sight—Hearing—Smell—Taste—Touch.

HAVING hitherto spoken to you of the disease termed insanity, and the
pathological condition of an insane man, I now proceed to consider the
symptoms, the things which insane people do and say, which, taken together, constitute their conduct—conduct by which we judge them to be
of unsound mind. Many acts and many ideas at once reveal the state of
the mind, though there may be no fixed line of demarcation between acts
or ideas, sane and insane.

If we take, first, the beliefs of insane persons, we shall have to examine the meaning of certain words met with in treatises on the subject.
These are *delusions, illusions,* and *hallucinations*—words used by different
authors in various senses, with some of which, at any rate, you must be
acquainted.

A *delusion* is a false belief in some fact which, generally speaking,
personally concerns the patient, of the falsity of which he cannot be per-
suaded, either by his own knowledge and experience, by the evidence of
his senses, or by the demonstrations and declarations of others. A man
thinks his head is made of brass, that he has a fire in his inside, that he
is a beggar or a prince; and no amount of proof convinces him of the
contrary.

Hallucinations are false perceptions of the senses, the eye, the ear, the
nose, and so on. The hallucinated patient thinks that he sees in the
blackest darkness, or hears a voice through any number of thick walls,
whereas his seeing or hearing is entirely subjective, taking place altogether
within his own head, without any excitation conveyed to his organs from
the outer world—when, in fact, he would hear and see the same were he
deaf or blind.

An *illusion* is also a false perception of the senses, or rather a mistaken
perception. There is something to see and something to hear; but that
which the patient thinks he sees is not the real thing, but something else.
He sees a chariot in the sky, when every other person sees a cloud; he
hears a voice, when others hear the noise of a carriage or a distant footfall.

Such are, I believe, the commonly received meanings of these words,
but such interpretations are necessarily arbitrary. And by some the word
illusion is used as synonymous with *delusion.* Prichard, in his well-known

work, nowhere speaks of *delusicns,* but uses *illusion* instead.   I advise you to avoid the word *illusion,* as it has been used in various senses by different authors.

There are various questions which arise in connection with these fancies of the insane—*e.g.,* how do delusions and hallucinations arise ? what is the relation between delusions and hallucinations? what is their significance in the diagnosis and prognosis of insanity ?

We know too little of the pathology of the brain to be able to put our finger on the seat of an hallucination; we can only approach the study thereof by comparing analogous phenomena in people not insane, and by contrasting the delusions and hallucinations of one insane person with those of another.

Hallucinations bear to delusions the same relation that the simple perception of objects does to judgment and reasoning founded on the perception, and both hallucinations and delusions illustrate the growth of ideas and intellect in the mind.   For, as I have said already, sensations, the stimulation of the organs of sense and the resulting feelings, and also the feelings experienced by the organism generally, whether of pleasure or pain, are the origin and material out of which ideas and intellect are developed.   These simple sensations, as they are linked together by memory, grow into complex ideas and complex feelings—feelings which we term emotion—but at the root of all are the bodily sensations caused originally in the majority of instances by stimulations coming from without, at any rate from without the cerebrum.

It would appear that delusions and hallucinations are false interpretations of morbid feelings and sensations occurring in various parts of the system, the falsity of which the disordered brain is not able to appreciate.

In the case of delusions this is not difficult to trace.   And thus we can explain how it is that the delusion of the insane man almost always refers to himself.   Many sane persons believe in absurdities of all kinds—in charms, witches, ghosts, spiritualism, and the like—yet they may hold such beliefs without having been personally concerned with any of these things. They entertain a mere abstract belief in them.   But the delusions of the insane man have reference to himself, just as in dreams we are always present and see and hear what is going on: we do not dream abstract notions or facts which wholly concern others, in which we do not in some way participate.

Delusions are not the first indications of a change in the mental condition of the patient.   This fact is important as regards both the pathology and also the legal diagnosis of insanity.   The changed feeling may express itself in depressed manner, an unusual excitement, anger, restlessness, or in acts of an extraordinary or outrageous nature, without being translated at any time into what we know as delusions; and these when noticed have been usually preceded by a period of alteration, which may or may not have attracted attention.

There is, as I have already said, an acute stage at the commencement of every case of insanity, though you may not see the patient during its continuance. But as insanity implies a deviation from the normal mental condition of the individual, so it connotes a physical disturbance of the brain-function, with impaired sleep, possibly pain, heat of head, flushed or pale face, suffusion of eyes, throbbing of carotids, and such like symptoms of cerebral disorder. This is the period of emotional alteration visible to others, of which the patient himself may or may not be conscious.

In the majority of instances this change is one which makes the patient feel or think that something is amiss with him, as in truth there is. His consciousness, however, of something being wrong with his head or his system generally will vary much. He may be quite aware of it, and may seek advice and assistance like any other patient. He is more likely to be unconscious of his real condition, and to attribute the feeling he experiences to external causes. According to the feeling, its degree and intensity, will be the nature of the cause to which it is ascribed, and the means taken to get rid of it.

The defective condition of brain operates here in two ways: First, the lack of nervous energy brings about the feeling of there being something wrong, and this may vary immensely in degree; secondly, the disorder in the various portions of the brain reduces it to such a state that the patient is unable to see the absurdity or impossibility of the explanations which he gives.

A man or woman feeling great depression of spirits proceeds to account for it according to his or her views. One thinks beggary the greatest evil that can befall him, and straightway fancies himself ruined, his wife and children starving, and the officers at the door to hale him to prison. So imperfect is the action of his entire brain, that he fails to assure himself from actual inspection of his accounts that everything is going on as usual.

Another looks not at the things of this life, but at those of the future. His soul is lost, he is in the power of the Evil One, he is Satan himself, or Antichrist—there is no hope for him. And inasmuch as the feeling experienced is strange, unaccountable, mysterious, patients fly to the mysterious for the cause thereof. It is due to mesmerism, to electricity, to secret and loathsome disease hidden in the flesh and bones, destroying heart, stomach, and bowels, though not to be discerned outwardly. The ignorant man will think it due to witchcraft, or the devil. According to his stores of knowledge, his education, and experience, each will invent a cause for that peculiar condition of which he is aware, but which he cannot rightly explain. Having no ideas connected with his feeling, he expresses it in those habitually associated with a feeling of deep gloom, anxiety, or displeasure. Feeling himself peculiar and changed, and another man, he thinks that all men are looking at him, pointing at him, deriding him. The cabmen and omnibus-men beckon to and mock him;

the passers-by avoid him.   And in the same way all the newspapers write about him; all the mysterious advertisements refer to him.   If he cannot fix his annoyances on any one he knows, he thinks bands of unknown conspirators are plotting against him, and that these can, by occult and supernatural means, affect him, even when far distant.   Then, as he feels discomfort in this or that part of his body, he says his head is of brass, or he is galvanized, or his inside burnt with fire.   With all or some of these delusions he may vary greatly in his emotional display, being profoundly dejected and in a state of melancholia, or being irritated or angry, and inclined to act on the offensive, when we call his affection mania.   And yet he may be equally melancholic and equally maniacal without his ideas being perverted into delusions, though they will be tinged with his prevailing feeling.   Similarly, his delusions will range from possibilities, or even probabilities that require some examination before they can be pronounced delusions, up to the wildest absurdities and the most incoherent nonsense that a madman can utter.   The latter are indications of a much greater brain-disturbance, a greater disconnecting of the relations of the various portions of the brain, and impairment of the brain-force and brain-circulation; but the prognosis is not always on this account more unfavorable.   They bear the same relation to the possible delusions, that the dreams and nightmares of the fever patient do to the natural dreams of the healthy sleeper.

There is a proneness on the part of most people to seek and assign a cause for all ailments, and for everything which is new, abnormal, or of which the origin is not plainly visible.   Every one can tell us exactly where he caught his catarrh, what caused his diarrhœa, or what makes him "bilious."   And people are still fond of flying to the mysterious, and of looking for their "causes" among things they do not understand. Take, for example, the common notions about the weather, the influence of the moon, of comets, and the like; and the superstitions and vulgar errors concerning all manner of diseases, from which quacks of every kind reap their fortunes.   The only difference is that the delusions of the insane man, as I have said, have reference to himself alone, inasmuch as they are invented to account for a feeling of which he only is conscious.   The proneness to account for it in this supernatural way is not peculiar to the insane, but is common to all, especially to those of uneducated or weak and unreflecting mind.   Similarly, in dreams, cold feet make us imagine that we are walking on ice; an uneasy posture causes us to think we are in chains; or the general *malaise* of dyspepsia or a heavy supper calls up a succession of horrible fancies and vivid nightmares.   We look on the brain of the dreamer as awake and acting in portions only; when it is all awake the dream is at an end; but in the case of some dreams which are extremely vivid, many moments may elapse before the whole force of the brain and the whole of its related parts can be so brought to bear upon the subject as to convince us that it is a dream and not a reality. The brain of the insane man fails in this power of perceiving the whole case.

So far we have considered those feelings of ill-being which express themselves in ideas of misfortune, accompanied by melancholy or anger; but there are patients manifesting not only in idea and delusion, but in countenance, manner, and action, a feeling of well-being, a conviction that the change is all for the better, that they are stronger, healthier, richer, happier than ever they were in their lives; and these symptoms in many cases, though not in all, are coincident with the presence of fatal brain disease, which will go on steadily and rapidly to death. All this is difficult to explain; nay, explain it precisely we cannot: we can only conjecture that the effect of this disease, which manifests itself in bodily paralysis as well as mental, is to blunt the sensibility of the brain, so that everything causes pleasure instead of pain to the half-roused centres: an asylum appears a palace, and a dinner of roast mutton a banquet of rich and *recherché* fare.

In all patients alike there is the same inability to see that the thing is a delusion. They see it when they recover—see it without the demonstration which failed to convince them while insane. Nothing, I think, can be more certain than this, that during the insane state the brain cannot act as a whole, cannot by means of one part correct the ideas which arise in another. These ideas are the concomitants of strange and altered feelings which have a real existence; and until the latter pass away, they are not to be removed by demonstration or argument. When the feeling subsides—the feeling of depression, or the excitement and elation which cause the grand and exalted fancies of some—the ideas in the majority of cases vanish also, especially if no long time has elapsed. The patient is said to have lost his delusions, and their gradual disappearance or occasional reappearance coincides markedly with the restoration of the general health and strength, of sleep, digestion, uterine or other functions. But it sometimes happens that the delusions remain after the feelings have gone, and we behold in the patient a confirmed monomaniac. The ideas which were at first the explanation to the patient of his altered sensations are stored up as facts of experience in a damaged brain, which never recovers from the injury it has received, and never resumes its entire working power; remaining permanently unable by means of one part to correct the false notions of another, it retains forever the dream that arose in its half-waking period.

The delusions most frequently met with amongst the insane may be arranged in comparatively a small number of classes. All, as I have said, are connected with self—the selfhood of the patient; all are supposed to indicate some change that has taken place with regard to himself. There is a change for better or worse; a change affecting his worldly or spiritual interests, his bodily condition or his surroundings; a change which has already happened, is happening, or is about to happen at some future time. The extravagance of the ideas will depend on the amount of brain-disorder, and we may often see this marked out by the delusions, as it rises to its climax, and then falls again to where it began.

The delusions presented by a patient who thinks that things are amiss with him, may be connected, as I have said, with his worldly or his spiritual interests. In the extreme melancholic condition which accompanies excessive depression and prostration of nerve-force, there will probably be delusions on both these points. The patient is ruined and a beggar, and has also committed the unpardonable sin and is doomed to eternal perdition, or he may fancy one of these things without the other. What do we learn from these delusions? How do they affect the diagnosis and prognosis? We learn from them that the patient's condition is one of melancholia, and they ought to warn us that in all probability he is, or will be, suicidal. Thinking that he is doomed, and that life is insupportable, incapable of reflection, and impelled by the ever-present horror of his position, he tries to shuffle off his mortal coil, and unless extreme precaution is used, he will certainly succeed. As regards diagnosis, the most valuable lesson taught by these delusions is, that the patient, being suicidal, will very likely try and escape, either that he might be free to commit suicide, or that he may wander over the face of the land to escape the evils that encompass him at home. The delusions may have reference to the past or the future; he may be in a state of profound remorse for imaginary crimes, or may shrink in terror from tortures and torments which are to come upon him unjustly hereafter; but practically there is little difference in condition of the two varieties; frequently we find either kind of delusion in the same individual: either may impel a man to suicide, or to running away from the scene of his past sins or expected torments.

As for the prognosis, it is not very unfavorable in such cases, provided that the bodily condition is not too much reduced by disease or starvation: patients recover from melancholia in large numbers, and after long periods of time; in fact, if they do not succumb to the disorder and die at an early date, we may have great hopes of restoring them to their family and the enjoyment of life.

Some may labor under a delusion that very much is amiss with them; yet their feelings may be not those of depression and melancholy, but rather of alarm, and restless anxiety, or anger and fury. So far as the delusions are concerned, it is evident that it is optional whether we call such person melancholic or maniacal; but the general deportment and feeling of many of them is far from being melancholic, and is unquestionably maniacal; while a certain number may be ranged with equal propriety under either one or the other of the two classes.

Their delusions resemble, to some extent, those of the melancholic class. They imagine that some evil is going to happen to them. They do not, however, think they have deserved it, but that they are unjustly treated—that wicked men are conspiring to ruin them or their family, to blast their character, or put them to death.   Here suicide will be rarer: more frequently we shall find attempts to escape, or avert by some means or other the impending catastrophe.   Consequently you will understand that the diagnosis of such a patient involves the belief that he may be very dangerous to those about him.   In order to escape he may set the house on fire, may try to obtain the keys from the attendants by force or bloodshed; may conceive the notion that those around him, especially strangers, are about to do him some evil, and murder them in supposed self-defense. A vast number of the homicides perpetrated by lunatics are done in fear and panic, especially those committed by patients suddenly waking out of sleep.   Murders are committed by those who imagine that their victim has accused them of foul and unnatural crimes, and who suffer from hallucinations in which they actually hear a voice repeating these slanderous words.   Or a man thinks that his food is poisoned, his clothes poisoned, the furniture and room tainted or filthy; he will say that his food contains blood or human flesh; and all these fancies make him refuse to eat, and very violent and dangerous without being melancholic.

The prognosis in such cases is not to be determined by the delusions alone, but must depend on the time they have existed, and on other circumstances which cannot be discussed here.   Generally speaking, however, where such delusions as the foregoing have existed for a twelvemonth, and all the symptoms of acute disorder have subsided, the prognosis is bad, and the patient is likely to remain through life a dangerous homicidal lunatic.

A number of delusions are presented by those who fancy that the change they feel within them is all for the better.   In their bodily or spiritual state, or worldly position and fortune, they fancy they are much better off than before.   All these are said roughly to be suffering from mania with elation; but among them are included those afflicted with the most fatal of all the forms of insanity—general paralysis.   When a man between thirty and fifty-five years of age is full of delusions that he is of great strength, rank, and wealth, we may suspect that his malady is general paralysis, and test him by the rules I shall lay down in a future lecture. But frequently the same delusions appear in men and women who are not paralytic.   They think they have, or are going to make, a great increase of income, that they are going into Parliament, are about to rise to the highest place in their profession, whatever that may be.   All their speculations, however venturesome or absurd, are to turn out very profitable, or they have invented, and are about to patent, new contrivances, which will be a source of endless wealth.

If the diagnosis in such cases leads us to conclude that the patient is suffering from general paralysis, the prognosis is summed up in one word—

the end is death; but we must not conclude too hastily from such delusions that paralysis is present: if it be not, patients often recover from this elated mania, always provided that the duration of the case is not so great as to make recovery hopeless.   In fact, when there is no paralysis, no hallucination, and when the attack is recent, the prognosis is favorable. Neither are they suicidal nor dangerous; suicidal they never are, except by accident; they have too good an opinion of themselves and their position.   They may be dangerous when thwarted, but it will be merely to escape from those who wish to confine them and curb their dignities and projects.   They do more harm to property than to their own lives or those of others, and in their elated condition will squander a fortune in a few days or involve themselves in endless liabilities.   And where recovery does not take place they often spend their existence very happily under a restraint which does not check their fancies, while it provides for their safety.

I now come to the subject of hallucinations which are met with daily amongst the insane, but, unlike delusions, are found also in people who it must be conceded are sane.   Difficult as it is to try to explain with anything like exactness the origin of delusions, it is still more difficult to account for hallucinations.   We derive some assistance, however, from the sane and insane persons who can give an account of the rise and nature of these false perceptions of the senses; and I may say, I believe, without fear of contradiction, that although hallucinations do not of themselves prove a patient to be insane, yet they accompany and indicate a disordered condition of brain.   They come and go as the brain health and force fail or improve, and point to a defective state of brain circulation and organization no less than delusions.   The difference is, that being mere perceptions—auditory and visual sensations—they are not necessarily compounded into judgments, and may exist and be recognized as hallucinations by the rest of the brain; whereas a delusion implies a judgment formed out of more than one perception or sensation—a proposition which necessarily consists of more than one term, the falsity of which the remaining portion of the brain is not able to understand.   In other words, an hallucination implies a certain disturbance of nerve-centres causing the false sensation; if the hallucination is acted upon, if something is done because of it, a proposition or judgment is implied, and the patient may then be said to act because of the unsound condition of his entire brain and mind.

Hallucinations are closely allied to, nay, are the same thing as frightful visions and nightmares which occur in children and adults, but do not come in perfect health, and point to some disturbance which ought to arrest the attention of a medical man.   Writers on children's diseases have especially noticed these night-terrors, and all who have seen them will agree that a child thus affected is out of health.   They often are witnessed in children when ill of acute disorders; in fact, hallucinations constitute the chief phenomena of the delirium of the young.

Authors formerly differed widely in their opinions as to the origin and seat of hallucinations, many supposing that they were due to lesions of the external organs of sense, others that the morbid seat was in the supreme centres of the brain.   Those holding the former view pointed to hallucinations perceived by persons suffering from cataract, or other disorders of the eye, which departed after the cure of the ocular defect; others who believed that they are due to disturbances of the highest cerebral centres cited the fact that they are regulated by the prevailing feeling and intelligence of the patient—the melancholic and maniacal man hearing words which correspond to his gloomy thoughts, the elated having his own peculiar visions and "voices."

The researches of Hitzig and Ferrier have, it is probable, solved the question in dispute.   The sensory centres in the cortex cerebri, as described by them, are generally looked upon as the seat of hallucinations. There is so great a resemblance between them and delusions, that it seems likely that the cerebral centres are the seat of both.   For both there must be a seat, and a want of proper relation between it and other centres.   We undergo hallucinations of sight in many ways in our daily life; we think the railway carriage we are in is moving when another train passes alongside, but we can immediately correct this by comparison.   So the hallucinations of ordinary illness are recognized as such, and corrected by reasoning, or by other senses.   But the hallucinations of the insane, like their delusions, are not corrected, and the entire brain does not co-operate, the relations of one part to another are interrupted or distorted, and normal mental activity is suspended.   An interesting paper on hallucinations has been written by a physician in Moscow, Dr. Kandinsky,[1] who was himself insane for two years, and affected with hallucinations of all the senses, with the exception of that of taste.   In the first month of his illness there was no hallucination, but an irregular mental activity, a race of delusions and involuntary thoughts.   The hallucinations began after the brain was exhausted by the rapidity of thought, and an anæmic condition produced through voluntary abstinence from food.   He thinks his observations confirm Meynert's theory that hallucinations are no proof of excitement of the cortex, but rather a proof of the abatement of its activity.   This theory appears to be borne out by the phenomena of hallucinations which occur in the sane, but are recognized as hallucinations.   A lady of my acquaintance, when out of health, always saw a cat sitting on a particular stair.   She was not averse to cats, nor afraid of the spectral cat, but it was to her an index of the state of her health; tonics and wine removed it, to return when next she fell into a weak state.   There was no defect in this

---

[1] Journal of Mental Science, xxvii. 430.

lady's eyes, nor in her mind, yet some disturbance in her brain apparatus brought up the image of a cat, which her other brain centres, being sane, recognized as an hallucination. If the latter had been disordered, and in a state of delirium, they would have been unable to perceive the falsity of the spectral appearance, and she would have been in the condition of those who, suffering from delirium tremens or other like disturbances, see snakes, rats, or birds in the room.

Patients may dispel, or fancy they dispel, hallucinations of hearing by stopping their ears, or putting cotton-wool in them. The relief may be due to association of ideas, or the exclusion of sound may preserve the morbid centre from stimulation. It is a fact that by some the "voices" are only heard when the patient is lying in a horizontal position, the brain circulation being probably modified thereby.

A good deal may be said about hallucinations with reference to diagnosis and prognosis, but it will be well to examine the hallucinations of the various senses separately, for in some respects they differ essentially.

The hallucinations which arise in the state of greatest disturbance are those of sight; consequently, in acute cases of insanity, as in acute disorders of the sane, we find them more frequently than any other kind. They rise and vanish with the acute stage: in chronic cases we find them less often than hallucinations of hearing.

Hallucinations of sight may be simply flashes of light, shadows, colors, or fires, or they may be objects: sometimes they commence as the former, and merge by degrees into the latter. I have already mentioned the hallucinations found in that disease which we term delirium tremens. In it by far the larger number are hallucinations of sight, visions of birds, animals, or snakes, in the room or on the bed. So, in fevers and other acute diseases, the wanderings and fears of the patient relate to what he sees rather than to what he hears, till at last he picks the imaginary flies from off the bedclothes in that stage which only precedes by a little the time when sight altogether fails, and he declares the room to be dark, though the sun is shining, or candles are burning in it.

Hallucinations of sight amongst the insane are frequently visions of the supernatural, especially in non-acute cases; patients see angels or visions of the Deity in some form or other, or spirits floating in the air in the shape of birds; they may see the form of departed friends or heroes, or of the absent: they may also see fiends, spectres, or the devil. But, on examination of my notes of a great many cases, I find that the hallucinations of sight which most strictly deserved the name were observed far more frequently in acute than in chronic cases—in cases where there was at the time great cerebral disturbance with violent emotional display, heat of head, and want of sleep. Epilepsy in the insane is constantly followed by hallucinations both of sight and of the other senses.

Certain phenomena, termed by some hallucinations, ought rather to be looked upon as delusions: conspicuous among them are the mistakes of identity so common amongst the insane. A patient declares a stranger to be a relation or friend, or declares a near relation is not the person, but somebody else—says her husband is not her husband, but a stranger, yet, possibly, asks after all at home—says the men in attendance are women, or the women men, or calls the medical attendant by the name of some former friend. Very curious are many of these assertions; but when they are made by patients who are free from all acute symptoms and can talk calmly on most points, we must look upon them as delusions of idea, and not as hallucinations, or even illusions of sight. They partake of the general change of feeling existing in the individual, which is projected outward, and extended to all he sees. He thinks his wife and children are changed, just as he thinks himself changed into a most miserable or most exalted person. The mistake is in his idea-region, and not in his organs of sight or sensory centres.

As hallucinations of sight occur in the acute rather than in the chronic stages of insanity, they do not warrant an unfavorable prognosis, inasmuch as there is more hope of recovery during the acute than during the chronic period. The prognosis of the case will depend, not upon these, but upon other symptoms, and the diagnosis will, by the same rule, be attended with no difficulty. In chronic cases I have generally found that where hallucinations of sight existed, others, especially of hearing, existed also. Of all the hallucinations caused by such substances as haschish, opium, and the like, those of sight are the majority.

Hallucinations of sight often occur in the dark, in which patients see not merely flashes of light, such as might be attributed to irritation of the optic apparatus, but actual objects—men or animals. This proves that the external apparatus is not the seat of the mischief, which is corroborated by the fact that blind persons are subject to hallucinations as well as those who can see. Curiously enough, many patients can dispel these hallucinations by closing the eyelids or covering the eyes. This is easily explained by the association of ideas. Being habitually unable to see anything with the eyelids closed, they do not see the phantoms of their diseased imagination.

In non-acute insanity, hallucinations of hearing are the most common and most formidable. They are difficult to eradicate, and while they exist they render him who hears them the most dangerous of patients. In chronic cases, or those in which insanity has revealed itself gradually and insidiously with few acute symptoms, hallucinations of hearing form at least two-thirds of all that we meet with. And when a patient of this class tells us that he or she hears "voices," not mere sounds, whistlings, humming, or the like, but words and sentences, we augur unfavorably of the case; our prognosis is gloomy, and our diagnosis is that such a one requires close watching and restraint. I have known patients lose the

fancy of mere sounds.   One gentleman used to hear "blowpipes" whistling down his chimney, and whistling at him in the street, and these by
degrees vanished, but his mental health was never fully restored.   He
leads in solitude the life of a hypochondriac, and when he is a little more
nervous than usual, he hears singing in his ears, and shows symptoms of
his former hallucinations.   Patients hear voices either of those they know,
or of unknown persons, natural or supernatural.   As no one is to be seen,
they generally imagine that the speaker is in the next room or house, or
in a cupboard or chimney.   Of course, if it be a supernatural voice, it
may come from the air inside or outside the house.   One lady was so annoyed by voices coming through the wall that she purchased the adjoining
house to compel them to cease.   I need not say she did not so get rid of
them.   Since the invention of telephones patients often think that by this
method the sounds reach them.   The whole life of many is regulated by
the commands they receive from "the voices."   They eat, drink, walk,
and sleep according to the commands they hear, and if compelled to act
contrary to them, they tell us that they will suffer for such disobedience.
They obey implicitly, holding the voice responsible, and so will commit
frightful crimes without looking upon themselves as guilty or responsible.
And as many patients will not reveal what the voices say, it follows that
the whole class is eminently dangerous and uncertain.   It often happens
that there is great difficulty in extracting from a patient the confession
that he hears voices, or that which the voices say.   He appears afraid to
tell, and seems bound down by some kind of compact not to do so.   He
thinks it a point of honor to conceal what passes, and it is only by our
overhearing him answering an imaginary conversation that we ascertain
the fact.

We may sometimes detect hallucinations of hearing in patients who
have never revealed them, by noticing while we talk to them that they
from time to time are inattentive, and appear to be listening to some one
else.   On being pressed, they will probably confess that they hear some
one speaking.   There is often the closest connection between the delusions of a patient and that which he hears.   The latter is, in fact, the delusion done into audible words.   Thus, a patient who hears a voice accusing him of various crimes, unnatural lust, and the like, entertains these
delusions at all times.   But there is great variety in what the voices say,
and sometimes it appears to have no connection with other well-known
delusions to which the individual is subject.

In some patients these hallucinations seem to deserve the name of illusions, for though they take the form of voices and intelligible words, they
are not heard in perfect stillness, but only when there is a noise going on;
which noise, whatever it be—footsteps, the rattling of a door or window,
or the wind—is converted into "voices."   On the other hand, it is often
difficult to distinguish between such hallucinations and mere delusions,

to say whether the patient imagines that he is falsely slandered and accused, or hears voices repeating the words. And yet our prognosis may be materially affected by the one or other of these symptoms.

Hallucinations of smell belong to the acute states of insanity rather than to the chronic; and when the patients get better, they vanish. Some will tell us that they smell fetid and noisome exhalations, the scent of the dead, or of vaults and catacombs, or say that their food or drink smells offensively, and for this reason refuse it. More frequently, however, they assert that an offensive odor proceeds from their own bodies, rendering them horrible to all about them, and contaminating the chairs, sofas, or beds they rest upon. This was the case with a gentleman, the subject of melancholia, who would only sit on a cane chair, because he thought that he polluted a stuffed one. A young lady always presented me with her smelling-bottle, because of the odor she believed to exhale from her, which must be disagreeable to me as I stood and talked to her. Both these patients recovered from somewhat acute attacks of melancholia.

Hallucinations of taste are so interwoven with disordered sensations, depending on the state of the tongue or alimentary canal, and with delusions concerning the food, that it is difficult to lay down anything with precision on this head. The patients who say that baby's flesh, human blood, human excrement, arsenic, or other poison, is put in their food, do not fancy they taste these substances. The idea is a delusion, not an hallucination. From my own experience, I am inclined to believe that true hallucinations of taste are uncommon, that they rarely exist alone, and that, depending on a disordered state of digestion, they generally are transient.

A number of curious phenomena may be grouped under the title of hallucinations of the skin and muscle. It is not uncommon to find a patient who calmly tells us that he feels himself touched on various parts of his body by little raps or shocks. A young gentleman felt these long after every other symptom of insanity had disappeared. They latterly gave him little concern, though at first he attributed them to supernatural causes. He had no other hallucinations. Others feel electric shocks, but these are often the subject of delusions rather than of hallucinations. Some feel snakes in their inside. One lady had a dog in her head, and complained of the sensations it caused. This I looked upon rather as a delusion. Some patients declare that they have felt persons have sexual connection with them. These are onanists, or patients with morbid sexual desires. Others constantly declare themselves to be with child, and say they feel it moving, or labor approaching. These are all delusions. Disordered muscular sensations, the feeling of being bound and incapable of moving, are akin to that experienced in nightmare, when we are pursued by a lion and cannot run away, or are falling from a precipice.

Patients may experience hallucinations sometimes or always. One gentleman told me he could always hear the voices if he listened for them. Some only hear voices when there is a noise of some kind in or out of the room. In all who are subject to hallucinations, we find them most frequent and most distressing when the bodily strength is lowest. They point to an exhausted nervous state even more than do delusions, especially the hallucinations of sight, which generally exist in acute diseases, as fever and delirium tremens, or can be produced by such poisons as haschish, stramonium, or belladonna.

LECTURE VIII.

The Acts of the Insane—Stripping Naked—Indecent Exposure—Fantastic Dress—
Eating and Drinking—Habitual Drunkenness—Suicide—Self-mutilation—
Talking to Self—Squandering Property—Homicide, for various Reasons—Py-
romania—Erotomania—Kleptomania.

BY the discovery of false beliefs or delusions, we are led to the conclusion that patients are insane. Let us now consider insane *acts;* for by
them, no less than by delusions, we may be guided to the same opinion. As
we find insane persons who have no delusions, so we see others whose unsoundness of mind is not displayed in acts, or, at any rate, in such of them
as are observable by us; for we should learn a great deal more of many
patients if we could watch them without their knowledge. Without discussing in this place the diagnosis of insanity, and its recognition in those
who commit insane acts without delusions, or have delusions without displaying anything insane in act or conduct, I wish to review the chief acts
which arise from the insanity, and are the best evidence of it. As insane
beliefs range themselves under certain heads, so we shall find that similar
acts are perpetrated by many insane people, emanating without doubt
from a corresponding mental condition. As the discovery of certain delusions leads us to inquire whether the individual has done, or attempted
to do, certain things, so the acts help us to discover the delusions. Many
acts which, taken by themselves, would not prove insanity, are, when explained or justified according to the insane ideas of the individual, valuable
evidence of his state of mind, and often afford a clue which otherwise
would be wanting.

We may roughly class insane acts under two heads—those which affect
the person or property of the patient, and those which affect the person
or property of others. Under the first head, we shall consider such acts
as stripping off all clothes, indecent exposure, fantastic dress, self-mutilation, starvation, suicide, dipsomania, squandering of property; while
under the second we find homicide, arson, rape, and acts of violence or
mischief of innumerable kinds. Every one of these may become a subject of investigation in a court of law, and it is as well to examine them
by themselves, just as we examine and consider delusions and hallucinations.

Stripping stark naked is not unfrequent amongst the insane. And
in this condition they will run out of their room, or the house, regardless
of decency. Putting aside all the cases where patients have got rid of

their clothes in a struggle, we shall find that they strip themselves either from a desire to destroy everything within reach, or from a wish to get rid of the feeling of heat or restraint engendered by clothing. In the latter case they may remove without destroying it; this they may also do thinking it filthy or poisoned, or may ruin it from pure mischief and wrong-doing.

In acute delirium, acute mania, melancholia, or general paralysis, it is common to find patients stripping off their clothes, and tearing them to pieces to get them off. The feeling, whether of restraint or of heat of skin, is one due to the physical condition, and they accomplish their end without assigning any cause whatever. Such patients are beyond the reach of argument or expostulation. The symptom, like the disorder, is of a temporary nature, and must be met with measures best suited to the case, by fastening blankets round the patient, or by placing on him a suit of strong material laced up the back, which he can neither tear nor remove. It is essential that many of these persons should be kept warm, as also those affected with general paralysis, who are given to stripping themselves if left alone at night.

Chronic maniacs who are noisy, mischievous, and destructive, but perfectly conscious, destroy their clothes from pure wantonness, just as they smash furniture and windows, or befoul their beds and rooms. Blankets or strong suits are of little use for them, as they are quick and ingenious enough to pick these to pieces by dint of nails and teeth, and, if left alone, each morning discloses a scene of rags and pieces, the result of a diligent night's work. These patients are good illustrations of the nonsense talked about the knowledge of right and wrong being a test of sanity. They know as well as any one that what they do is wrong, and they delight in doing it, because it is wrong. I have never found any expedient of use but the presence of an attendant, or two, if necessary. When the patient finds that he is not allowed to destroy, he gives up, and in time loses the habit and desire.

In a case where the clothes are stripped off or destroyed in conformity with certain delusions, such delusions will commonly be assigned as the reason. These must be met like any others, and will pass away as the patient improves. They generally belong to an acute stage, and are found in connection with hallucinations of smell, or sensations of the cutaneous surface.

As a matter of diagnosis little need be said in connection with this subject; the insanity of such patients will be patent in many other ways; at the same time, sane people do not strip themselves naked except for justifiable reasons, and such stripping may be adduced as evidence of insanity along with the other symptoms which are sure to be observable.

The same cannot be said of the next topic I shall mention, indecent exposure of the person. This is constantly practiced by sane people, as the records of our criminal courts prove. It is done by the insane both in the acute and chronic state. It is a not unfrequent symptom in the early stages of general paralysis; it is often done by patients, men and women, who are passing from the stage of chronic mania to that most hopeless condition, chronic dementia. We are not to infer insanity from such an act unless there be other corroborating symptoms; but when these leave no doubt in our minds that the patient is insane, our prognosis of the case, whether it be acute or chronic, will be unfavorable. I am speaking, of course, of a deliberate act of indecent exposure, not of the accidental exposure which occurs when an excited patient strips himself or herself entirely.

Open and shameless masturbation is a common occurrence in patients both in the acute and chronic state.

We shall constantly see something bizarre or extravagant in a patient's dress and general appearance, which, if it does not of itself prove insanity, may at any rate reveal the delusions which prompted it. The hair, the condition of face and hands, the clothes, articles attached to the dress as ornaments, the state of the nether garments, shoes, and stockings, may all betray singularity, and corresponding singularity of ideas. One might fill a volume with anecdotes of the extraordinary appearances presented by patients under the influence of delusions. There are few who dress just as they ought, if they have the opportunity of giving the rein to their fancy. The man of exalted ideas wears a crown or coronet of paper or straw; orders and bits of ribbon adorn his buttonhole; he winds shawls, rugs, or sofa covers round him for robes. The melancholy man thinks it is not worth while to wash his face or body, or brush his hair; unwashed and unkempt, in ragged or untidy clothes, he mourns his destiny. The apartment in which a patient habitually dwells frequently presents appearances which correspond to his personal dress and demeanor, and may vary from the height of eccentric tidiness to the extreme of filth. Constantly, if it be one which the patient has tenanted for some time, we shall find signs of singularity of conduct and idea. Many choose to keep all their provisions and do all their own cooking in their living-room, from fear of poison: others sleep in strange fashion; others keep numberless animals. In every case where we notice oddity of appearance in an individual or his surroundings, we are to bear the facts in mind, and question the patient as to their meaning; but we are not in every case to infer insanity, for eccentricity may exist to a great extent without insanity, and is often displayed in outward adornments. If, however, we are called to a person who has never been eccentric in demeanor or dress, and who suddenly decks himself in a strange and unwonted garb, we may do more than suspect insanity.

The degree of dirt compatible with sanity is a question which will vary according to the experience of the observer. Those who are familiar with the beggars of the south of Spain or Italy will think this test very uncertain; but here, too, the same rule is applicable. If an educated and hitherto respectable gentleman appears in a state of deplorable filth, we may well question his sanity.

Our prognosis will not be determined by such appearances until further information is obtained. If the case is recent, the oddity of the appearance is of little moment; if chronic, some very slight token may point to a delusion which may have existed for years, and is not likely to be removed.

A patient's eating and drinking may arrest our attention. He may eat voraciously, or very little, or absolutely nothing. He may eat only such things as eggs, from fear of poison, only the food cooked by himself, or may require others to taste everything before he partakes of it. He may eat filth of all kinds: in fact, there is nothing too nasty to be carried to the mouth by the insane in all stages of the disease. Such acts are evidence of unsoundness of mind, generally of dementia or idiocy. Absolute abstinence from food would also warrant our coming to the conclusion that the individual was unable to take care of himself. Between these extremes, singularities of diet and mode of eating, like eccentric personal appearance, point our way to the detection of insanity, but are frequently not in themselves evidence thereof. Cases are on record where patients have swallowed such articles as lancets or a pair of compasses, and have habitually eaten nails, coals, rags, tobacco-pipes, etc.; but we know, from the post-mortem examinations of our general hospitals, that these depraved and perverted tastes are not peculiar to the insane, but are indulged in not unfrequently by sane persons. The eating of hair by girls, and the chewing of slate-pencil, are by no means uncommon—and I have heard young ladies descant on the pleasures of the latter.

It may be as well to mention in this place that craving for alcoholic drink, which has been called by some a monomania, under such titles as dipsomania, oinomania, methyskomania, and by others is merged in the class of *moral insanity*. Perhaps, for the practitioner and the medico-jurist, this is the most perplexing of all varieties of unsoundness of mind, if—and this is a moot point—it is of itself unsoundness of mind. I have already, in my third lecture, said something concerning the pathology of alcoholism. I have here to inquire whether drinking to the extent of what is called dipsomania is an insane act. There can be no question that there are hundreds of habitual drunkards who are in no sense insane. The very fact that a certain proportion of them abandon the habit, reform, and for the rest of their lives live temperately, shows that they are not suffering from chronic insanity. To them inebriate asylums are a boon—nay, a necessity, not because they are insane, but because they require assistance to enable them to break through a bad habit, like an inveterate smoker, snuff-taker, opium-eater, gambler, or masturbator. There is,

again, a number of insane people who drink, their insanity being shown
not merely by drinking, but by their whole history and insane acts and
ideas of various kinds. But in addition to all these, there is a class of
patients who are insane and suffer from a dipsomania or oinomania, which
is in truth a form of insanity to be distinguished from mere habitual drinking, though the latter may be the chief symptom, and may so overshadow
the others, that they almost, if not quite, escape notice. Such people are
generally members of an insane family; they show the first symptoms at
an early age, and in a majority there is a periodicity in the attacks wherein
they resemble other insane patients, and do not resemble the ever thirsty
and habitual drinker. This symptom, however, is not constant, and some
oinomaniacs whom I have known had no remission of the disorder, but
were, like other lunatics, in a state of perpetual drink craving. To distinguish them from mere habitual drunkards is not easy, and in many
cases where the family history leaves little doubt, we may still shrink
from pronouncing them legally insane. The cure is well-nigh hopeless,
though the periodical attacks may leave them free for a longer or shorter
time. The habitual drunkard may be cured by prolonged abstinence,
and for this reason I greatly wish that the machinery for legally controlling
him may become the law of our land.

Both the habitual drunkard and the oinomaniac may drink himself or
herself into brain disease—acute or chronic alcoholism, the latter more
rapidly attacking those who have a strong tendency to neurotic disorder.

We may notice in the next place the acts of self-mutilation and self-
destruction so often committed by the insane. If we walk round the
wards even of a small asylum, we shall rarely fail to find some one or more
patients who have tried, with more or less success, to do damage to themselves. They may be laboring under suicidal melancholia, or other forms,
as that which is called suicidal monomania or suicidal impulse. Critics
and writers on the subject vary greatly in opinion, some thinking that all
who commit suicide, are insane, others that delusion must be ascertained
before we can pronounce any suicidal or homicidal patient to be of unsound mind. I have coupled suicidal and homicidal patients, for the condition of the one is often closely allied, or even identical, with that of the
other class. The same patient often commits both homicide and suicide,
or at one time is homicidal, at another suicidal; and so, in reducing to a
number of heads the patients who are homicidal, we find that we can
range in almost identically the same divisions those who are suicidal. It
is of the latter that I shall first speak.

That sane people commit suicide is a fact that must be apparent to
every one who exercises common sense in looking upon the subject. The
hundreds of poor creatures who are rescued from the Thames, or brought
to our general hospitals half poisoned or with throats half cut, are not
insane in any medical sense of the word. Putting these aside, let us look
at the insane who are suicidal.

I. First, we have the melancholic patient, who has been noticed by his friends to be a little low-spirited, but nothing more. They have not heard of any delusions; he has not done or said anything that could warrant their calling him insane. He has only appeared changed in spirits and capacity of enjoying himself, and this they have thought it better not to notice; so he blows his brains out, or jumps from the top of the house, and then they are extremely anxious that he should be called *insane*, and not *felo de se*. This is pure suicidal melancholia, insane *tædium vitæ*, where, without any marked or overwhelming delusions, the whole feeling of the individual makes him look on life as not worth the keeping. He is perplexed and annoyed with everything and everybody:

> "Weary, stale, flat, and unprofitable
> Seem to him all the uses of this world."

And so he ends them. We see the insanity of the man in that he is entirely changed from what he was; there is no cause for his depression, but perhaps there is cause for his insanity, his suicidal melancholia.

II. The malady may still be fitly called suicidal melancholia; but the desire to commit suicide may be directly prompted by delusions, *e.g.*, that he is going to be horribly tortured; or he hears a voice commanding him to kill himself; or thinks that by this he shall gain heaven; he is ruined, and shame and poverty are staring him in the face; or he sees visions of the departed beckoning him to come to them. This form is easy of diagnosis, and the prognosis is favorable if the general health be not much broken. Of this I shall have to speak when I come to the subject of melancholia.

III. In almost any case of acute insanity—in delirium tremens, in epileptic mania—suicide may be committed in fear, in a paroxysm of rage, or a general outburst of destructiveness; or in attempting to escape a man may jump from housetop or window without any definite idea of self-destruction. In all cases of acute insanity, as acute delirium, acute mania, and the like, this must be borne in mind, and opportunities of self-harm removed from a patient. Many at this time are subject to paroxysms of ungovernable fury, and will then try and hurt themselves as well as others, will dash their heads against a wall, bite their arms, or even do more serious mischief if they have the chance. We cannot say that they are suffering from suicidal melancholia or suicidal mania. Suicide is like breaking the windows, or tearing in pieces their clothes or furniture—a mode in which their vehement destructiveness finds vent. Of the insanity of such persons there is, of course, no doubt, and the prognosis is not affected by the fact that suicide has been attempted.

IV. Besides these, we find true suicidal mania and suicidal impulse. Patients both in an acute and unmistakable state of insanity, and in one that is hardly recognizable as being insanity except in this one feature, have a violent desire and longing to commit suicide. According to the capacity of self-control, they may make the attempt all day long, so as to require an attendant to be constantly within reach; or may keep it to themselves for days, weeks, or months, till they see a fitting opportunity of satisfying the desire. Although in a state of marked insanity, their delusion may have no connection with suicide, and their feelings and demeanor may be the reverse of melancholic. The most suicidal patient I ever saw was a friend who was perfectly free from everything like low spirits, and whose delusions, of which to me he made no secret, had no reference whatever to any of those things which usually prompt patients to suicide. One might have conversed for hours with this gentleman, and given an opinion that he was a patient not likely to be suicidal, yet he would do himself harm in every way he could—swallow scalding beef-tea, try to strangle himself on a five-barred gate, in short, in some way or other hurt himself. He went on in this way for some two years, the suicidal impulses occurring in paroxysms every two or three months, and then died epileptic.

Suicidal impulse occurs in children, who, when they commit this act, do so usually not from delusions or from settled melancholy, but from pure motiveless impulse. There is an uncontrollable desire to gratify the morbid idea which is present in the mind, as in dipsomania there is the desire to gratify the physical craving of drink, regardless of all consequences. Just as, in the case of a patient suffering from some foolish and absurd delusion, the rest of the brain is not sufficiently at work to correct the falsity thereof; so here it is not sufficiently at work to correct, by ideas of duty, prudence, and self-love the feeling and idea which urge to self-destruction. Probably, in many cases both of murder and suicide, the insane man or woman experiences in an insane degree a feeling of anticipated pleasure or curiosity akin to that felt by the spectators of the horrors of the amphitheatre or auto-da-fè of past times, or the bull-fights of our own days—a morbid delight, nay, a keen enjoyment of a ghastly spectacle.

Besides those who try in every way to put an end to themselves—who refuse food for the purpose, and lose no opportunity of doing to themselves bodily harm, however trivial—there are patients who, for definite reasons, will damage or mutilate some one or other part of their bodies.

Some, following the literal precept, will pluck out their eye or cut off their hand. I had a lady under my care whose right eye was destroyed under these circumstances, and have recently seen a gentleman who attempted to do the same. Both recovered their reason. Frequently the organs of generation—the penis, or testicles, or the whole—are cut off, because they are supposed to have " offended," to have incited the patient to wicked lust, or to have been the means of gratifying them through masturbation. Patients will set themselves on fire, place their hands in the candle or coal, to inflict upon them the torments of hell, and will display a fortitude or an indifference to pain in so doing that is truly marvelous. All such acts are done from delusion, not from mere impulse, as the acts of suicide. Whether it be from one or other motive, such patients demand constant and unceasing vigilance on the part of those about them. They are safe nowhere save in an asylum, and even there they are the source of endless anxiety. Being for the most part in full possession of their faculties, they never let an opportunity escape of secreting a knife, a nail, a piece of broken glass, or a bit of cord. They will do themselves mischief under the bedclothes with an attendant sitting by their side, and not indicate it by a muscle of their countenance; and when all weapons are taken away, they will gouge out their eyes, or pull down the rectum with their fingers.

Besides desperate injuries inflicted on themselves by patients, either through strong delusion or expressly from suicidal motive, many daily do themselves smaller mischiefs without any motive at all. This one picks his face or hands till he covers them with sore places, from restless fidgetiness; another bites his nails down to the quick; another plucks out hair after hair from his head till he is nearly bald. Such tricks, less perhaps in magnitude, are common among nervous people who are not insane. Most nervous persons have some trick or twitching, and, like the insane, twist and fidget their dress, or pick and scratch their hands, face, head, or limbs: the extent to which they do it at any given time is often a valuable criterion of their nervous condition.

There is another common habit of the insane which must be noticed, that is, talking to themselves. Many people who are sane talk to themselves: they automatically convert into audible words those they are combining in thought, so that this talking to self is not to be reckoned as a sign of insanity. But it is often done by the insane in a way and in a tone that is strongly symptomatic of the disorder; and frequently, if we listen, we find that they are holding conversations with imaginary beings, and answering imaginary voices. And by so listening valuable information may be gained as to what is passing in the patient's mind—information which he will not give us when questioned.

Hitherto I have spoken of what is done by a patient, so far as it relates to his own person. But he may also squander, destroy, or give away his property, and measures will be necessary to protect this no less than himself. But the man who squanders or destroys his property is not the same as he who harms himself. The latter is commonly a melancholic, who thinks he has no property, or, in the overwhelming fear and anxiety that possesses him, neglects it entirely. The man who squanders it is he who in his exalted notions thinks himself a millionaire, or, imagining he can find a source of endless wealth, beggars himself to discover his *El Dorado*, or makes presents to every one he meets, and orders silk dresses or jewelry to be sent to all he knows. In every asylum we see such patients. They spend their days in writing letters to every shop whose advertisements they see in the *Times,* and ordering goods to the extent of hundreds or thousands of pounds. And before they are put under legal control such orders are often given, and great trouble is experienced in getting them annulled. Then there is the senseless squandering of money by the imbecile class, who have not brains enough to know its value. We may see the same thing done by patients in the first stage of paralytic, senile, or other dementia. This, together with debauchery and indecent conduct, is often the earliest symptom of the decay of mind.

I now proceed to consider insane acts committed upon the person or property of others. In this category we shall have to examine the so-called homicidal monomania; pyromania, or arson; erotomania, or rape, and kleptomania, or pilfering. And first of homicide, the gravest question of all, as an excuse for which it has been said that the homicidal monomania has been invented. Here, as in the case of suicides, some writers and physicians hold that the crime may be committed from moral obliquity equivalent to insanity, which they term "moral insanity;" others affirm that murders are committed from sudden insane impulse; while many, especially lawyers, require that delusions shall be found before they admit that the perpetrator is insane, and even then do not allow that in all cases the insanity is a bar to punishment. This is not the place for the consideration of moral or impulsive insanity. I shall confine myself at present to the homicidal act, as it is committed by this or that patient, assuming that impulsive insanity has at any rate a recognized existence.

1. Homicide may be perpetrated without any impulse or delusion by a patient enraged at being restrained, simply to free himself from detention. Many of the attacks to which attendants are subject are of this nature.

II. It is frequently done under delusions of various kinds. The patient thinks that he is himself about to be tortured, murdered, or led to prison, and he slays the man who is his fancied enemy. He will kill his wife or children to save them from worse ills which he thinks are to befall them. He fancies that he has a Divine mission to murder some one, or hears a voice urging him on, or the voice of some one he knows insulting him—and this will probably be the voice he hears oftenest, his attendant's or doctor's. The delusions vary indefinitely; but where the crime is committed from such a motive, it is generally confessed, and the insanity is manifest. It is not possible in every case to connect the delusions of a patient with the act of murder, yet we are not on that account to assume, as do the lawyers, that there is no connection between them. Whence delusions spring, and whither they lead an insane man, is what no one can tell. To attempt to trace a madman's ideas from their source to their outcome in act is at all times an unprofitable task.

9

III. Murder is committed by an idiot, imbecile, or demented patient, from wanton mischief or folly, from ignorance of the act or disregard of its consequence, or from mere imitativeness. An idiot may kill a child because he has seen a fowl or a sheep killed.

In all the preceding classes the unsoundness of mind of the homicide is patent. There is no question of diagnosis. Lawyers may quibble about the responsibility, but medical men will detect the insanity, which is not left to be inferred from the homicidal act alone.

There are also acts of homicide committed by persons whose soundness of mind was never called into question before the murder, and they may be arranged in three classes.

IV. Homicide is not unfrequently committed by epileptics. Epilepsy is not the equivalent of insanity. Many persons suffer from epileptic fits for years, yet mix in society and discharge the duties of sane and responsible people. But it has been asserted, and not without good grounds, that the mental condition of an epileptic is not thoroughly sound. "All authors are agreed," says Baillarger [1] "in admitting the fact that epilepsy, before leading to complete insanity, produces very important modifications in the intellectual and moral condition of certain patients. These sufferers become susceptible, very irritable, and the slightest motives often induce them to commit acts of violence; all their passions acquire extreme energy." An act of murder may be committed by an epileptic in the furious mania with hallucinations and delusions which follows a fit or succession of fits. It may also be done in the period which precedes a fit, a period during which especially some patients show strangeness of mind and manner, or in the state of unconsciousness which takes the place of, or follows an attack. And the convulsion may be exploded, as it were, in the act of violence, the fit not occurring as it otherwise would have done. Murder may also be committed by patients rendered weak-minded by fits. In 1869, Bisgrove, an epileptic, was tried and found guilty of murder, but afterward removed to Broadmoor Asylum. He saw a man lying asleep in a field; he did not know him, but took a big stone that was lying near, dashed out his brains, then lay down by his victim's side and went to sleep. From epileptics generally we may expect acts of sudden and unaccountable violence, whether they occur in close connection with fits or take the place of, or follow them.

V. Murder may be committed during a paroxysm of insanity, brief, but furious, which, from its duration, we may call transitory mania. Here, while it lasts, the insanity will be recognizable; but as it rapidly passes off, and possibly has never occurred before, the difficulty will be to discover any traces of unsoundness of mind a few days after the event. Such attacks are really for the time paroxysms of acute mania. The

---

[1] Ann. Medico-Psych., Avril, 1861.

patient afterward may not be conscious of what he has done, and by his expressions at the time may or may not indicate the feeling or delusion that prompts him. Frequently mania occurs in patients suddenly waking out of sleep, and is of the character of a nightmare, probably a continuation of some horrible dream. The act is often committed in a state of panic rather than rage, and the committal may thoroughly bring the patient to his waking senses, and the contemplation of what he has done.

VI. Lastly, homicide, like suicide, may be committed from "homicidal impulse." We are told that a man is impelled to commit murder by a craving impulse, like the craving thirst for drink. It is in cases of homicide that this variety, impulsive or instinctive insanity, is chiefly alleged, and criticised; and, merely mentioning here that homicide may be committed in this fashion, I postpone the discussion of this insanity to a subsequent period.

Besides the homicidal monomania, we hear of others, as *erotomania, kleptomania, pyromania.* Having already said that I do not consider homicidal monomania to be a specific disease, I still less acknowledge that the acts which these terms indicate proceed from special disorders. They are committed by insane patients of various kinds, but the insanity is not likely to be confined to one of these acts, and is sure to be noticeable in other ways, if it exist at all. To take the last mentioned, *pyromania,* we might as well erect into a special form the window-breaking mania. Patients set fire to the house out of an insane impulse to destroy, as they break windows or anything else. They will do the same thing in order to escape in the confusion. This has happened several times within my knowledge. They may do it from suicidal motive, or from commands received from hallucinary voices. But a sufficient examination of such patients and their history will reveal insanity other than the mania for incendiarism.

Neither is erotomania a special form of insanity; there are numberless patients who in their mental disorder present marked erotic symptoms. The majority of young women in states of acute mania do so, and will assail the physician with words or embraces if he is not on his guard while visiting them; but all this is a result of their insanity; it does not constitute it. Those who have exalted erotomania into a special monomania, say that it is not the same thing as nymphomania or satyriasis; that the disorder is not in the reproductive organs, but in the head, and that it is an error of the understanding, an affection of the imagination, in which amatory delusions rule.[1] If this be so, *a fortiori* we may deny that it is a special variety of insanity; it is a form of disordered intellect, accompanied by delusions. I have known girls conceive an insane love, not for men, but for other girls, and fancy themselves influenced by them,

---

[1] Esquirol, Malad. Ment., ii. p. 32.

and in their power; and I presume this would come under Esquirol's *erotomania;* but in one case these preliminary fancies culminated in an attack of acute mania, and there was nothing about the case which removed it into any special category. In the acute stage, it might have been called *nymphomania* rather than *erotomania.* We may be justified in describing forms of insanity of which the origin is in the sexual organs, or there is marked disturbance of the latter. Such a division would be in accordance with pathology; but a class distinguished by intellectual disturbance, in which love is the chief feature, is not really separable on this account from other forms of insanity where there is disorder of intellect, with or without delusion.

I now come to *kleptomania,* another form of insanity supposed to arise from so-called perverted emotion or impulse. And of this I would say, as of the foregoing, that it cannot be exalted into a class. Lunatics, undoubtedly, steal both in and out of asylums. When the lunacy is beyond a question, little is thought of their stealing; but when it is not so easy to detect, then the stealing is set up as a proof of insanity, if the thief is in a social position to make us astonished at the act.

When we hear of a theft committed by a person supposed to be insane, but not clearly proved to be so, we may suspect one or other of several things.

1. If the thief be a man between twenty-five and fifty-five, we may suspect him to be in a very early stage of general paralysis, in which acts of foolish and unprofitable theft are not unfrequently committed. Here, of course, we should look for the early symptoms of general paralysis, an incipient stutter, loss of memory by which sometimes the so-called thefts may be explained, the occurrence of epileptiform attacks, a general condition of exaltation, and ideas of being able to afford anything.

2. We may meet with that form of insanity called *moral,* when the patient. is either on the high road to general intellectual insanity, or, stopping short of delusions, shows his malady in insane acts which he justifies, in general alteration of character, and intellectual defect.

3. The individual may be congenitally idiotic, or imbecile, or may become demented gradually, after reaching adult age, from brain disorder. This introduces us to a class of half-witted boys and girls, who go sometimes to jail, sometimes to an asylum, according to their good or bad fortune, friends, and circumstances. To lay down the exact amount of responsibility of each of these is no easy task; but one thing is certain, each case must be examined and considered by itself, in view of the general mental capacity, bodily disorders, and history. We shall gain no assistance from constituting out of these weak-minded people one special class of thieves. Whether their propensity is thieving, drinking, squandering their money, or frequenting the company of the lowest of the low, they must all be judged by their general capacity of intellect, and not by one special class of acts alone.

4. With regard to acts of theft committed by well-to-do women in a pregnant state from supposed longings, I confess I should look upon any such with the greatest suspicion, unless there were other corroborative evidence of insanity. Stealing in shops is not very uncommon amongst ladies, if we are to believe what we hear. The impulse to appropriate an article, if it appears that it can be done with safety and secrecy, is one that is not seldom felt by many ill-regulated minds, and to erect this into insanity would be fraught with the greatest danger to society. Acts of violence are common amongst the insane; act of secret and systematic theft are, in my experience, much less frequent, and other evidence of insanity should always be looked for.

LECTURE IX.

The two Extremes of Insanity—Acute Delirium and Stupor—Early Symptoms of Derangement—Insanity with Depression—Treatment, Medical and Moral—Prognosis—Acute Melancholia—Symptoms—Treatment—Prognosis—Mental Stupor—with Melancholia—with Dementia.

In this and the following lectures it is my intention to describe clinically certain patients who may come under your notice in ordinary practice. I shall portray their symptoms, and, so far as is possible, indicate the treatment. I shall also endeavor to connect the symptoms with the pathological varieties of the disease previously enumerated. Concerning some, however, you will be consulted, not so much with a view to cure or treatment, as to elicit your opinion upon the question of their sanity or insanity. Of these I shall speak later; but first I propose to enter upon a description of the forms of insanity which urgently demand medical treatment, and which interest us as practical physicians and pathologists. As a rule, there is no difficulty in perceiving that the patients are of unsound mind; the question is, are they curable, and how are they to be cured?

In what order am I to treat of these? After examining the pathological phenomena presented in the full development of the disorder, I think that there are two forms which may be placed, one at either end of the scale, and that between these all the rest may be ranged.

The patients I put at one end of my list are those suffering from acute delirium—acute delirious mania—which runs a rapid course to death or amelioration in a week or fortnight. Here we see, for the time, entire sleeplessness, incessant action of brain and body, with the evolution of great heat, speedy emaciation, a quick pulse, tongue coated and soon becoming brown, symptoms pointing to an excessive decomposition of every tissue, and a general excess of brain action, which, if it does not cease within a certain time, leads to death by exhaustion without morbid change of structure. No other kind of insanity is so rapidly fatal or calls so imperatively for medical treatment.

At the other end of the list of cases I place patients suffering from what was formerly called "acute dementia." The name has been objected to, and the disorder is better described as one of the forms of mental stupor. In such a patient we see almost the opposite of the former. There is not excess but defect of action and oxidation. The skin is cold, the hands and feet in hot weather are blue with cold, the patient sits motionless, lost, answers no questions, does not appear to understand them, looks idiotic, sometimes almost comatose; the pulse is very weak and slow, the tongue pale and moist, food is taken passively, and sleep is not absent. Such a case often resembles one of chronic dementia, the result of long-standing insanity, old age, or brain disease, and without a history we shall find it difficult to form an accurate diagnosis. However, I am not going to describe it in this place. As the very opposite of acute delirium I place it at the end of the scale, and between the excess of cerebral action and this extreme defect we may range all the varieties of disorder, and try to ascertain, both for the purposes of pathology and treatment, how far their departure in one or other of these directions corresponds to the mental symptoms exhibited. We shall, unquestionably, meet with cases that some would call melancholic, while others will think they ought rather to be termed maniacal—cases that one person will call acute delirium, another acute mania. There are such on the border-land of all diseases—cases of rheumatism that can hardly be called acute—cases which one may term rheumatism, another gout; cases of fever which one calls typhus, another typhoid. Yet, speaking generally, we employ the names acute rheumatism, gout, typhus, and typhoid, not to *describe* a disease, but to denote a number of symptoms which usually coexist in the individual said to be suffering from it. In the same way we may use such terms as mania, melancholia, and dementia. These names do not, as some object, merely denote varieties of mental symptoms and delusions: stupor with dementia is pathologically a different disorder from acute delirium, occurring at a different age and running a different course. It is a malady of youth, while acute delirium attacks patients in the prime of life, and melancholia is most common in those whose vigor is on the wane. Each variety may be due apparently to a similar cause, as a fright or mental shock. Each may be idiopathic, that is, may come on without assignable cause, owing to the inherited predisposition combining with some constitutional disturbance, but when developed they are different diseases, requiring different treatment.

I shall not, however, commence by a description of either of these extreme forms, but shall put before you the earliest indications of mental disorder, and trace their progress and development in the various directions. In the exanthemata, in fevers and allied diseases, there are premonitory symptoms threatening approaching mischief, but not clearly indicating what it is to be, so that we say the patient is "sickening" for something, but cannot definitely say what; similarly, many signs of mental disorder make us apprehensive of coming insanity, but we cannot always say with certainty what form the malady will assume. Speaking generally, the more rapid the onset, the more acute are the symptoms, and the shorter the attack: that which gradually and insidiously comes on, gives us, it is true, a better chance of arresting its progress short of actual acute disorder, but this, if reached, will probably be of considerable duration.

It often happens that a patient is conscious of there being something amiss with him for a very considerable time before he says anything about it, or any of his friends notice it.   Frequently friends, and relatives also, if they be not near relatives, are afraid of mentioning what they observe, even to the patient himself.   This period of *alteration* precedes many forms of insanity, and I shall have again to allude to it.   It is of the greatest consequence that treatment should be at once adopted, but often a long time elapses before the doctor hears of it—time which would have been most valuable in trying means of relief.   Patients are sent to asylums whose insanity is stated to be of a week or a fortnight's duration, but who have been thought odd by servants or others perhaps for months.   Many complain at first of confusion and dullness of head, of disinclination or inability to do their day's work; often they suffer headache; almost always they sleep badly; periods of depression alternate with periods of excitement.   If there be less and less sleep the advance will be rapid, the depression more marked, the excitement more irrational; and now fancies, at first transient and recognized to be fancies, afterwards permanent and unmoved by the arguments and demonstrations of friends, vex and torment the patient, and drive him to acts of insanity.

I propose in this lecture to speak of patients whose insanity is characterized by depression, who are suffering, to use the technical term, from melancholia.   I do this for two reasons: first, because the slighter forms of melancholia are those which you will most frequently meet with in ordinary practice; secondly, because depression is the first thing noticed in many other forms of insanity, though the depression may afterward be converted into symptoms of a totally different character.   It has been said by distinguished physicians that every variety of insanity commences with depression.   My experience enables me to say that to this rule, there are many exceptions.   Nevertheless, it is witnessed in many cases, and I cannot impress this upon you too strongly, for serious dangers and difficulties may arise if it is overlooked, and if, instead of a mild case of melancholia, it turns out to be one of very acute mania.

After the early symptoms already mentioned, which are physical rather than mental, and are called the period of alteration, well-marked mental depression may be noticed.   This depression will vary in degree, being slight or very intense, and yet distinct delusions may be absent.   Most men know what it is to feel depressed at some times of their lives.   Their friends attribute it to dyspepsia, or liver, or weather, or may set it down to difficulties or overwork.   Some feel it frequently and periodically; in some it lasts for a few hours only, in others for days or weeks; some may require treatment in order to disperse it, in others it passes away of itself.   If it is of any duration and at all intense, you may be consulted about it, and it will demand your attention, but many shake it off without the doctor's help, and even when you are consulted, you will often recognize the malady as being transient and constitutional, so to speak, one that will yield to time without any very active measures. But if the physical symptoms increase and the mental state gets worse, it will be your duty to tell the patient's friends that the matter is not to be treated lightly, that the sufferer is in a state of insane depression, which will, if not arrested, go on to still more urgent melancholia, or may possibly be converted into an active mania.

There may be a very considerable amount of melancholia without any absolute delusion. It is a morbid feeling due to a physical condition, and the patient himself may be quite unable to account for it. He does not know why he is so sad, and if occasionally toward evening the cloud lifts for awhile, he is apparently himself again. But though he cannot account for his melancholy, and may have no delusions with regard to it or any assignable cause, it may be very intense, and shows itself in great prostration of bodily strength and energy, in complete disorder of the visceral functions, and rapid emaciation.

Let us first consider the milder forms of melancholia where depression without delusion is the most prominent symptom. If any of you devote yourselves to the study of mental disorders, you will encounter such cases constantly. They are the out-patients, so to speak, of this special practice. A considerable number may be treated as out-patients, and recover without the restraint of an asylum. Not unfrequently we find persons who suffer from periodical attacks of melancholy, lasting from six to twelve months, and passing away, perhaps without treatment, to return in two or three years' time, with no apparent cause beyond an inherited constitution and a periodical tendency to recurrence. For these treatment— at any rate medicinal treatment—does little, but in ordinary cases it is necessary; and unless it is special and appropriate, the patient will drift into a more advanced stage of melancholia, or the existing malady will become confirmed and obstinate, and require months rather than weeks for its cure.

Depression is the chief feature of mild or simple melancholia. The sufferer has no pleasure in life; life itself is one long pain, hence the wish to end it. After a short and restless night, with little and unrefreshing sleep, he wakes in the deepest gloom, with all his morbid thoughts intensified, without hope in this world. Bear in mind that the morning is the time when all melancholics are at their worst, and most likely to do themselves harm; you will see again and again accounts of suicides which have taken place at this time. Patients in a state of despondency, whether they have plain and unmistakable delusions or not, are always disposed to suicide, and ought not to be left alone. But it often happens that an attendant, who has slept in the patient's room, gets up in the morning and goes to another room to dress or breakfast, or some such errand, leaving the patient quietly in bed at the very time of all the twenty-four hours that his presence is most needed. And the patient takes the opportunity to throw himself out of the window or over the banisters. The number of suicides committed yearly by patients suffering from mild and curable melancholia is enormous, and truly lamentable they are, for the commonest care on the part of friends would prevent four-fifths of them. The depression may be general, ranging from excessive nervous trepidation or irritability to quiet sorrow and despondency, and it may or may not be connected with some real event. You will constantly be met with the argument, raised either by the patient or his friends, that it is only a state of grief arising from some adequate cause. He has lost a relative or a sum of money, or suddenly discovered that he is a sinner, or has some bodily ailment about which he is anxious. In short, say they, he is not suffering from insane depression, but from real grief. Your diagnosis here will be aided if you compare the way in which people in general are affected by such matters with the case before you. If a lady mourns the death of a daughter which happened several years ago, and accuses herself of having caused it, you will hardly think it ordinary sorrow. If the money lost is small in amount, causing no real difference in the patient's circumstances, you will not see in it a reason for his folding his hands and utterly neglecting his business, family, and friends. When a man suddenly discovers, without any process of introspection, reflection, or counsel from others, that he is wicked and beyond the hope of salvation, the chances are that he is suffering from religious melancholia, not that he is converted. In short, you will sometimes find that there is no assignable cause for the gloom, or that if there is, it is one totally out of proportion to the mental distress you witness.

There is another question of diagnosis which you will not find so easy of solution. Is the malady which begins by depression likely to continue as depression, or will it turn into violent and dangerous excitement, as I have told you it sometimes does? There is no certain rule for your guidance. A patient may be dull and depressed for several weeks, or even months, and then maniacal or even melancholic excitement and violence may set in very suddenly, and an asylum will be imperatively demanded. The longer the melancholy lasts, and the slower and more gradual its access, the more likely it is to remain melancholy. An insanity which progresses rapidly, advancing quickly from depression to delusions, and from delusions to hallucinations, will very probably terminate in violent excitement, call it what you will. As you cannot arrive at certainty on this point, it is important that you should be prepared for it. The patient must not be sent away out of reach or abroad. Those about him must be warned of the contingency and must be sufficient in number, or must have the means of adding to their number. Friends when they first consult a medical man cannot bring themselves even to contemplate the possibility of an asylum. Yet it is as well to remind them that such a thing may be necessary, and in a few days the development of the disorder may materially change their views on this point.

When you see a patient in an early stage of melancholia, you will probably find not excessive rapidity of pulse, but slowness. Very likely it will not be more than fifty or sixty. We may possibly hear of headache, or pains in the vertex or back of the head, together with constipation. I am now assuming that the sufferer is in that condition that he can inform us or his friends of what he feels; but not every patient in this state will see a doctor. If a woman, there is in all likelihood some irregularity in the catamenial function. You inquire the cause of all this. There may be one, plain and palpable, as overwork, mental anxiety, a bodily illness, or pregnancy, lactation, or parturition. The cause may be removable or irremovable. And very probably you are informed of something which is not really the cause, though it is put forward as such by the friends; you will have to inquire further before you come to a conclusion on this point. The cause may have already passed away—as parturition: you only have to deal with the resulting condition. If it be lactation, you can end it; but if the patient be pregnant, you will have to nurse her through the period of gestation and parturition, or will have to consider the question of inducing premature labor. Yet when the cause is removed, we do not always at once arrest the disorder; the mischief is lighted up in the brain, and must run its course. Frequently, however, we may have the good fortune to restore the balance.

The two things to be kept in view may be called, in concise terms, moral and medical treatment. We must inquire into, and if need be correct, the patient's external surroundings, and, by diet and medicine, try to restore his physical health. In many cases, perhaps in most, it will be advisable to send him away from home—to produce an entire change of ideas and objects—to remove by this means painful subjects of thought constantly presented by the sight of home, or wife, or children, subjects already, it may be, distorted by fancies, and incapable of being regarded in their true aspect. Then comes the question, where is he to go, who is to accompany him? Here difficulties arising out of the patient's circumstances, pecuniary means, *impedimenta* of all kinds, may prevent us doing exactly what we wish. One thing is certain—he should not go alone. Morbid fancies come thick and fast to a man who has no one with whom to interchange ideas. I once saw a gentleman who had a transient attack of mania, and recovered quickly in his own home. Six months after he went away by himself on a sketching tour, and though he was quite well at starting, his old fancies all came back to him in his solitude in about ten days, and he fled back to his friends to have them dispelled, and happily this was done. Who is to be the companion? Is the wife to go with the husband, the husband with the wife? Who is to go with single people? As to this there is no rule. The companion must be a person of sense and tact, and devoid of fear. Of this you must judge as best you can. If no one can be found fit for the post, the patient had better stay where he is. Where is he to go? Not abroad: a trip to the Continent is all very well at the end of an illness, to give the finishing stroke to the cure; but at the beginning, when we are uncertain what is about to happen, a patient, as I have said, should not go out of reach of assistance and immediate restraint and treatment. The place to which invalid Londoners and many others are chiefly sent is the seaside, but I do not find that sea air is beneficial to those threatened with insanity. I have seen so many get rapidly worse after a few days' sojourn at Brighton, that I cannot help coming to the conclusion that there is something about the seaside which tends to convert the preliminary stage of confusion and depression into wild excitement; and for this reason I prefer to send patients for change to an inland place. What can we do in addition to avert the evil that threatens our patient's reason? First, with regard to diet: assuming that the incipient symptoms of insanity are those of deficient nerve-force, I inquire closely into the eating and drinking of the patient, and constantly find that, whatever may be the proportions of the latter, the former is in defect. Mental trouble or bodily ailment, fears of dyspepsia, anxieties about liver and constipation, have caused the amount eaten in the day to fall below the normal standard, often to a very considerable extent. Thus, for lack of nourishment the brain becomes more and more exhausted, just at the time that it ought to have an extra supply. Regularity in meals, by which I mean the eating an adequate quantity at regular intervals, and not allowing a very long period, even at night, to elapse without food, often does much good; for we constantly find that before any mental symptoms have been observed, the patient's friends have noticed that he has grown thinner. Malt liquor, wine, and the morning beverage, rum and milk, may be given as you see 'fit; but I attach more importance to the administration of good and wholesome food in this stage than to stimulants, which in many produce heat and pain of head, or undue excitement.

If sleep be insufficient and irregular—and you will rarely find it otherwise—are you to give medicine of any kind? At this time I think that we derive valuable assistance from the hydrate of chloral. I have known more than one threatened attack of insanity warded off by its administration in the premonitory stage of sleeplessness, and I would make trial of it in preference to opium where we are uncertain as to the form which the disorder will finally take. For opium, as you well know, not only procures no sleep for some people, but absolutely prevents it, and by raising the bodily temperature causes the very symptoms we are endeavoring to dispel. At a later period it is often of the greatest service, but at the commencement it is hazardous to try it, whereas chloral is not attended with the same disadvantages, and in a mild attack is far more sure in its action. Should this fail, you may try bromide of potassium alone or combined with chloral, or extract of henbane, or tincture of digitalis; but assuredly chloral should be given first, in doses of twenty or twenty-five grains. Tonics may be required—iron, quinine, arsenic, or strychnia. There is nothing special to be said concerning these, which must be left to your discretion and judgment. Patients at this time are not capable of great fatigue, mental or bodily. If sent into the country, very hard exercise—as boating or very long walks—must be interdicted. I have known it produce suddenly a very acute attack. Neither must they be exposed to great heat of sun. Amusement they require, not work, and this must be regulated by the companion, without whom, as I have said, they are not to be trusted, and who is to have supreme authority in everything. In some such fashion many a slight attack of melancholia may be cut short.

I am now about to consider those cases which have not been cured by the treatment already spoken of—change of scene, the companionship of a friend, fresh air, and good diet. I assume that the depression has become undoubted insanity, leading to ideas and acts of an insane character, and requiring special interference and constant medical care. What are the most prominent symptoms of the disorder?

The patient is in a state of general depression. An utter lack of energy is exhibited in all his ideas, feelings, and acts. Not only has he many insane delusions, but he takes a desponding view of everything that happens around him, and connects his own position and fancied misery or wickedness with all the disasters that he reads of in the papers, or hears in his own circle. The most distant events, earthquakes abroad, shipwrecks at sea, battles and murders, all depend on his evil fortune, or have happened for the express purpose of making his lot more wretched than before. He would give himself up to the police for the committal of all the murders he reads of, or would flee away, and hurry from place to place to escape those who are accusing him of crimes that have never been committed. Or he has committed sins so black, so unpardonable, that he dreads not human, but Divine justice; his soul is lost beyond chance of redemption; he is without hope in the world. Again, he is beggared in fortune, his wife and children are going to the workhouse, he is to be arrested for debt. Then his health is in as sad a state as his affairs. He is eaten up with syphilis, his inside is all gone; he has in him a burning fire consuming his vitals, and reducing his excreta to cinders; he has the leprosy of the Old Testament; a loathsome smell emanates from his body, and contaminates every chair or sofa on which he sits. No amount of argument, no demonstration, however plain and undeniable, shakes his conviction or banishes his fears. These are to be removed by medical treatment, not by any method of moral persuasion; in fact, argument is often hurtful.

If we examine his appearance and bodily condition, what do we find? His aspect is dejected, dull, and heavy, or woe-begone to an extreme degree. He sits or stands in one place for hours, or constantly tries to wander away. He is, in all probability, much thinner than usual. He sleeps badly, and eats little; the tongue is foul, coated, and creamy; the bowels obstinately constipated, the breath offensive, the pulse slow and weak.

Now, among which of our pathological varieties are we to look for these melancholic patients? Chiefly we find them among those who are passing into the decline of life, whose insanity has been termed " climacteric." We do indeed see melancholia in patients of all ages, and see it accompanying all causes and conditions; but it is the exception to find it in the young, the rule to find it in those whose vigor is beginning to fail, whether at forty, fifty, or sixty years of age. Of 338 cases of melancholia admitted into St. Luke's Hospital, only 9 patients were below the age of twenty. Occasionally it occurs after parturition in women who have been much weakened by their confinements. The insanity which appears some weeks after confinement generally takes this form, and often rises to a very acute state, with sleeplessness and obstinate refusal of food. The weaker the patient the more urgent are the symptoms, and the greater the need for active and immediate treatment. There is nothing of the sthenic character which marks the wild excitement of mania. It is not usually found in phthisical patients, who are commonly excited and maniacal, but we sometimes see it thus associated. It is an old belief that it is connected with the abdominal organs, as the liver; or, according to Schroeder van der Kolk, the colon; but there seems reason to doubt this. The disorder of the liver and the loaded state of the colon are as likely to depend on the general derangement of the nervous system as to be the cause of the mental disorder. The question is not set at rest by their vanishing together. The *propter hoc* in such cases is very hard to come at.

Whatever be the cause of the insanity, whether we call it idiopathic or sympathetic, phthisical or sexual, or even paralytic, the melancholia is the effect and indication of the condition of the sufferer at a particular time. He is generally depressed, his slow and feeble circulation imparts little force to his brain centres, and the supply there is always in defect. Not only is it in defect—for this appears to be the condition of almost every phase of insanity—but there is little action going on. The metamorphosis is not rapid; there is no immediate danger to life, emaciation does not occur rapidly as in acute mania. The patient is not absolutely sleepless, though he may sleep little. The depressed state may last for years and then pass away; and, so soon as the feeling is gone, all the delusions and fancies bred of it vanish too. We are far from understanding the exact pathology of that state that gives rise to melancholic feeling.

I have known it to exist in a gentleman who ate heartily, who was stout in body and florid in face, who was free from all bodily disease. In him it appeared without assignable cause, and in process of time vanished; and all that we can say of such a case is, that by some concurrence of conditions beyond our recognition, the nerve-power of this man's brain was insufficiently produced.

We find patients whose melancholic delusions are attended with so much excitement that they may rather be called maniacal. This only indicates that their condition, though one of depression and defect, is at the same time one of greater disturbance of the brain and more rapid metamorphosis; and, as we shall see, when it attains a certain height, it becomes as formidable and dangerous a disorder as any other acute form of insanity.

But here I wish to point out the treatment of an ordinary case of melancholia, attended with great depression and melancholic delusions, disinclination to take food, little sleep, and obstinate constipation.

The first thing you are to remember is, that every patient of this kind is to be looked upon as suicidal. Never mind whether he has, or has not as yet, made attempts, or shown signs of such a disposition. He may have had no opportunity, or may never yet have felt the particular idea or impulse. Where is the treatment of such a patient to be carried out? An asylum is not absolutely indispensable, if the patient's means will afford him what he requires elsewhere. If a poor man, there is nothing for it but to send him to an asylum. For he must not be left for a moment where he can do himself harm, or make his escape. He requires the companionship of some person his equal in education, as well as of attendants; must be removed from home to a house, airy, light, and quiet, and should have facilities for taking exercise without going into crowded thoroughfares. All this implies some considerable expense. If, as I say, his means suffice, such a plan often works a cure more rapidly, in my opinion, than the asylum, with its depressing influences and lack of sane companions; but if funds are scanty, the latter is a necessity, for the other plan is impracticable unless carried out completely in all its details.

Having removed your patient into a suitable abode, and having arranged that he shall never be left alone for a moment, what are you to do by way of treatment? Your object is to restore the defect of brain by means of food and sleep, and you will find that in many cases a most satisfactory result follows the treatment, and this in no long time. I have seen some very bad cases recover perfectly in two months, and recover in a manner which was clearly due to the medical treatment, and not to mere change of scene and surroundings; for this had been already tried, and tried in vain.

Besides the mental symptoms, there are three things specially to be attended to—the want of sleep, the tendency to refuse food, and the constipation.

Chloral will produce sleep in these cases, as in others, but is better suited to the slight than the severer forms of melancholia. It is a sleep-compelling agent; beyond that its effect seems of little import. It does not appear to have such a *healing* influence as opium where the latter is beneficial. In sub-acute melancholia, the preparations of opium are of great service, whether given by the mouth or subcutaneous injection. I have very rarely been obliged to discontinue them, and have almost invariably found the patient mend after their administration. The preparation which, according to my experience, has succeeded best is the *liquor morphiæ bimeconatis,* for it does not cause sickness or constipation, such as too frequently follow the administration of the acetate or hydrochlorate of morphia. As the patient is already inclined to refuse food, often on the plea of nausea or loss of appetite, and as his bowels already are obstinately constipated, it is important that we do not increase this state of things by our remedies. Dover's powder, in some cases, or solid opium, or Battley's solution, we may give, and give freely, in full doses at night, and in smaller doses two or three times a day. It is of little use to give at night by the mouth less than the equivalent of half a grain of morphia.

We now come to the food question. We read that patients refuse their food because of dyspepsia, and that the latter is indicated by the foul, coated tongue, fetid breath, and loaded bowels. I am obliged to say that I think all the symptoms of dyspepsia are the result, and not the cause, of the depressed nervous condition; that the tongue is covered with old dead epithelium, which, for the same reason, is not thrown off; that the fetid breath is caused by this, or is due to actual starvation; and that the loaded bowels must also be ascribed to the want of general power. And I say this with some confidence, having treated a very considerable number of these cases, and having removed all the symptoms by means which were in no degree directed to cure dyspepsia. This is the kind of diet which I have frequently given for the purpose. Before getting out of bed in the morning, rum and milk, or egg and milk; breakfast of meat, eggs, and *café au lait,* or cocoa; beef-tea, with a glass of port, at eleven o'clock; and a good dinner or lunch at two, with a couple of glasses of sherry; at four, some more beef-tea, or an equivalent; at seven, dinner or supper, with stout and port-wine; and at bedtime, stout or ale, with the chloral or morphia. We give less stimulant now than we did twenty years ago, but a certain amount is useful in almost every case. This allowance I have given to patients who were said to be suffering from aggravated dyspepsia; who, I was told, had suffered from it all their lives; who had never been able to take malt liquor, or eat more than the smallest quantity at a time; who, in fact, had constantly been living on about half the quantity requisite for their support, and, through chronic starvation, had come to this depressed condition. I need hardly tell you that the patients and friends were aghast at the quantity ordered to be taken; but improvement has taken place immediately, the tongue cleaned, the constipation given way, and the depression diminished; and I have known patients themselves become so convinced of the necessity of this augmented diet, that after recovery they have continued to take about twice as much as before the illness. How dependent these melancholic patients are upon food has often been proved. Some, when nearly well, if they were out for their walk or drive longer than usual, or from any other cause postponed their meal, felt at once a return of the depression and delusions, which vanished again after the reception of food.

I am speaking now of patients who would refuse their food if left to themselves, or protest against it, but who take it when told to do so, or allow themselves to be fed without downright resistance. Wherever you can, give solid food; do not be content with beef-tea and stimulants, which are supposed to be the sole diet suited to invalids. The best proof that dyspepsia has nothing to do with these patients' condition is, that even with this enormously increased diet, I never knew any reject it from the stomach, except one lady, who did it willfully; when made to take another supply, she kept it down. You may vary the diet to any extent, for every kind of good, plain, nourishing food may be given—poultry, game, fish—not merely mutton and beef. But even turtle-soup and champagne are not to take the place of solid food, which is a far better sedative than mere liquids.

The constipation will often be remedied by the stimulus of the increased amount of food; but when the patient is first subjected to treatment, the colon is often clogged by hardened masses of fæces, which nothing but enemata will remove. You may use for this the *enema terebinthinæ* of the Pharmacopœia, with or without castor-oil, and the operation will probably have to be repeated more than once. Afterward you may promote regular action of the bowels by giving a daily dinner pill of the watery extract of aloes, or by a daily teaspoonful of castor-oil, or some such laxative. Active purgation only makes matters worse, and should be avoided. I have also found great benefit from giving such patients bran bread. This has been called a cure for melancholia, and it certainly brings about a regular action of the bowels in persons who for years have never been relieved except after medicine.

Melancholic patients are almost invariably better toward evening, and worse on first rising in the morning, owing, in my opinion, to the long abstinence from food. When a patient habitually wakes after three or four hours, and cannot go to sleep again, some food, as a sandwich, glass of milk, or some of his matutinal rum and milk, will often bring sleep back to him. If this fail, he may take a small dose of his sleeping draught.

These people all suffer much from cold, are generally worse in the winter, often lingering through the spring, and waiting for hot weather
10

to thoroughly restore them.   They will derive benefit from the hot-air or
Turkish bath, if they can have it regularly, and in all respects they require warmth.   Their rooms should be sunny, and their clothing sufficient.

Of course, in such cases moral treatment is not to be lost sight of; and
although no precise rule can be laid down on this subject, yet the recovery
of a patient may be greatly aided by the judicious care of those about him.
Every one must be struck by the intense self-feeling of the melancholy
man.   His egotism exceeds even that of the paralytic or maniac.   He
thinks that everything is centred in him, that he has committed the greatest sins, or is to endure the greatest torments.   His superlative misery is
a theme on which he loves to descant as much as the paralytic loves to
describe his wealth and greatness.   His depression is great, but he mag-
nifies it in the recital of his woes.   Therefore it is necessary to lead him
away from self-contemplation, and to awaken in him an interest in others;
and it is curious and interesting to see the gradual improvement in this
respect.   By degrees he will listen to news told to him or to what is mentioned in his presence, will furtively look at letters or the newspapers, and
in this way, little by little, return to his normal state.   Many a patient
has been suddenly cured of melancholy by some event which called for
immediate action.   Thus, a lady's only son was seized with dangerous illness, and she was obliged to go and nurse him.   In her work and anxiety
she forgot her own melancholy, and when he recovered she too was well.
Melancholy is banished in this way when the patient is on the road to recovery; but the commencement of an attack, before the strength is restored, we are not to expect such a sudden termination.

What is the prognosis in these cases of sub-acute melancholia?   Generally very good.   In my experience almost every case of this kind, if it
does not run on to acute and excited panic-stricken frenzy, with desperate
determination to resist food, and total want of sleep, progresses to a favorable termination in a longer or shorter time, whether in or out of an asylum.   Of former patients now at large and well, more have suffered from
melancholia than from any other form of insanity, and some of these
were inmates of asylums for long periods.   In the second volume of the
St. George's Hospital Reports, I have given an account of three patients
who recovered after long treatment in asylums.   One was a gentleman
who thought he had committed the unpardonable sin—nay, that he was
himself the devil.   He also thought himself ruined and afflicted with
leprosy, but did not refuse food.   He went on in this way for seven years,
till at last affairs necessitated action on his part, and he woke up out of
his melancholy, and has now for twenty years keenly enjoyed life and its
pleasures.   Another was a lady of fifty-six years of age, who had all the
worst symptoms of melancholia, refused food, did not converse, but paced
her room, ejaculating " My God, my God," and picked and rubbed her
hands in terror and panic till they were sore.   After five years she began

to mend, gradually improved, and in six months was discharged quite well. Another was a gentleman, aged thirty-one, who had been in an unhealthy tropical climate. He had all the symptoms of melancholy, was suicidal, tried to avoid food, would not converse, but muttered to himself, and thought he was going to be put to death for murder and forgery. He too recovered perfectly, after being in this state for five years. Two others, ladies, recovered one after nine, the second after no less a period than thirteen years.

I believe that depression is the only form of insanity in which we may expect recovery after such periods, and we may perceive in the recovery an indication of the pathological state of the patients. It would appear that during the whole of the period the general nerve-energy is in a state of defect, the result of which is, first, the feeling of intense melancholy; secondly, ideas and fancies—in other words, delusions—growing out of the feeling. All the ideas, in fact, are tinctured by the prevailing gloom. But if the nerve-force is restored, if the physical condition of the brain is raised again to its normal level, all such delusions vanish, the abnormal feeling passes away, and the mind resumes its proper work unimpaired by what it has gone through.

In dealing with cases of this kind, it is important, for many reasons, that we should keep before us the possibility of recovery after a long period. For our opinion will be asked by those who, having the disposition of property, may regulate their wills by what they hear of the chances of the patient's recovery. I always advise that even an inquisition in lunacy should in these cases be deferred as long as possible, to give time and opportunity to see what the probable duration of the disorder will be. But a commission in lunacy can be superseded, and the patient restored to the management of his affairs. A will, however, is another matter; the testator may die, and soon after the lunatic may recover from his melancholy to find himself disinherited.

I have been speaking hitherto of patients suffering from what may be called sub-acute melancholia, who require constant watching, but are not violent, do not resist feeding, and are, in fact, sufficiently tractable to be kept till cured in an ordinary house, if there be means adequate to their necessities. Such generally recover; but I am now going to speak of some who do not stop in a quiet stage of melancholy, but go on from it to an acute condition, of which the prognosis is the very reverse of favorable.

This form of acute melancholia demands as much as any the care of an asylum. It is hardly possible to keep a patient in safety in any ordinary house, or to treat him with any but the large staff of officers which an asylum supplies. He is not in a state of mere depression or mild melancholy, nor in silent stupor, but he is panic-stricken. In violent frenzy and terror he paces the room, dashes at the doors or windows, eager to escape from the doom that waits him, from the police, who are on his

track.   He will not sit on a chair or lie still on his bed, but is incessantly
running about, exclaiming that he is going to be burned or tortured, that
the room is on fire, the floor undermined, and everything ruined and
lost.   He is suicidal in an extreme degree, and may try not only to put
an end to himself, but also to harm himself in every way he can, to gouge
out his eyes, cram things down his throat, swallow nails or bits of glass,
or break his legs or arms in the furniture.   Though he will not attack
others like a dangerous paralytic, he nevertheless resists with the utmost
violence all that is done for him.   He will take no food, wear no clothes,
will not be washed, neither will he remain in bed.   This is a condition
very different from that last described.   It implies a much lower degree
of nerve-force, and a much more serious disorder of brain, and the hopes
to be entertained of cure are but small, if the patients remain in this state
beyond a very few days, for they are generally broken in health before
they reach it.   They are advanced in life, or enfeebled by other diseases,
and melancholia is in them the commencement of dying, the later stages
bringing not unfrequently gangrene of the lungs or bronchitis, or a condition closely resembling scurvy, indicating a gradual termination of the
vitality of the whole bodily frame.

Such patients refuse food, not with a passive resistance like the former,
who allow themselves to be fed with a spoon, but with all their might,
ejecting it from their mouths, even after we have managed to place it
there.   They drive us to the adoption of forcible alimentation, about
which I shall have something to say in a subsequent lecture.   We are not
to wait long before we have recourse to this.   They sink rapidly if not
fed, and the more they are weakened by lack of food, the more are all
their symptoms exaggerated.   It is of no use to let them go for a day with
a mouthful or two, neither can we afford to let them wait till they are
hungry, or in the humor to eat.   The case is too serious, and too much
is lost by abstinence, even for twenty-four hours.   In the case of some insane persons there is no occasion for hurry.   When they are tolerably strong
and vigorous, and when the refusal of food is due to mere opposition or
whim, we may wait, and frequently the humor will change, and they may
be coaxed into taking food.   But coaxing does little with the terror-
stricken melancholic.   Nevertheless, it must be tried.   It is an old suggestion that the persuasiveness of the opposite sex may prevail upon
patients to eat; that female friends and nurses should try to overcome the
fears and reluctance of men, and males that of women.   In some forms
of insanity this plan unquestionably is of service; but here, I fear, we
shall find it of little use.   The sufferer is determined not to eat, and will
die rather than do so; in fact, the refusal is generally from a suicidal
motive.   In spite of our feeding such a one may sink, for, as I have said,
this acutely melancholic state is often only the last stage of a melancholy or
some other disease which has been reducing the strength of the individual

for months, and which has gone to a point where cure is impossible. Nevertheless, we must not let a patient die of starvation, and as a long and exhausting struggle is not to be thought of, we must feed with the stomach-pump twice a day or oftener. These are the cases where Dr. Clouston's "sheet anchor," milk, is of such service. No one more strongly advocates abundant alimentation in melancholia. "Milk," he says, "in very many cases, is my sheet anchor. I have given as much as sixteen tumblers a day with surprising benefit. The nervous diathesis does not put on fat naturally, therefore we must combat the tendency to innutrition by scientific dieting. Adipose tissue and melancholia I look on as antagonists; therefore, when we want to conquer the latter, we must develop the former."

Chloral may be given to these patients, but unless specially contra-indicated, I always combine it with some form of opium. Opium here, as in all the severer forms of melancholia, is most valuable, not only to procure sleep, but to nourish and stimulate the brain. No mode of administering morphia is so suitable to these violent patients as the subcutaneous injection, for they will no more take medicine than food, and in the struggle to give it by the mouth much may be wasted, and we may not know exactly how much has been taken. Never give pills; patients will hold them in their mouths till your back is turned, and then spit them out; you can never be certain that they are or are not swallowed. The morphia may be mixed with the food, but this is not nearly so satisfactory a method as the subcutaneous injection. Other medicines are not worth the struggle of getting them down. In so great debility, such drugs as bromide of potassium, tartar emetic, hydrocyanic acid, and digitalis are, in my judgment, out of the question. The warm bath may promote sleep, and great warmth of rooms and clothing will be necessary. Such persons are not to be allowed to lie on the floor of an ordinary room all night, which they are very prone to do. There is a tendency in all to fatal pneumonia and gangrene of lung, and rather than run any risk of exposure, I would employ mechanical restraint, and fasten them in bed. Suicide is their one end and aim; and, at the suggestion of the Commissioners in Lunacy themselves, I have employed mechanical restraint at night for such reasons as I have stated.

In addition to the ordinary warm bath the hot-air or Turkish bath may be tried in these cases of acute melancholia, if opportunity enables us to do so. I myself have not been able to apply this mode of treatment to any except those who could be sent out of the asylum; but in some of our larger asylums it is systematically adopted, with proper rooms for carrying it out.

The prognosis in this extreme form is, as I have already said, unfavor-

---

[1] Clinical Lects. on Mental Diseases, p. 133.

able. It occurs chiefly in persons debilitated by age, disease, or child-birth. And the obstinate refusal of food, and the struggles with which its administration is attended, add greatly to the danger of the disorder and to the difficulty of dealing with it. This form of melancholia is rapid compared with the other. Very rarely can we give sufficient food, if the refusal is persistent. Hence many of these patients gradually sink. If the disorder becomes chronic, and the bodily health improves, recovery may take place as in other forms of melancholia. Our power of prognosis will, of course, be aided in these acute cases by the thermometer and the sphygmograph. The chances of life or death will be indicated to some extent by the temperature and character of the pulse. But I need not say that it is difficult in the extreme to take observations of such a kind in the case of patients who resist all that is done to them. Our fingers must be our guides, and by them we may learn the heightened tempera-ture and rapid pulse which are of such evil omen in acute melancholia and acute delirium.

Instead of advancing to the wild delirious melancholia just described, the subacute disorder may take another form. The patient sinks into profound dejection, and sits all day speechless and motionless, lost to everything around him, and apparently regardless of what is said or done. This is one of the forms of stupor, a morbid mental condition which now ranks as a variety of insanity. The French long ago gave it the name of *melancolie avec stupeur*. I prefer to call it *stupor with melancholy*, to distinguish it from other forms of stupor wherein melancholy is not the predominant feature. All are marked by a lethargic and torpid condition, indicating a brain whose working and nerve force are at the lowest possi-ble point. It is an advanced stage of a disorder which commenced with symptoms of melancholia, and progressed not to the violent, frenzied, terror-stricken form last described, but to the motionless state which de-serves the name of stupor. Of the treatment little need be said. Such patients must be fed, and fed largely. Some are passive and manageable in the hands of those about them, but many will resist and oppose what is being done for them with silent, dogged, persevering resistance. They may even require to be fed by force. They require to be led about for exercise, to be washed and dressed. Though they sit motionless for hours, they are not to be trusted alone, for they may eagerly avail themselves of an opportunity to commit suicide. They demand even more food and stimulants than those who suffer from the milder forms of depression, and must be kept as warm as possible, for their motionless condition does little to circulate the blood stagnating in their vessels.

Writers have speculated upon what is passing in the mind in this state of stupor, on the absence of will, on the hallucinations and illusions that torment the patient. Our information must necessarily be derived from the patient himself after recovery, and it must needs be that such testimony is very fallacious. In fact, it is just as trustworthy as an account of the whole psychological state of a man in sleep derived from what he himself recollects of his dreams. There is, however, a very close resemblance between the dreaming condition and profound melancholy. That which is going on around the patient is observed through the medium of the depressed feeling, and is altogether unreal and illusional. All people are changed, and places, and things; and, pathologically, the states of the dreamer and the melancholic are probably nearly akin; for in dreams it would appear that the brain is only partially at work, that the whole of the idea-organs are not in a state of activity, but only a few, therefore there is no correction, but everything seems real. Similarly the nerve-power in melancholia is so low that the entire idea-faculty of the brain cannot be employed. The depressed condition influences the ideas that are formed, and these are not corrected by the entire brain as they would be in health. The sensations experienced from affections of the skin or viscera are converted into flames, tortures, snakes, and so forth, just as in dreams cold feet make us think we are walking on ice, or dyspepsia originates ideas of legions of devils. These melancholic patients, though the present the appearance of stupor, sleep very little, and require narcotics as much as the last mentioned class.

After speaking of these patients, I come to a second class which has been by some confounded with them. This form of stupor has received various names, and has been brought under one or other of the varieties of insanity according to the views of different writers. It was described by Esquirol, and called by him "acute dementia," and this name has been more often applied to it than any other. Pinel confounded it with a kind of idiocy, and named it "stupidité." M. Baillarger pronounced it to be a form of melancholia, and, owing to the torpor and inactivity of the patients, thought their malady identical with *melancolie avec stupeur.* Dr. H. Newington and Dr. Clouston propose to call it "anergic" stupor, a bad name, as all stupor must necessarily be "anergic;" an energetic stupor would be a contradiction in terms. Dr. Monro applied to it the term "cataleptoid" insanity, and the word not inaptly expresses the lost automatic condition so often witnessed in those suffering from it. They are not, however, lost in woe, like those buried in profound melancholia, in *melancolie avec stupeur.* Rather are they lost in vacuity; they look utterly idiotic and silly, like the chronic demented people seen in asylums and workhouses. Hence the name "acute dementia," "acute" being used in contradistinction to chronic, meaning a curable disorder lasting comparatively a short time. The term has been objected to, because it is said that we cannot call such a passive disorder "acute," and because the minds of such patients are not really demented. However, I will try to describe to you the disease, to which you may affix what name you like, provided you clearly understand the symptoms. Dementia it is while it lasts, and if we are to give up the old name, acute dementia, we had better call it *stupor with dementia*, as we called the other form *stupor with melancholia*. The patients are all young, from fifteen to twenty-five years of age. When we hear of a man or woman of fifty being thus affected, we may presume that the malady is *melancolie avec stupeur*, or something of the kind. It is not stupor with dementia. A young man, then, or a young woman, after some shock or fright, some appalling sight or intelligence, is frightened "out of his senses." He is horror-stricken, paralyzed in mind, not merely deranged, not depressed or excited, but deprived of feeling and intellect; his movements, if there be any, are automatic, but frequently he is motionless, standing or sitting, staring at vacancy for hours and days. As I have said, this may come on suddenly after a fright or may be developed more gradually after some slight cause, so slight that it has been unnoticed or forgotten by the friends. In the case of one young man, it was a fall from a scaffold without injury beyond the fright and shake. Such are always persons of weak nerves, boys and girls who have outgrown their strength, and whose nervous condition is still further weakened by delicate bodily health. They do not converse: their answers are those of an idiot or demented patient. More frequently they give no reply. They are not always motionless; often they are in incessant motion. One girl used to snap her jaws together for days at a time, and then change to wagging her head from side to side. This action was truly automatic, for no effort of will could have kept up muscular motion for so long a period. They do not stop when spoken to, but if we give them a shake, they may perchance direct their attention to us for a moment, and then begin again. They do not always remain the same. After being for hours in a complete cataleptic or trance-like state, so as to make those about them think they are in a fit or comatose, they will commence to laugh, chatter, or grimace in a silly and idiotic fashion. There is apparently a complete mental blank, and sometimes after recovery we find that nothing is recollected. Sometimes they recollect all that has passed, though at the time they had no power or control over their actions, and it is going much too far to say, as some do, that these patients are unconscious. One young man, who was apparently so unconscious that his attendants thought him dead or dying, told me afterward every word that had been said about him. They may, however, resist violently when fed, dressed, or moved, resisting, like the chronic demented, without any reason, except that their quiet or automatic action is interfered with.

The physical condition is peculiar. The circulation is so feeble that in the hottest weather hands and feet are blue with cold. In winter they are covered with chilblains, and there is great difficulty in keeping these from becoming sore. The heart's action and pulse are proportionately weak. Such patients do not lose flesh, as those in acute mania or melancholia, but are flabby and pale. They do not eat, but can generally be fed without difficulty. They are wet and dirty, or do not pass water unless at long intervals, neither will the bowels of some act without enemata. Sleep may be irregular, but is seldom absent, and frequently the amount is normal.

You will have to distinguish this form of insanity from other varieties of stupor. There is the stupor with melancholy already mentioned. The patients suffering from this are seldom young. They have been melancholy from the commencement of the attack, are suicidal, rapidly emaciate, refuse food obstinately, and sleep little. In all of which they differ from those whose malady is primary dementia. The latter in the beginning are maniacal rather than melancholic, and rarely suicidal; and when they get better, and can tell us something about themselves, we see that depression is not the leading characteristic of their state, but that it is mainly extreme weakness of mind, consequent for the most part upon mental shock or bodily debility, or both combined. How can we distinguish it from chronic dementia? In other words, how can we say whether the patient will recover or not? I confess that this is not easy. I have seen patients of whom I had hopes who have remained permanently in a demented condition. Everything will depend on the history. The appearance of the sufferer may be identical in primary and secondary dementia. You are shown a young man or a girl in a state of fatuous imbecility, grinning idiotically, lost, and dirty. Nothing can appear less promising. But if you are told that this condition came on almost suddenly, especially if it followed a mental fright or shock, which seems to have been the origin of it, and if you observe the physical symptoms I have mentioned, indicating great weakness of the system and circulation, you may pronounce hopefully as to the result. But if the patient has gradually and imperceptibly drifted into this imbecile condition without assigned cause, then you may set it down that the cause is hereditary taint in the first place, and masturbation in the second, and that though improvement may take place, recovery is impossible.

What is the treatment of these patients, and where is it to be carried out? I do not think an asylum absolutely necessary. In many the mental shock they have undergone would be intensified by such removal. If they are passive and tractable, they may be treated in a family, or even at home. And as they improve, change of scene will be of infinite service. They require nutrition and stimulation by wine and stout, and also such stimulants as shower-baths or galvanism. In these cases especially, I believe shower-baths to be useful, a short sharp shower, and then plenty of friction to restore and promote the circulation. Above all, they require warmth; warmth which to us would be excessive heat will not do more than warm them, and you will find that many of them will continue in the same state through the winter, and wake up and recover when the

hot summer comes.   Tonics we may give, especially steel and small
stimulating doses of morphia.   Menstruation is sure to be absent, but we
need not direct any special efforts toward it.   As the general strength
returns, it comes again, and meddling in this direction does more harm
than good.   We shall soon see, when a patient is subjected to treatment
of this kind, whether there is an awakening of attention, and a return of
mental strength.   If there be, if we are conscious of an improvement, we
need not despair, even if it be slow.

As a result of great nerve prostration after an acute attack of mental
disorder there may be for a time a state of lethargy or stupor, which will
pass away if the patient is to recover, or may merge into chronic dementia.
The history here will prevent any mistake of diagnosis, and the same may
be said of states of stupor occurring after epileptic or epileptiform attacks
or the excessive use of alcohol.   In short, various causes may bring about
a condition of stupor amongst the insane, as in the sane who are suffering
from some grave cerebral disorder, but the only two forms which can be
ranked as varieties of curable insanity are the stupor with melancholy and
the stupor with dementia, formerly known as acute dementia.

I said in my last lecture that patients attacked with primary dementia
might be ranged at one end of the scale of the insane, for here we see the
minimum of cerebral action and metamorphosis, the negative side of mind
disorder, manifested in silly idiotic vacuity rather than in depression,
excitement, or delusions.   The bodily phenomena correspond; the circulation is lowered, the surface cold, there is but little change going on,
and no waste, nothing but automatic movements or torpid inactivity.
In my next lecture I shall present to you certain patients who are the
very reverse of all this, whose insanity is shown in furious delirium, with
incessant violence and entire sleeplessness, with so great waste and exhaustion that life may come to an end in a few days, or, if recovery take
place, the sufferer may be found to have lost flesh to a great extent, even
in a very short period.   This form I shall term acute delirious mania,
and it may be fitly placed at the other end of the scale.

LECTURE X.

Acute Delirious Mania—Diagnosis of Transitory Mania—How to Arrest it—Treatment of Prolonged Acute Delirium—Food—Nursing—Medicines—Baths—Purgatives—Prognosis—Diagnosis.

INSTEAD of a patient feeling something wrong for weeks or months, or the same thing being noticed for as long a period by those around him, very acute symptoms of mania may arise in a few days or even hours. The premonitory stage of an attack of acute delirium, or acute mania, may be extremely brief; in fact, a patient may awake out of sleep, and at once become delirious. The more sudden the invasion, the shorter will be the duration of the attack in the majority of instances, but to all such rules there are many exceptions. Sudden outbursts of delirious mania frequently have their origin in a mental cause, as the death of a friend, a suddenly announced misfortune, a violent quarrel, a disappointment or cross in love. Any such circumstances occurring to a person of weak nerve, hysterical, and by nature predisposed to mental disturbance, may bring about very acute delirium in a few hours.

The same thing may proceed from a cause clearly physical. It may arise in the course or during the decline of acute disease, as pneumonia, measles, or fever. It may come on in a patient who is phthisical, or has acute rheumatism, or who has undergone too great fatigue, as a very long walk. It may succeed a paroxysm of epilepsy, or take the place of one. It may come on quite suddenly after childbirth, exposure to the sun, or indulgence in drink.

Now, we cannot say in every case whether an acute delirium will last a few hours or days, or whether it will run the ordinary course of an attack of delirious mania, and require special care and treatment for weeks or months. But it is all-important that we should arrive at some conclusion before we move a patient to an asylum. The prognosis in such a case is of the greatest consequence, and at the same time extremely difficult.

Much information may be derived, first, from a consideration of the character, constitution, and past history of the individual; secondly, from the cause of the attack; thirdly, from the symptoms observed.

Some patients there are whose organization is so unstable, who are so prone to violent disturbance of the nervous system, that attacks of delirium may in them supervene upon an occurrence comparatively trifling. It is reasonable to hope that such an attack will be transient. If the individual has been hitherto unknown to us, we must discover, so far as is possible, what his or her temperament and character ordinarily are; above all, whether there have been previous attacks of a similar nature, and whether they were of long or short duration.   If the attack has been gradually developing during a week or longer, without any assignable cause, it is not likely that it will suddenly terminate.   Observation of the patient will teach us something.   If amidst the paroxysm of delirium there are intervals of calm, during which the patient is rational, we hope that the attack will be brief, also if there are intervals of sleep.   If there is no sleep for three or four days, except perhaps short snatches of half an hour, and if during the whole of this time the patient is becoming more and more maniacal and delirious, we conclude that an attack of acute mania has begun, through which he must be nursed, and which cannot now be arrested.   We shall also be aided by observing the physical condition, noticing if there be symptoms of hysterical delirium, copious voiding of pale urine, indications of amatory feeling, pretended inability to speak, and other pecularities denoting that even with apparently very violent delirium there is complete appreciation of all that is going on, and considerable exercise of volition in what is done.   We constantly see cases which would be described by one as mere hysterics, by another as an attack of transitory delirious mania, and it is not easy to distinguish one from the other.

What can we do to bring this condition to a conclusion?   It is necessary above all things to determine whether the attack is going to be transient or prolonged before we remove the sufferer to an asylum, for the terror inspired by the removal would be very likely to convert one of the short attacks into a prolonged and obstinate mania; and if this were not the result, we should yet regret that we had placed a person in such a position for an insanity of so passing a character.   Three or four days will set our doubts at rest, and then if the patient cannot be managed out of an asylum, to an asylum he must go.

There are doubtless many patients, who, if treated at home by friends and among friends, would rapidly recover, but who, when removed by force and placed among strangers, experience a much more prolonged and severe attack.   Yet such a step is often unavoidable.   Suppose that a man becomes acutely maniacal in a seaside lodging: he cannot remain there three or four days till it is decided whether his disorder will suddenly terminate or not.   He would require to be violently restrained, most probably by strangers, and he might not be safe in such an abode: removal becomes imperative, and we can advise nothing else under the circumstances.   But, as I have said, when it is possible to wait and watch the case, let it be done.

A short time ago I saw two most acute cases of delirious mania commencing in seaside lodgings.   One was that of a gentleman aged thirty-one.

He had had two previous attacks of mania within a twelvemonth. The first occurred in his own home. He was then full of fancies, which at times he recognized as delusions; but once or twice he woke in the night in a paroxysm of terror, accompanied by great excitement, and violently attacked a friend and his brother, who were his nurses. This great excitement, however, passed off in an hour or two, and in a few days he was well. The second attack merely consisted of a number of the old delusions, which reappeared and disappeared again after change of scene and visit to friends. The third, as I have said, commenced at the seaside, after premonitory symptoms of sleeplessness and religious delusions of about a week's duration. He was inclined to run away; his wife called in the police to her assistance, and he then broke out into violent delirium, was brought to London, and at once taken to an asylum, where I found him. He perpetually heard voices, talked of religious subjects, refused food, had paroxysms of violent fury, in which he attacked all about him, and intervals of rationality, in which he washed and bathed himself, and ate a hearty meal. He was treated with drachm doses of bromide of potassium. The first night he slept three hours, the second two, and the third four, and after that he slept naturally, and by the end of a week he was perfectly himself, and resumed his usual work and occupation in the asylum, where he stayed a fortnight longer. In six months he had another attack, which also began at the seaside. In this he was even more violent and desperately suicidal. He was now treated with chloral, which quickly subdued the sleeplessness and delirium. He was one of the many examples met with of patients who hear voices. Whether these are chronic or acute cases, the prognosis in all is unfavorable. Here, though the patient seemed perfectly to recover, the symptoms returned every six months; and at last he committed suicide while at home with his family.

The other was a young lady of twenty-one, in whose family, as in that of the gentleman, insanity existed. She was not removed from the sea for a fortnight, her violence making it impossible. She experienced a most acute attack of delirious mania, recollecting nothing afterward of what occurred. The contrast in these two cases was evident to one watching them. In the girl's the attack had been coming on for months: there was no sleep beyond the briefest snatches for a week, and there were no intervals of quiet or rationality. Her tongue, her odor, indicated from the first the serious nature of the disorder; but of this prolonged mania I shall speak hereafter. It ran a course of violence lasting nearly a month, and then came great weakness and prostration, and gradual but perfect recovery.

What treatment are we to adopt when this acute mania first breaks out? We cannot say for certain whether we can cut it short; but we hope to do so.

We have at this time a remedy in the hydrate of chloral, of more value

than anything that ever was given before its discovery. Opium in anything like a sthenic case generally made matters worse. Bromide of
potassium was of greater value, but it did not surely produce sleep. Hyoscyamus, digitalis, cannabis indica, were uncertain remedies, on which we
could not depend. But many of these cases are cut short and cured, like
delirium tremens, if we can procure one long and sound sleep, and I believe that chloral, or chloral combined with bromide of potassium in half-
drachm doses of each, will generally be found to cause sleep of a longer
or shorter duration. I have seen a very violent maniac sent to sleep by
such a dose, and wake clear of everything like delirium, though he still
had delusions. Chloral does not cure insanity: it is given nightly to the
chronic insane; they sleep, but are not cured. In them the disorder of
the brain is fixed and incurable; but in the early stages of acute insanity,
after one sleep the sufferer often wakes restored, and this sleep we can
bring about by chloral in a way we never could formerly.

By the side of this drug it seems scarcely worth while to mention the
modes of attaining sleep which were once adopted. But I may say that I
have seen the violent mania of an approaching attack subside more than
once after a brisk purgative. Whatever be the action of this, whether
we are to call it, in the language of a bygone age, a derivative, a revulsive,
or a counter-irritant, its effect is often marked and evident, and the
patient recovers.

There is another remedy sometimes tried, viz., packing in a wet sheet.
A sheet dipped in water, hot or cold, according to circumstances, and
wrung out, is laid on a mattress protected by a mackintosh sheet and a
blanket, and the patient, placed on this, is wrapped so as to include arms,
legs, and, in fact, the whole body, with the exception of the head. The
blanket is then tucked over him, and other blankets laid over all. Dr.
Lockhart Robertson, the great advocate of the cold-water wet sheet,
recommends that when the patient has been in the sheet for an hour or
an hour and a half, he should be taken out, rubbed thoroughly with a
dripping cold sheet, and replaced in another wet sheet and blankets, and
that after each change of sheet two pails of cold water should be poured
over him. He says that in some cases of recent mania he has pursued this
system throughout the day, or three or four times in the day, or less frequently. We must be careful, however, how we employ it, for like the
prolonged warm bath, the prolonged shower-bath, and other desperate
methods of cutting short mania, it is not unattended with risk. I
believe, however, that merely packing up in a sheet wrung out of hot
water gives us without danger the chief advantages of the system. It is a
powerful sudorific, and promotes sleep by reducing to the minimum the
power of motion. There can be no question that when the latter is
taken away, patients will often fall asleep. This is one of the chief arguments used by the advocates of restraint, and I have no doubt that in their

experience they saw this effect produced. Now, the wet sheet, as a medical appliance, has advantages over the strait-waistcoat. It will, I presume, be denied by those who use the wet sheet that its chief good arises from its being a form of mechanical restraint; but that it is the latter, for good or evil, there needs no argument to prove.

If a patient sleeps three hours at a time within twenty-four of the commencement of the delirious outbreak, and awakes calmer, and sleeps again in the next twenty-four hours,—and if his attacks of violence are paroxysmal, with comparatively lucid intervals, his mind not getting more and more lost and confused,—we may hope that the malady will soon terminate in long and healthy sleep, after which he may awake comparatively well, like one awakening from the sleep which terminates delirium tremens. But if the snatches of sleep become shorter, the patient waking in a paroxysm of rage or terror—if the delusions grow wilder and more senseless, and he takes less notice of those about him, violently resisting all that is done for him—we may make up our minds that he has entered upon an attack of acute delirious mania which cannot be cut short, but through which he must be nursed for days or weeks.

I need not take up your time by a lengthy description of the disorder, for in truth it is hardly possible to mistake it for any other. The name "acute delirious mania" sufficiently describes it. I shall have a few words to say on the diagnosis and distinction between it and some other forms of delirium; but however alike these may be for a time, the history and progress sufficiently distinguish them from this form of insanity.

I have already spoken of the premonitory symptoms of acute delirium. The oncoming of a transitory attack is usually more rapid than that of one more prolonged. In the latter case the change may have been noticed for a long or short time, from a week to three or four months, presenting the ordinary features of mental derangement, such as in one may terminate in melancholia, in another in mania, while in a third it may be fixed in a quiet monomania. Generally speaking, after a brief period of *alteration*, some casual opposition, heat of weather, or accidental circumstance, lights up the violent stage, while gradually lessening sleep precipitates the attack. When any such symptoms are apparent, it is right to bear in mind that the night is the time when the acute stage is most likely to commence. This is important, because a patient is apt to be less guarded at this hour than at any other. Wives are alone with their husbands, out of reach, it may be, of all male assistance; or a man may be in a room at the top of the house by himself, possibly locked in. Friends are afraid to place attendants with a person in this condition for fear of irritating him, and will often keep them out of sight till mischief is done. The most violent stage is not usually at the commencement of the malady, though a patient may be violent for a short time, paroxysmally. As I have said, this may terminate in sleep, or he will go on get-

ting worse and worse, more full of delusions, more unconscious, and utterly
sleepless, till all hope of speedy termination is past.

We have thus arrived at a stage when the patient, man or woman,
cannot any longer be treated in an ordinary bedroom, or nursed by
relatives and ordinary nurses.   Whether he must be removed to an asylum
will depend on his means, on his house and the inmates of it, or of
adjoining houses.   An asylum is not absolutely necessary to his recovery.
He does not require amusement or occupation, or grounds and garden.
He is going to be acutely ill for a fortnight or longer, and to be confined to
one room for that time.   Upon the room much will depend, but little on
its locality.   He cannot be nursed by near relatives; it is therefore essential that very skillful people shall be about him, in whom full confidence
can be placed.

The room, in or out of an asylum, should be large enough to be airy
and cool; the windows must be out of reach, or protected, yet capable of
being opened, and also of being darkened, and the darkened state must
be kept up during the height of the attack.   The patient will not lie
quietly on a bedstead, and attempts to compel him to do so will end in
many bruises, if not in broken ribs.   So the bed must be made up on
the floor, a considerable surface of which may be covered with mattresses.
Few patients will allow any clothes to remain on—they will strip them off
and tear them in pieces.   They must not, however, go naked: a strong
suit, consisting of jacket and trousers, or petticoat, fastened together in
one piece, and laced up the back, may be securely fastened on them, and
underneath it may be placed the requisite body linen; or a patient may
have a blanket fastened round his neck and shoulders so as to form a kind
of poncho.   If this be well fastened around his arms, they will not be
very available for mischief, and yet there will be no irksome restraint.
He can also be easily held without the infliction of a single bruise.   The
scantiest furniture should be left in the room—utensils will be little
needed, for such patients are almost invariably wet and dirty—and for
drinking a horn-cup is better than glass or china.   Near relatives must
keep out of sight, for their presence will not be tranquillizing, but the
reverse, to the bewildered mind; yet the occasional glimpse of some one
he knows—an old servant, friend, or doctor—may reassure and make him
think that the strange faces he sees are not those of enemies, persecutors,
or fiends.   Many are aware of what is going on, far more than is suspected;
though they make no sign, they take in all that passes, and no better
advice can be given to those in charge than to be careful that they say
nothing in presence of a patient that he is not to hear.   If we in his
presence order medicine to be mixed with food, we may run the risk of
its being refused, even if he is apparently in the most unconscious
delirium.

The old writers used to describe in glowing terms the violence and ravings of these maniacs; and in the days of chains, manacles, stone floors, and straw, doubtless their appearance was terrible. But they are not the really dangerous lunatics, and their violence is for the most part temporary and paroxysmal, though they may pass hours, and even days, in singing, shouting, and perpetual motion. We see in such cases an excellent example of a *reduction* of the higher brain-centres, and over-action of the lower, owing to the removal of the control of the former.[1] The whole of the ideas and movements are more or less automatic. There is incessant talking or singing, such as no effort of will could keep up. The patient rhymes, repeats page after page of poetry, repeats the same phrase in a perpetual sing-song, talks some language he has at other times nearly forgotten, in short, brings out of the recesses of his memory scraps of all kinds, not by an effort of wish and mind, but by the unaided, uncontrolled, automatic action of his lower centres. Dr. Crichton Browne, in an interesting paper,[2] has drawn attention to the movements of patients in this delirious mania. Some will run about night and day in a purposeless manner. These, he thinks, may be suffering from irritation of the postero-parietal lobule of the brain, in which Ferrier has localized the crural movements. Others will toss their arms about incessantly, or busy their hands with the bed-clothes; in these the irritation may be concentrated in the ascending frontal and parietal gyri, in which the brachial and manual movements are localized. "Some talk vociferously and jargonize. May we not infer that in them there is an irritative lesion of the oro-lingual region in the third frontal convolution? Others are resolutely silent, but shake their heads from side to side without intermission. May we not suppose that in them the cortex of the superior temporo-sphenoidal gyrus is hyperæmic or inflamed?" From the motor symptoms Dr. Browne passes to the consideration of the sensory, and conjectures that hallucinations of hearing, sight, taste, smell, or cutaneous sensibility may be indicative of lesions of the various portions of the cortex which correspond to these senses. Similarly, there are grounds for believing that the prefrontal lobes are the centres for the highest mental functions, and lesions of these may produce either loss of motor power, so that the patient stands fixed and motionless, being, as it were, over-inhibited; or this may pass away, and the higher centres may, instead of being stimulated, have become paralyzed. All control being now withdrawn, the subordinate or motor centres are let loose, and we witness the incessant automatic movements executed from their stimulation. These suggestions are interesting, but require close observation of many cases. So incessant is the motion, and so great the exertion of many patients, that the wonder is that they do not more frequently sink from exhaustion.

---

[1] *Vid. ante*, p. 31, *note*.
[2] A Plea for the Minute Study of Mania. Brain, iii., 347.

11

Yet for the renewal of their automatic centres little sleep is required.   A certain amount is absolutely needful, and without it death ensues; but that which is requisite to support life is not large, and if we can procure it, the sufferer may be kept alive till the violence of the brain-storm subsides; and by ever-increasing sleep the health not merely of the automatic, but also of the highest brain centres, is restored.   For continuous work and effort of the latter a considerable amount of sleep is required, as we see in our daily life.   Without sufficient sleep work requiring deep thought and concentrated attention is deficient in quality or exhausts the brain, and so ceases; but for routine and mechanical tasks, less sleep is demanded.   There is not much to say about the mental symptoms; commonly the speech is an incoherent jumble of sentences, or the constant repetition of a word or phrase, beginning in a low tone and rising in a *crescendo,* till the room rings with their piercing cry.   Such patients in many cases can hardly be said to have any delusions—at least these are not to be made out in the confused jargon, but the prevailing feeling is markedly shown in their expression, tone, and gestures.   This may be fear, in which they will call on some familiar name for assistance, will scream "fire!" or "murder!" and in every way indicate the terror they are in. Or they may be filled with religious dread, and refuse food, remain incessantly on their knees, and see the horrors of hell before them.   This is, in my opinion, a very unfavorable symptom, and this variety is closely akin to, if not identical with, the worst form of acute melancholia.   They may be furiously angry, and attack those in charge, applying opprobrious names in accordance with their ideas; or they may be gay and hilarious, laughing and shouting with glee and mischief.

I believe our prognosis may be materially affected by a careful consideration of the emotional state.   On the one hand we find the panic-stricken, on the other the hilarious.   When I see a patient laugh and frolic, however noisy and outrageous his or her delirium, I augur favorably concerning the termination.   I believe that the gayety indicates a reserve of force which does not exist in the other; we may often observe during the disorder that the spirits of the patient will rise or fall as he loses strength or regains it from food, or a period of quiet or sleep.

What are the bodily symptoms?   Entire absence of sleep for days together.   One young lady, in the heat of an extremely hot summer, only slept one hour in eight days.   She then slept five hours, and began to mend.   The time at which sleep first comes varies much, and the time also varies during which patients can go without sleep without fatal exhaustion occurring.   Women can go much longer than men, and unless debilitated by some such cause as parturition, generally battle through an attack of mania; we have not the same need for alarm as in the case of the other sex.   Young women from twenty to thirty years of age are often the subjects of this disorder, and almost always get well—at any rate in the first attack. Patients will exist a long time without becoming exhausted by want of sleep if they are well kept up with food, and are kept from wearing themselves out with incessant motion and fatiguing exertion.

The tongue of a patient in this condition is generally furred and coated with a thick sticky layer of dead epithelium, and in proportion to the heat, feverishness, and shouting of the patient, it will become brown and dry, and the teeth will be covered with sordes. As exhaustion, and that condition we call typhoid, come on, all this will of course become worse. In some cases, however, particularly in women, the tongue keeps wonderfully clean and moist, even if the attack be a long one, and this is a good sign. Taken in conjunction with the pulse, the state of the tongue is a valuable indication. A very quick pulse is a bad symptom. During the paroxysms of violent struggling and incessant motion the pulse may become very rapid, and then fall again in an interval of comparative quiet to a normal rate. Such a condition of things is hopeful. If, however, the acceleration continues, and, with the temperature, is kept up at a high rate, then the case is likely to do badly. Taken generally, the pulse of patients who recover is in frequency below that which we should expect when we see their great muscular efforts and profuse perspiration. The latter is very offensive, and in females we often perceive an overpowering odor of sexual excitement.

The temperature in these cases, though it may be heightened, rising to 99° or 100°, is not usually that of fever. If we find a high temperature, 104° or 105°, it is a most grave symptom, supposing the patient is suffering from uncomplicated delirious mania. We should suspect some complication, or that the delirium depends on a febrile condition due to such causes as typhoid or pneumonia. I lately saw a gentleman, said to be suffering from acute mania, whose friends proposed to remove him to an asylum. He was maniacal beyond all question, and most hostile to doctors. Careful examination was impossible, but the nurse managed to get a thermometer under his arm, which rose to 105°. I said that in my opinion he ought not to be removed.

The water is generally scanty and high colored, and may be contrasted with that of patients who are suffering from transitory or hysterical attacks, when it is often copious and pale. The bowels rarely act without medicine, and very powerful aperients may be required; the stools are usually dark and very offensive. Both urine and fæces are passed about the room or bed-clothes, and often daubed about the patient's person or surroundings.

The appetite is very capricious, but as a rule these patients do not refuse their food. At the commencement of the attack they may labor under the delusion that it is poisoned, but as the state of excitement increases they seem to forget this, and will take a large amount, if it is given to them judiciously.

What is the treatment and what the prognosis of these cases of acute delirious mania? Our object is to nurse a patient through the violent and sleepless stage, and to support his strength, so that he shall not die of exhaustion during it. It is exhaustion that kills; we do not find by post-mortem examination any lesion of the brain or other organ sufficient to cause death. It would seem as if the disorder is one which has a natural tendency to subside after a longer or shorter time, if only he can hold out till this time arrives; and so we see that the young and the strong pull through, while those debilitated from any cause succumb. Everything, then, depends on the amount of support we can introduce by dint of feeding, and the extent to which we can prevent a patient from exhausting himself by violent struggles and want of sleep. When he is quiet—when he does not take much out of himself—our anxiety is comparatively small, even if sleep is long in coming. Now, with regard to food: skillful attendants will coax a patient into taking a large quantity, and we can hardly give too much. Messes of minced meat with potato and greens diluted with beef-tea, bread, and milk, egg and milk, arrowroot, and so on, may be got down. Never give mere liquids so long as you can get down solids. As the malady progresses the tongue and mouth may become so dry and foul that nothing but liquids can be swallowed; but reserving our beef-tea and brandy let us give plenty of solid food while we can. I never knew a patient in this state vomit his food, or suffer from diarrhœa. He will require plenty of drink, also, especially if the weather be hot; and we may give cooling mixtures, as lemonade, in any quantity. If the patient is young and vigorous he will want no stimulants—that is, brandy and wine—at any rate at the commencement of the attack. These, like opium, will only increase the excitement. We may, however, give bottled ale or stout, a glass at a time. The waste of tissue in this disorder is enormous, and patients rapidly lose flesh, and we learn hence the necessity for an increased quantity of food. It is quite possible by patience and tact to get down a liberal allowance: without actual forcing, they will permit themselves to be fed, and a female nurse will often be of great service in this respect among male patients. In the same way a female will often take food from the doctor's hands rather than from the nurse's.

Good attendants will not only feed a patient—they will also nurse him. A man or woman in this condition is to be looked upon essentially as a sick person, and not as a mere violent and noisy lunatic to be allowed to run about, or to be shut in a padded room and occasionally looked at and fed. Such a patient should not be allowed to be on his legs the whole of his time till he falls on the floor exhausted. At times he can be made to lie quiet enough by the side or in the lap of an attendant, especially if the latter fans his face or applies cold cloths to his head, for his violence is, as I have said, paroxysmal, with frequent remissions, and he can either be allowed to walk about at his quiet times, being held on his back during a paroxysm till this subsides, or he can be allowed during the violent stage to go at large, being made to rest during the intermediate period. In this we must be guided by his strength, the degree to which he exhausts himself by his fury, and by the staff of attendants we have at hand. A man kept in the recumbent posture is far more likely to drop off to sleep than one who is perpetually pacing the room.

There remain for consideration the various methods by which we are to procure or promote sleep. For a sleep of some hours' duration will be the turning point of the illness, and this we look for anxiously till it arrives. Probably, nowadays, chloral will be the remedy first thought of, and none is more likely to succeed. Before its discovery those who had most experience were disposed to lay aside all drugs, as doing, upon the whole, more harm than good, and to trust rather to baths and such appliances to bring about sleep. But chloral has come to our aid, and by it, or the combination of it with bromide of potassium, sleep may be procured in the majority of cases, and in my experience, deaths from exhaustion have been greatly diminished. I lately treated one of the most violent cases of this disorder which occurred in a young lady after confinement. There existed the most violent sleepless delirium, complicated by total refusal of food. Chloral alone did not bring sleep, but combined with bromide of potassium in half-drachm doses it never failed, and her recovery was most rapid and satisfactory. If chloral and bromide fail to procure sleep, I would lay aside drugs, and trust to other measures.

I have found that it is better not to give chloral and bromide for any great length of time. In the height of the delirium I give it at night to procure a certain amount of sleep, and prevent the patient sinking and dying of exhaustion, as patients did five and twenty years ago. I never give chloral to produce quiet. There comes a time when sleep will take place without drugs: at first it may be of short duration, but yet enough to preclude danger to life. When we have arrived at this stage it is better to leave off chloral and bromide. We may procure by it six or seven hours' sleep, but the patient next day wakes as maniacal as ever, and this condition, I am sure, is often prolonged by such means. The temptation to give it for the sake of nurses and friends is great, especially when we are treating a case in a private house, but the effect on the patient is often marked when these drugs are omitted, even when for the first week the amount of sleep is greatly lessened. I am as strong a believer in the virtues of chloral as ever, but, like all powerful medicines, it must be used with judgment, and I refrain from giving it for a great length of time to any patient. It is a medicine for acute cases, not for chronic. And this is a reason why patients should never take it at their own discretion.

It may occur to some that we ought to try opium or its preparations. I have said something of this already, and have only to add that in prolonged delirious mania I believe opium never does good, and may do great harm. We shall see the effects of narcotic poisoning if it be pushed, but none that are beneficial. This applies equally to opium given by the mouth and by subcutaneous injection. The latter, as it is more certain and effectual in producing good results, is also more deadly when it acts as a narcotic poison. After the administration of a dose of morphia by the subcutaneous method, the patient will not improbably at once fall asleep, and we congratulate ourselves that our long-wished-for object is attained. But after half an hour or so the sleep suddenly terminates, and the mania and excitement are worse than before. Here you may possibly think that had the dose been larger, instead of half-an-hour's sleep, you would have obtained one of longer duration, and you may administer more, but with like result. Large doses of morphia not merely fail to procure refreshing sleep: they poison the patient, and produce, if not the symptoms of actual narcotic poisoning, at any rate that typhoid condition which indicates prostration and approaching collapse. I believe there is no drug, the use of which more often becomes abuse, than that of opium in the treatment of insanity. Among the ancients the disorder was treated invariably by hellebore; by our fathers by bleeding and tartar-emetic; now, till lately, by opium. Do not be led away by the fatal facility with which you can administer it by subcutaneous injection. Inject it in the case of a melancholic patient, if you like, but in this furious delirium you must abstain from the administration of opium in all its forms. We may, however, promote the access of sleep by other means. And, first, I must speak of baths, which are largely used here, and still more abroad, as sedatives in cases of excited mania. The baths recommended for this purpose are always, so far as I know, hot; even mustard is to be added, according to some, to promote the determination of blood to the surface. This is the *rationale* of the treatment; cold is to be applied to the head by cloths and ice-bags while the patient is in the bath, and so the blood is to be "derived" thence.

The French employ these baths for very long periods. M. Brierre de Boismont recommends that a patient should be kept in one for ten, fifteen, or eighteen hours; but there must be considerable risk in so doing, and I should be sorry to recommend it. Unless the bath can be given effectually, it had better not be attempted. Merely holding a man in hot water for a few minutes, kicking and plunging, is not likely to quiet his mania, or lower his circulation down to sleeping point. To be of any use, it should be given for half an hour at least, with cold to the head. It should not be too hot: one at 90° or 92° is more likely to be cooling than one at 98°, and is less likely to cause syncope. After such a bath a patient may go to sleep. I have never seen a cold bath given; but I should like to see the effect of placing a patient in a tepid bath, and allowing him to remain there till the water cooled. I cannot help thinking that a great general cooling of the body would thereby be produced without any shock, and that by the skin much would be absorbed to allay the fever and thirst, and replace what is lost by the profuse perspiration.[1]

Shower-baths are wholly inadmissible here: to produce a sedative effect they would require to be given for a considerable length of time, and would be productive of very great danger.

Concerning purgatives: if we see the patient at the commencement of the outbreak, we may give a brisk purgative with the hope of arresting it. Later we cannot effect this by any means, and we do not wish to lower his strength by strong purgation. If it is necessary to interfere at all, we may give a few grains of calomel, because they are easy to administer, or a dose of castor-oil, if we can get it down. The latter is perhaps more likely to act than anything else. We must beware of giving pill after pill, or dose after dose, when those taken previously have not acted; for if the stage of reaction sets in, we may have violent action from their combined effects.

Are we to bleed in this acute delirium? A few years ago such patients were invariably and largely bled. Nowadays general bleeding in insanity is in this country entirely abolished. Local bleeding by leeches or cupping is to some extent advocated and practiced, but of it I myself, having had no experience, can say nothing. There would be considerable difficulty in accomplishing it in most cases, and, in my opinion, it is not needed, for the only patients whom we should think of bleeding are the young and strong, and these almost invariably get well without it. The cases which succumb to this disorder are those in which we should never think of abstracting blood.

I have seen two girls to whose nape and spine blistering fluid had been applied. They passed many days in furious mania, and the sore places became dreadful wounds, depriving them of both sleep and rest, and, being followed by a crop of boils and pustules, retarded greatly their recovery. I advise you to abstain altogether from blisters, and from everything which is likely to make a wound or excite suppuration, which is very prone to occur. Surgery under such circumstances is most difficult and unsatisfactory, and nothing should be done which is likely to call for surgical interference or dressing.

What is the prognosis? In my own experience I have not found this a hopeless disease, though it is most formidable to treat. A well-known French authority, M. Marcé, says that of these patients (delire aigue) scarcely one in there or four recovers. I should say scarcely one in ten dies. I have found that young unmarried women, who frequently are the subjects of this disorder and of hereditary insanity, generally recover; that the women who die are those weakened by childbearing or phthisis, or ad-

---

[1] This was written before the administration of cold baths in cases of hyperpyrexia.

vanced in age; that young men seldom die in the first attack; that the men who die are those who have had repeated attacks, or are advanced in life, or weakened by such causes as phthisis, hæmorrhage, or want of food. The disease, as a rule, occurs in the young, in patients between the ages of twenty and thirty. Modifications and varieties of it occurring in older persons are more akin to delirium tremens, or the transitory delirium of wasting and acute diseases. But this is not the true acute delirious *mania* of which I have just been speaking.

The terminations are almost invariably recovery or death. If the delirium continues and no sleep comes, the patient becomes weaker and weaker, the pulse more uniformly rapid, the mind more unconscious, not recognizing at all those in the room. The tongue is brown and dry, profuse sweating bedews the forehead, emaciation rapidly advances; and when at last the sufferer becomes quiet, and lies on his bed without attempting to rise, instead of sleep we see coma and collapse supervening, and an exhaustion which rest does not remove. When coma sets in it generally increases quickly; there are evident signs of serous effusion taking place within the cranium, and death soon follows. When the termination is in recovery, sleep is the first harbinger of it. Sleep comes either naturally or the result of medicines; at first it may be a sleep of two or three hours only, but on awaking there is manifestly an improvement, evidence that this short sleep has had an effect. Probably in no long time there are other sleeps, then a long one of six or seven hours, and we consider the crisis past; and although a very maniacal condition may last perhaps for weeks, there will not be again the same absence of sleep, or danger to life. The time occupied in recovery varies indefinitely. A patient may be perfectly well in a month from the commencement of a very acute attack with a long period of sleepless delirium. On the other hand, he may remain for weeks or months in a condition of nervous prostration and dementia, and then slowly recover. We find here and there a patient who never recovers, but remains permanently in a state of chronic mania or chronic dementia; but such instances are, I think, rare, and even those whose families are saturated with insanity, and who have an attack of acute delirium without other apparent cause, get well in a most satisfactory manner, at any rate in the first attack. There is one variety, however, which is the most formidable of all to treat, and which does not in general recover, but terminates fatally. This is a class of patients who, with all the violence, sleeplessness, and delirium of genuine acute delirious mania, obstinately persist in refusing food, as obstinately as others afflicted with acute melancholia. They do not make the passive resistance of the melancholy; they waste in their violent condition far faster, and require much food, yet none can be got down, except by force. As the utmost we can give in this way is inadequate to their need, and as it always implies a struggle, whatever method of feeding we adopt, they gen-

erally sink rapidly and die.   Among them we do not find the gay and
lively; but, on the contrary, they are filled for the most part with religious and other horrors.   In fact, the malady merits the name of acute melancholia rather than acute mania.

I need say but a very few words concerning the diagnosis of this disorder.   Those most likely to be confounded with it are various forms of
delirium.   1. I have known a gentleman sent to an asylum as suffering
from acute mania who was raving drunk.   He had been drinking for
days, but he had not even delirium tremens; he was simply mad drunk,
and so soon as he had slept off the liquor he was well, and could not be
detained.   Here a grave mistake was made, and it should be a rule with
you never to sign a certificate for a patient who is at the time under the
influence of drink.   2. A man may be in a condition of delirium tremens.
If it can be helped, we do not send such patients to asylums, knowing
that the attack will be brief.   You will observe the tremor, the busy restless fumbling, and the terrified aspect which is not generally seen in maniacs.   Even without a history you can hardly mistake the two, and delirium tremens occurs in men advanced in years much more frequently than
acute mania.   The delusions and hallucinations also are very different.
Maniacs seldom see snakes, or rats, which so often torment the drunkard.
3. The patient delirious from fever, acute rheumatism, or pneumonia, is
a very different being to the maniac.   There is no difficulty in keeping
him in bed, though he be noisy and incoherent.   His aspect is quite different, his temperature is different, and of course his malady was not at
its commencement mental.   4. Still less need I mention patients suffering
from acute disease of the brain, meningitis, etc.   We have other symptoms to guide us—rigors, squinting, vomiting, coma, convulsions, a totally
different pulse, and complaints of excessive pain, which last is rarely complained of in mania.

LECTURE XI.

Acute Mania—Symptoms—Treatment—Medicines—Prognosis—Terminations—Alternating or Periodical Insanity—Mania and Monomania—Treatment of various Insane Patients—The Insanity of Puberty—Masturbators—Puerperal Insanity—Insanity after a Blow—Coup de Soleil—Syphilis—Rheumatism—Epilepsy—Phthisis.

THE next form of insanity of which I have to speak to you is very different. It is called by many acute mania, as is also the last described. You may, if you like, call it "acute mania without delirium," as the other is "acute mania with delirium;" but it is a very different disorder. Whereas the former is a disease running a rapid course to recovery or death in a week or two, this may last for months without much danger to life. The patients are little less violent or noisy, but they know what they are about—their violence has in it far more of design; and sleep is not absent as in acute delirium. This mania may come on like the latter, almost suddenly, or with premonitory symptoms of some duration. In two cases lately treated, the disorder was stated to have lasted in one for three days, in the other for a month. These persons are mischievous in an extreme degree, wet and dirty, not from unconsciousness, but to give trouble, abusive, filthy in language and habits—presenting, in short, all the worst features of the insane. Having their wits about them, they know very well how they can annoy, and are extremely ingenious in provoking the attendants, and complaining of them afterward. They may be, and often are, full of delusions, or their insanity may consist chiefly of outrageous conduct and language. They are perfectly unmanageable out of an asylum: acute delirium can be treated in a suitable room in any house, and the sufferer may recover, as so many do who are said to have had "brain fever;" but I do not think that many of these noisy maniacs could be treated with a view to recovery in a private house. To keep them there at all would probably entail very rough handling.

What is the mental, and what the bodily state of these people? They may talk quite coherently for some time, or a few minutes only, rambling off in incoherent nonsense, or may commence singing or shouting, or will abuse us violently with full consciousness of who we are. Their mental state varies from a power of keeping up a conversation and concealing the insanity so as almost to baffle us in signing a certificate, to a degree of wild incoherence which might be taken for delirium; their acts and deeds vary also, but at the height of the disorder they almost all destroy clothes, bed-clothes, furniture, and windows. They are abusive and inclined to fight, and rack their vocabulary for opprobrious and obscene terms. Often they are given to self-abuse, which they will practice with open and shameless audacity; at other times they expose their person, to cause annoyance and disgust.

The bodily health generally is tolerably good, and suffers less than we might expect from the severity of the disorder. They seldom die of it unless their health is much broken at the commencement. It comes on, however, occasionally in persons whose strength has been much impaired, and then, if they do not recover shortly, they may gradually wear themselves out and sink. This was the case with an officer who had suffered severely from wounds and exposure in India. He never recovered, but sank after being many months in confinement. He was in a state of extreme emaciation when the mental disorder began.

Such patients are not all young, like the majority of those who suffer from acute delirium. Many are men of forty years and upward, yet, as I say, few die. They generally eat heartily, nay, voraciously, if allowed, and although in time they get thinner, do not waste rapidly. Sleep is deficient. They pass a good night, sleeping six or seven hours, and then go on for days with perhaps only one or two hours' sleep, making night hideous with shouting, laughing, and singing, and disturbing all within hearing. They seem to acquire an ability for doing without sleep, which may last for a couple of months. The tongue is not particularly foul or coated, and may after a short time become quite clean, while the bowels will act without much trouble. In short, the bodily health is not a matter of grave anxiety: our chief object is to calm their unquiet mind. Now, although medicinal means are not to be neglected, much may be effected by moral treatment, moral control, and discipline. Not by stripes and chains, as was recommended by Cullen and others, but by a system of moral rewards and punishments, their violence is to be checked, and a wish for a return to civilized life to be roused. Here no rule can be laid down which is applicable to every case, or even to a majority; but your own common sense will probably suggest to you how a patient is to be encouraged, how repressed. There are many things which he will delight in and look for, as tobacco, snuff, or wine; various indulgences, as walks or games. They may be granted for good behavior, and withheld for bad. The noisy maniacs are, in many respects, very like children; their turbulent behavior, and mischief committed often for mere bravado, their dirtiness and untidiness, and the utter senselessness and folly of their proceedings, remind one constantly of willful and naughty children, and they are often capable of being influenced by analogous processes. We must devise something which shall help them to restrain themselves—something which they will attain to if they behave properly—something which they will lose if they continue in violence and mischief. I once had under my care a most outrageous patient, the terror of attendants and the pest of the asylum, who committed great damage and destruction to property. He went on in this way for four months. I then placed two men with him constantly, so as to give him no opportunity for mischief, and entirely stopped his tobacco, of which he was greatly enamored. This had a marvelous effect, and in a month's time he was discharged, and continued well for some years. Some patients who tear their clothes to pieces will abstain if they have a new suit given to them, or one of a different kind or color; others will require the constant surveillance of attendants till the habit is broken through. For, like pulling out the hair, or picking the face or fingers, tearing-up is often a habit and employment for the hands, which must be kept from it, and, if possible, employed in something else. Nothing is of so much service in these cases as prolonged exercise in the open air. Let the patient be in the open air as much as possible, and let him be well walked between two attendants; his exuberance of spirits will be checked, and he will sleep all the better. As the last-mentioned patients were to be considered sick, and kept in one room and nursed, so these are to be looked upon as in fair bodily health, are not to be kept in bed or in one room, but, on the contrary, are to have as much exercise as we can give them, unless it be contra-indicated by some diseased condition.

Are medicines which procure sleep of service here? In some they are, in very many they are not. We may procure sleep by means of chloral and bromide, longer sleep than the patient would have without their aid, but he awakes refreshed in body, yet not improved in mind, and recommences, with renewed vigor, his course of noise, violence, and mischief. As I have said, we are not anxious about the lives of such patients, and sleep is not to them of vital importance. In these cases soporifics do not always produce the beneficial effect which we anticipated some years ago, and which friends and relations think must be produced if they procure the sleep for which they are given. When we attend maniacal patients out of an asylum, whose friends are with them, the latter naturally attach the greatest importance to the amount of sleep enjoyed, and they cannot believe us right if we cast aside drugs, and prefer a small amount, perhaps a very small amount, of natural to a larger quantity of sleep produced by medicinal means. In all such cases we have time for trial, and trial we can make. Some patients in a subacute condition improve rapidly when put to sleep by drugs; without them they waste, become more maniacal, more lost and demented, and dirty in habits. A week's good sleep works wonders, and recovery follows. And in the administration of such drugs, I often find that, given every other night, or every third night, they are more beneficial than if taken nightly, and that small and diminishing doses are often of more service than large ones. Sometimes in the less sthenic cases opium is of service, and procures sleep at night; we cannot always say beforehand whether its effect will be beneficial, but here we can try it: it is not a matter on which hangs life or death. We may administer morphia by subcutaneous injection, if there is a difficulty in getting medicines swallowed. Comparing this mode of administration with the ordinary plan, I have found but little difference in the effect produced by the drug; but a less quantity of the salt is required if it be subcutaneously injected. If it is of no use, we may give digitalis, bromide of potassium, extract of henbane, cannabis indica, or a combination of them, e.g., the bromide and cannabis. In the sixth volume of the West Riding Asylum Reports Dr. Robert Lawson speaks very highly of hyoscyamine, and its efficacy in subduing the wild excitement and destructiveness of patients of this class, whether the cases are recent or chronic. From Dr. Lawson's account of the effect it would seem to be specially suited to the conscious mania and willful outbursts which I have related. The dose given ranged from $\frac{1}{18}$ to $\frac{1}{3}$ of a grain. The smaller dose may be given every two hours till the physiological effect is produced, or the larger dose may be administered once; the latter was frequently found to be effectual, and a marked change followed. Beyond doubt it is a most powerful medicine, but it is not unattended with danger. Dr. Clouston [1] says, "Hyoscyamine is an admirable quieter of motor restlessness, and often does no harm, but I have seen dangerous coma produced by it." I shall have to speak of it again hereafter. Two other drugs I may mention: one, which was formerly used almost universally in the treatment of insanity, is the potassio-tartrate of antimony—*antimonium tartaratum*, as it is now called; the other is hydrocyanic acid. The former has been largely given for years in many asylums, and frequently acts beneficially, quieting the patients I have described, and also chronic maniacs. Here it produces no nausea in small doses ($\frac{1}{3}$ or $\frac{1}{2}$ gr.), neither does it cause any aversion to food. One young lady to whom I gave it took food better than before. It might be given subcutaneously, but of this plan I have no experience. Its small bulk, solubility, and absence of taste and color, enable it to be mixed with facility in any kind of drink.

The other, hydrocyanic acid, is, like digitalis, an old remedy revived. Dr. Burrows, in his "Commentaries," tells us that Dr. Balmanno, of the Glasgow Asylum, used to give from 15 to 30 drops of a diluted solution of it, preparatory to giving narcotics, but that he himself had found no good results. In papers in the "Medical Times and Gazette," March 1863, Dr. Kenneth M'Leod, of the Durham Lunatic Asylum, strongly recommends its use as a calmant in cases of acute mania and melancholia, in doses of from ℥ij. to ℥vj., to be given by the mouth or subcutaneously, and repeated every quarter of an hour till an effect is produced. This may be worth a trial in these cases; but if the dose, say ℥v., which

[1] Clinical Lectures, p. 178.

Dr. M'Leod recommends, is to be repeated every quarter of an hour, it
must be done under your own supervision. I should think that the drugs
of modern times have altogether superseded these two.

What is the prognosis of the disorder? As in most forms of acute insanity, it is favorable, if there be no complications of other disease. Our
opinion as to recovery will be founded on the consideration of the following facts:—1. How long has the attack lasted? This, of course, is a matter of some importance, for if the disorder is becoming chronic, our hopes
will be less; but these patients often continue their violent conduct for a
long time, and yet at last recover. I have known one recover after having been in an asylum nearly two years, and it was his fourth attack. 2.
The character of the mania. If great noise and turbulent excitement are
the predominant features, with no very marked delusions, or with delusions ever changing and not fixed and immovable, we may have hope. If
the delusions do not vary, if they have reference to the unseen and supernatural, above all, if the patient hears voices, the cure is very doubtful.
3. The age of the patients. Many of these are not very youthful, as I
have said; the younger they are, however, the more chance they have of
recovering; and, as a rule, I believe men recover more frequently from
this form of insanity than women. 4. If the patient at the commencement of the disorder is greatly debilitated, or if there is other disease, the
violence and want of sleep still further reduce his strength, and interfere
with the chance of recovery, and if there be much difficulty in getting
him to take food, his prospects are still more gloomy. And this brings
me to another portion of my subject, viz., the terminations of the disorder.
First, the patient may recover, and that after a very considerable time.
And it is not to be forgotten that if medicines and the moral control of
one asylum and one set of attendants have failed, great good may be
effected by removal to another, to entirely new hands and surroundings.
It is often the only thing left for us to try; but no less often is it a remedy
of marvelous effect. I shall have to recur to this again; but I mention
it here as in these cases the good that is done is frequently very apparent.
Secondly, the patient may sink and die, gradually worn out by the disorder, or by intercurrent disease. Thirdly, he may lapse into a condition of
chronic mania or dementia; or, what is more favorable, he may quiet down
into melancholia. From this he may recover, or after a period of depression may again become excited and maniacal, his disorder becoming the
folie circulaire of the French.

This name has been given by the French physicians to a form of
insanity which goes through a certain cycle of states or stages with
extreme regularity. The patient suffers at one time from mental depression. After a period spent in the depressed state, which may be months,
or a year or two, he gradually emerges, shakes off his gloom, and then
either remains for a time in a condition which appears to be perfect

sanity, or passes at once into one of excitement and exaltation. This in turn lasts for a varying time, to be followed by another period of apparent sanity, or by depression without any intervening stage. And this alternation may continue for years, even to the end of a long life. In some the depression and excitement are so severe that control and restraint are necessary, but many do not require it, and a man may live his cycle in his own home with the same regularity, being at one time dull, inactive, unable to face the world, predicting misfortune in all things, incapable of effort and unduly irritable. Then the scene changes, and everything is rose-colored. He is keen to try new projects, to launch out into speculation, he jokes and tells a good story, and, as his friends say, "there is no holding him," and then as time advances there comes again upon him the cloud of gloom. The duration of the depression and excitement varies greatly in different patients, and on different occasions in the same patient. But the succession is uniform and regular, and the prognosis in such cases is decidedly unfavorable. Other patients there are in whom the respective attacks last a much shorter time, weeks, or even days; some have alternate days of wild excitement and quiet depression. Two I have known where one day succeeded the other year after year with such even regularity, that visiting them once a month I could always tell by reference to the date of my last visit whether I should find them quiet or excited. Neither of these recovered. A third, who alternated in this way for a year or more, gradually lost the regular alternation, becoming more uniformly depressed. This depression lasted some years, and has lately passed away, being followed by a slightly elated state, the duration of which remains to be ascertained. I have known several patients in whom the stage of excitement lasted from five to seven weeks, followed by a period of depression without any sane intervals. One lady now under my care has these alternate stages, and has had them, I believe, since she was eighteen; she is now upwards of seventy. I am told they commenced with epileptic seizures. They have occurred for twelve years since I have known her with undeviating regularity. Another, a gentleman, had attacks of excitement with constant hallucinations of "voices." These, too, lasted for about five or six weeks, and were followed by a period of depression, in which he was furiously angry at being detained, threatened every one, then gradually got more cheerful, and so lapsed again into excitement. Both these were gay, hilarious, and expansive during the excited stage.

Besides these patients who alternate from one phase of insanity to the other, from depression to excitement in a never-varying round, we see many others who suffer from periodical attacks of mania or melancholia, their intervening state being one of perfect sanity. That insanity should be a periodic disease is not surprising. All nervous disease is marked by periodicity. Look at the periodic attacks of epilepsy, of epileptiform tic, of neuralgia, of neurotic gout, of migraine or sick headache. Many have these attacks at intervals throughout a long life. I knew one who was in asylums more than thirty times. Another, now under my care, had his first attack thirty years ago, and has been under care and treatment every three or four years since. There is an extraordinary uniformity in the illnesses of such persons. They reproduce their old delusions, their old tricks and ways, their old language. Friends try and discover, or think they have discovered, a cause of each attack, and on subsequent occasions try zealously to ward off a repetition by extra care and precaution. Vain hope! Try what we will, as time rolls on the disorder reappears; many are aware of its advent, and tell us that they know they are going to be ill. Some come to us for succor, others try all they can to avoid the restraint they know will be their fate.

There are, however, many patients whose insanity cannot be called melancholy of any kind, neither is it acute delirium nor acute mania. It is something short of the latter, and, in ordinary parlance, it is called *mania*. Depression is not a marked feature, except, perhaps, at the outset. There is at times angry excitement, complaints of being detained, of delusions not being listened to and requests not being complied with. Such patients are capricious in temper, sometimes friendly, sometimes hostile, dangerous according to their delusions, seldom suicidal, prompt to escape, threatening vengeance and an appeal to a jury. They require tact and patience, are not to be subjugated by intellects inferior to their own, and are, as a class, to be cured by moral rather than medicinal measures.

We feel no anxiety about the life of these maniacs. When the disorder is recent, and the acute sypmtoms at their height, sleep may be defective, their eating may be irregular, and they may threaten to refuse food. But they seldom do; hunger asserts its claim, and they satisfy it. And a dose of chloral, repeated for a few nights, generally produces sound sleep. We may notice other bodily ailments requiring medical attention, or a general debility, to be met by tonics and generous diet; but many have little the matter with the bodily health, and seclusion from home and the world, change of occupation and amusement, and regularity of life work a cure with but little assistance from the Pharmacopœia. Frequently they refuse to take medicines, and none being specially necessary, none are to be forced upon them. If you search the case-books of any large asylum, you will see that a not inconsiderable proportion of the patients recovered are of this class. Probably they would be termed by certain writers cases of idiopathic mania in persons who upon some moral or mental disquiet, or even without assignable cause, suddenly or gradually show signs of non-acute insanity, and when properly cared for, get well again without any very special treatment beyond seclusion and restraint. The majority of them inherit the disorder. These, too, are the people so often allowed to run loose, without care or treatment, till chance of cure is gone. The symptoms are not urgent, the patient is rational and well-conducted in many ways: it seems a shame to call him a madman, and deprive him of liberty, and so months and years are allowed to elapse, till the friends become tired of his vagaries or expenditure, and he is then brought to an asylum, where it is expected that he will immediately be cured, because they have always heard that it was the proper place for him. Possibly at the commencement the cure might have been effected; but it is certain that chronic mania, or monomania, as it is often called, cannot be cured when it is of two or three years' standing, though by the restraint of an asylum the outrageous behavior of such patients may be greatly checked.

I am supposing that all these patients labor under delusions or hallucinations. When we speak of chronic mania this is implied. And the recent and curable mania which I am describing is equally characterized by these fancies. In a subsequent lecture I shall bring before your notice people whose insanity is not marked by delusions, but by insane acts and conduct. Now, however, you will bear in mind that in these patients delusions are a marked feature, and the prognosis will be determined in some degree by the character and continuance of them. Of this subject I have already spoken, and have endeavored to point out how the prognosis is influenced by delusions or hallucinations of one kind or other.

In addition to mania, certain authors describe a form of insanity which they call monomania, but are not agreed as to the symptoms which the term denotes. Griesinger, following Esquirol, who first introduced the word, would confine it to those patients whose mental disorder is displayed in expansive delusions, is over-estimation of self. Dr. Bucknill says that it is seldom primary, and is in the majority of cases, a transformation of melancholia, i.e., the self-feeling is one of depression, not of exaltation. Nothing, however, is to be gained by dividing patients into those affected by mania, in contradistinction to others whose disorder is monomania. Probably that which is most commonly called monomania is chronic insanity, when the patient is removed from deep depression on the one hand, and gay or angry excitement on the other, and when the bodily health has assumed its ordinary level, and all pathological marks have by time been effaced. The distinction between mania and monomania is for the most part verbal. Formerly all insanity was called "melancholy;" nowadays it is often spoken of as mania, and, if chronic, as monomania. There is nothing pathological in such a nomenclature, which only serves to draw us away from the due consideration of the pathology of the disease we have to consider and treat. We may retain such terms as acute delirious mania, acute melancholia, or general paralysis, because they connect a certain set of pathological symptoms occurring in individuals of various ages, requiring special treatment, and capable of

12

receiving a similar prognosis. We may, if we like, also retain the general terms mania and melancholia, but beyond this we need not go; nothing is to be gained by retaining such a term as monomania, to which no two persons attach the same meaning. Any further divisions should be made, not according to mental peculiarities, but according to the pathological causes or conditions of the case.

If we try to connect the symptoms with the pathological conditions mentioned in my third and fourth lectures, how are the treatment and chance of cure affected thereby? There is an insanity of pubescence, of pregnancy, after childbirth, at the climacteric, an insanity connected with masturbation or sexual irritability, a sympathetic or reflex insanity, an insanity depending on alcohol. How does the diagnosis of one or other of these affect our prognosis or treatment?

The insanity of pubescence is marked by violence, excitement, maniacal conduct, rarely by depression or melancholic features. Delusions there may be, and hallucinations, but not of a formidable or persistent nature. The majority of cases would be called by some "sthenic" rather than "asthenic." Sleep, though defective, is not altogether absent, and may be brought about by exercise in the open air, and by chloral, or bromide of potassium, rather than by opium. Stimulants are not needed; they produce excitement and a prolongation of the symptoms. In girls, we shall find that the catamenial function is generally disordered, the menses scanty or profuse. In the latter case, local treatment is preferable to anything given by the mouth, and pessaries, containing tannin or matico, may be applied *per vaginam*. In the former, iron and aloetic preparations are often of service. The majority of such patients get well, at any rate in the first attack; but as they almost all inherit the malady—which inheritance is, in truth, the cause of it—its recurrence is to be looked for at some period or other of their lives. This occurrence it is our business to prevent, if possible; and also to ward off insanity from other children of the family, or any children who inherit an hereditary taint. Such should be separated, and placed with others of robust nerve constitution; their diet should be plain and unstimulating; they should lead an outdoor life, and not spend their playtime indoors, devouring poetry and fiction. The boys should at an early age be warned of the evils and dangers of the habit of masturbation, and a strict watch kept on the girls. Such habits are acquired early; and, when acquired, advice on the subject is for the most part useless. These neurotic childern easily become the prey of any bad habit, and with difficulty break it off.

After this, I may fitly speak of the insanity of masturbation,[1] reminding you that patients whose insanity is caused by masturbation are generally by inheritance prone to mental distubance. That masturbation

---

[1] *Vide* page 41.

by itself is not a frequent cause of insanity, is a fact of which all must be aware: were it so, in many of our schools insanity would be an everyday occurrence. In some persons, already predisposed, it may light up the disorder, and may coexist with it in others without being the cause. These may, and do, get well, and when well often relinquish the habit. But the habitual masturbator, whose insanity has been gradually developing perhaps for years, is incurable. Thin and sickly-looking, he seems ever on the verge of consumption, and, though he may eat voraciously, his appearance is always a discredit to those who have to care for him. You have doubtless heard of various methods of preventing or curing the habit —of clitoridectomy, which is certainly inefficacious, the clitoris not being the only sensitive portion of the genital organs—of the application of ice to a female—of blistering the prepuce of a male—of the administration of antaphrodisiacs, of which bromide of potassium is supposed to be the most efficacious—and of various mechanical cages and contrivances for preventing the contact of the patient's fingers. After considerable experience, I am sure that it is next to impossible to prevent it in a very determined person. Slight confinement of the hands of a quiet or demented patient in what is called a pinafore or muffler, may suffice; but many women will boast that they can effect their object by the friction of their thighs, or the application of the heel. And a man whose hands are confined may do the same by friction against the bedding. Circumcision is in some cases beneficial. Blistering the prepuce can rarely be maintained for a sufficient time; and when the sore is healed, the man returns to his practice. I have certainly found bromide of potassium produce some effect; and one woman confessed that it took away the power of accomplishment. But bromide perseveringly given for weeks and months will make many patients wretchedly thin and weak, and it is by this effect that it appears to act. We are obliged to discontinue it, and with the strength the habit returns. Nothing but close personal watching will really stop it. A patient's clothes may be so constructed that he cannot by day carry it on in the presence of others; and at night, some slight confinement of the hands, and rigorous surveillance, may greatly check it. It will also be necessary to see that it is not practiced at the closet, whither such patients frequently resort on pretense of a legitimate desire. I lately saw, in the case of two chronic male patients, the experiment tried of confining their organs in a pair of leather breeches made in such a way that access of the hands, or any kind of friction, was entirely prevented, and masturbation at night stopped By day they were closely watched. In one, a case of chronic mania, the effect was very good. From an emaciated being with no appetite, wan and sickly, he became fat and strong; and the habit, though not entirely given over, has never since been carried on to the same extent. The other was a youth in a state of chronic dementia Here the prevention of the habit caused excitement to such an extent that he appeared likely to pass into a state of acute mania. The mechanical contrivance was removed, and he subsided into his usual quiet but demented condition. In many instances, while a prospect of cure exists, it is worth while to exercise every precaution, and to try if a total suspension may not only eradicate the habit, but also restore the mind. An examination of the linen will afford us information of the continuance or discontinuance of it in males: in females it is often difficult to arrive at a just conclusion, for questions upon the subject are better avoided. We gain little information, for by all but the most hopeless it is sure to be denied. We are not to suppose that every patient suffering from acute insanity who may be detected in this is therefore incurable. I have known many recover, even after a considerable period, who were guilty of such acts. But of their insanity masturbation was not the cause, but only the concomitant, and when reason returned, this, like any other insane and dirty habit, was abandoned. Yet I .have heard a most unfavorable prognosis pronounced, which was founded solely on the fact that such acts were perpetrated. In acute delirium occurring in young women, we may find it in the large majority of cases: it existed to an extreme degree in two lately under my care, yet both recovered rapidly and perfectly.

Turning to another form, viz., climacteric insanity, we see a number of patients whose malady and condition is truly asthenic, whose disorder is evidenced for the most part by deep depression, who are suffering, in short, from melancholia. The treatment I have sufficiently indicated in my ninth lecture.

Puerperal mania is a disorder often described in works upon midwifery, as well as in those upon mental disorders, but the term very inadequately describes the condition of the women who from this cause become insane; for they may show signs of aberration before childbirth, and both before and after it the symptoms may be those not of mania, but of melancholia. Dr. T. Batty Tuke, in an examination of 155 cases admitted into the Royal Edinburgh Asylum, tells us that 28 were of the insanity of pregnancy, and that the symptoms in these are, as a rule, of the melancholic type, the suicidal tendency being very marked. " The prognosis in this form of disease is generally favorable; nineteen cases recovered within six months; and of itself it is not fatal." My own experience confirms this. Such women are suicidal, refuse their food, and require to be treated in all respects like other melancholic patients. With care and good feeding they usually recover. The question will arise, whether labor ought to be induced prematurely. Looking at the favorable results of cases which have gone their full time, I should say that, except for very special reasons, such a proceeding is not necessary.

Dr. Batty Tuke examined the records of 73 cases of puerperal insanity, in all save two of which, symptoms appeared within one month of confinement. Where they commenced in the first sixteen days, the form of insanity was acute mania; later it was melancholia.

Bearing out my own observations as to the general treatment of the two types, Dr. Tuke tells us that in the melancholic women the administration of large doses of morphia was attended with the very best results, while he feels certain that where mania has established itself, sedatives in large doses will not be found successful in the majority of cases. In the very commencement of puerperal insanity, I know that the greatest good has followed the administration of chloral, and that an impending attack has been cut short; but at this critical period I should apprehend as much evil as good from a dose of morphia. As the period after childbirth is lengthened, the maniacal symptoms decrease, and those of melancholia become more prominent. Of 54 cases called by Dr. Tuke "insanity of lactation," 10 were examples of acute mania, 39 melancholia, and 5 of dementia. "The acute mania, as a rule, in this form of insanity is severe, but evanescent; it rarely lasts more than ten days or a fortnight, and is generally attended with hallucinations of the different senses, and delusions, as in puerperal mania, of mistaken identity. In almost all cases of insanity of lactation which have come under my notice during the last two years, exophthalmia and *bruit de diable* have been marked symptoms." I think these cases plainly indicate what I have so strongly insisted on, that the mental symptoms depend on the general pathological strength and condition of the patient at the moment of the outbreak, and not upon the nature of the exciting cause. Taken in conjunction with Dr. Batty Tuke's observations, those of Dr. Clouston at the Edinburgh Asylum are most-interesting. Of 60 cases of puerperal insanity no less than 43 were very acute in character and symptoms, 59 being generally maniacal in character, and 14 generally melancholic with motor excitement. In the mild cases the prevailing character was depression, 14 of the 17 being melancholic. "In at least 18 of the acutely maniacal cases, the mania amounted to absolute delirium. I know of no clinical form of insanity that would yield so large a proportion of very acute cases." Of the 60 patients, in 26 the temperature rose over 99°, and in 14 it was over 100°. All the deaths occurred in patients whose temperature rose over 100°, yet of 5 patients whose temperature was over 103°, 3 made excellent recoveries. Half the cases required forcible feeding. Of the 60, 45 patients recovered and 5 died. One died of phthisis, one of septicæmia, a third of simple exhaustion; in the other two cases no post-mortem examination was allowed.[1]

Of the remaining pathological varieties of insanity, some are incurable, and require little notice here. That which follows a blow on the head, and gradually and insidiously makes its appearance, is a form from which

---

[1] Clinical Lectures, p. 504.

recovery rarely takes place.   Treatment can do little, for, as a rule, such patients are not subjected to treatment till the disorder is chronic.   Many persons, however, who have received blows on the head, though not chronic lunatics, suffer from attacks of temporary insanity, if subjected to any exciting controversy or irritation, or if they partake even of a moderate amount of strong drink.   Precautionary measures and advice may be of benefit, and may ward off more serious evil.   We frequently meet with patients of this class who are in an incipient state of insanity—in what is termed moral insanity—and are very difficult to deal with legally, till they arrive at a stage when hopes of cure are small.   It is worth while to note that writers, and amongst them Drs. Bucknill and Robertson, have recorded the fact that a blow on the head has produced sanity in a patient previously of unsound mind.   With this form another—viz., insanity following sunstroke—has been coupled by Dr. F. Skac.[1]   It is, however, a mere conjecture to suppose that the pathological state produced by a blow is identical with that produced by *coup de soleil*.   In this country we find but few examples of the latter.   In my experience the insanity does not come on gradually as in patients who have had a blow, but quickly develops, runs the course of acute mania with acute hyperæmia of the brain, and terminates in death or recovery in a comparatively short time.   I have seen many persons who went through such attacks in India, were invalided home, and arrived in England quite well.   Some had subsequent attacks, the mania recurring as it might have done in any one who had suffered from an attack of acute insanity.   The prognosis here is very much more favorable than in cases where the disorder follows a blow.

The treatment of so-called "syphilitic insanity" must depend much upon the history of the case.   It is the fashion to call all insanity syphilitic which occurs in individuals who present any symptoms of recent syphilis. A gentleman contracts syphilis for which he is subjected to specific treatment.   In the course of this he becomes insane.   Is the specific treatment to be continued for the cure of the insanity after all syphilitic symptoms have disappeared, or are we to treat the mental affection as we should treat it in a non-syphilitic patient?   I think there are many cases, especially those of melancholia, where it is advisable to stop specific treatment when the mental condition is not improving, and to try food, tonics, and opium.   I have seen mental recovery take place even where secondary symptoms occasionally showed themselves, although no specific treatment was adopted and nothing but tonics administered.   In true syphilitic insanity we find evidence of the disease having attacked one or more of the intracranial tissues, the vessels, neuroglia, membranes, or bones.   The brain cells and fibres seem to enjoy a remarkable immunity, but they soon

[1] Edinburgh Medical Journal, Feb. 1866.

become secondarily affected through the disease of the arteries or other parts as the result of such inflammation and degeneration. We find mental symptoms of all kinds, maniacal and melancholic, often those of progressive dementia even simulating general paralysis of the insane. In most of them there will be evidence of some amount of impairment of motor power as well as of mental, showing that by the thickening and obliteration of the vessels, or pressure of syphilomatous deposits, actual organic change is in progress. About the majority of such patients our prognosis is most unfavorable, and they gradually sink into dementia or death; but here and there we find one, and there are not a few on record, where under specific treatment persistently continued recovery has taken place in a remarkable manner; even patients who presented all the symptoms of general paralysis have lost them under such treatment, and, regaining mental and bodily health, have been called cures of general paralysis. In view of such recoveries we are encouraged to persist in specific treatment in such cases for a long period, for if such treatment is not to work a cure, there is nothing else, I fear, that will.

If you have reason to suspect that an attack of insanity is connected with rheumatism or gout, and the patient has already suffered from either of these maladies, you may direct your treatment with reference to them. There is not, I fear, any method by which we can compel the disorder to leave the brain and invade other organs, but it may be worth while, by counter-irritation, or by poultices and fomentations applied to parts formerly affected, to invite it to the feet, ankles, or knees. And such medicines as potash, iodide of potassium, or even colchicum, may be administered. It is to be noted that the cerebral symptoms may precede the articular inflammation. There may be violent mental disturbance lasting some time, and subsiding quite suddenly, to be followed by swelling and pain in the limbs, varying from a slight affection of the great toe up to genuine acute articular rheumatism. The transition may take place more than once, the rheumatic symptoms vanishing more or less completely as the cerebral return. The prognosis is, generally speaking, favorable: recovery takes place after a time, but the patient is liable to a recurrence of the attack. Besides the gout there may be other predisposing causes of insanity; one gentleman, whose mania generally terminated in gout, had had one or two epileptic fits. He died, however, of general decay, neither insane nor epileptic.

Though insanity is found in connection with acute rheumatism, it is a variety differing from the delirium which I spoke of as making its appearance toward the decline of acute disorders, as fevers, pneumonia, measles, and the like. This insanity, as it appears suddenly without premonitory warning, disappears for the most part as suddenly, one long sleep terminating the attack. Though its unlooked-for advent may cause the utmost consternation amongst the patient's friends, our prognosis need not be unfavorable, for most cases do well, and only require to be carefully watched and guarded during the transient abberration; opium has been found to agree with them, as it does with the asthenic generally, and I have no doubt that chloral also will bring sleep.   Indeed, recovery takes place without any special treatment, if only plentiful nourishment is administered, for there can be no question that this condition is brought about by the debilitated state of the patient, and by the tendency to disturbance of the brain circulation, which probably all such persons possess by idiosyncrasy and inheritance.   Instead of an acute delirium we sometimes find a melancholy or demented condition coming on as the sequel of one of the acute disorders, more formidable as regards the prognosis, though to the bystanders the symptoms may be less alarming.   In the treatment of these forms of insanity you will proceed on the plan laid down in the lectures on melancholy and stupor with dementia, remembering the need of food and warmth in both of these varieties, remembering that melancholy is curable, even after a very long period, and that from the most lost and apparently hopeless dementia patients will emerge and recover, if they receive the constant care, watching, and feeding, they so urgently require.

Very violent is the mania which follows an epileptic attack, or series of attacks, in certain cases, yet we may reasonably expect it to subside in a comparatively short time, and measures for the treatment of the patient must be arranged in view of such a result.   It seldom happens that mania follows the first epileptic seizure.   According to the frequency of them we may infer that the individual will be a long or a short time free from mental disorder.   If the attacks are infrequent, and it is possible to guard him at home during the maniacal period, we may do so, as in a few days he may be in a condition of sanity; but if the epileptic seizures are, or are becoming, frequent, and the maniacal state continues more or less throughout the intermediate time—if in his mania he is violent and dangerous, and his means and surroundings do not admit of safe treatment at home—it will be proper to remove him to an asylum.   These persons are often very dangerous, sudden, and furious in their acts, and haunted by voices and other hallucinations.   We must try and bring about sleep by chloral or bromide of potassium, and with each sleep the symptoms will be mitigated; occasionally, when we find mere delirium instead of mania with delusions and hallucinations, the whole may subside after one long sleep.   Chloral has also been found very effectual in diminishing the frequency of the fits, and given by the rectum it will often arrest the constant repetition which is called the *status epilepticus*.[1]   Should this fail, the inhalation of chloroform should be tried.

Of mania in conjunction with phthisis little need be said specially.

---

[1] West-Riding Asylum Reports, v. p. 257; vi. pp. 67, 232.

Where the symptoms rise to the height of acute delirious mania, the tubercular complication renders our prognosis unfavorable, and the sufferer will probably sink. Where we find ordinary mania with excitement and little sleep, but with no very exhausting violence or deep depression, we may entertain great hopes of a recovery, at any rate from the mental disturbance; there may be very little cough, or complaint of chest affection, and our attention may be altogether drawn aside from the latter, if we have no previous history of hæmoptysis or other lung symptoms. Patients may gain flesh while in the insane state. When the latter has passed away the lung disorder may again become prominent; and some have thought that there is a kind of vicarious action between the two sets of symptoms.

LECTURE XII.

General Paralysis of the Insane—Discovery of the Disease—Three Stages—First
Stage, Alteration—Second Stage, Alienation—Mental and Bodily Symptoms
—Epileptiform Attacks—Terminations of Second Stage—Temporary Improvement and Apparent Recovery—Third Stage, Progressive Paralysis and Dementia—Sex and Age of Patients.

I HAVE now to lay before you the description of a terrible form of insanity, which is probably the most fatal disease that attacks man, destroying in a short period not only mind, but life itself—so fatal, that a well-
authenticated case of recovery is, I believe, unknown—so mmon, that
among 194 patients admitted during the year 1869 into the Devon Asylum,
43 were affected by it. More, I think, has been written concerning this
than about all other kinds of insanity; yet about the pathology and nature
of the disease there are still great doubt and controversy. It is best known
by the name of General Paralysis of the Insane.

To the French physicians unquestionably belong the credit of having
first recognized and described it as a special form of insanity. The credit,
however, must be divided amongst several. Esquirol recognized the incurability of insanity complicated with paralysis, but he looked upon the
latter as a complication, and did not consider the whole a distinct malady.
In 1822, Bayle, for the first time, noted that the mental disturbance and
paralysis were synchronous, and attributed them to chronic inflammation
of the arachnoid. In a complete description of the disease, which he divided into three periods, he calls it *arachnitis chronique.* M. Delaye, in
1824, thought that it was not always accompanied by insanity, and was a
softening or atrophy of the brain, with adhesions of the membranes.
In 1826, M. Calmeil gave a most complete account of it, and to him frequently is ascribed the merit of having been the discoverer. For many
years he has paid great attention to the disease, and in his "Treatise on
the Inflammatory Diseases of the Brain," published in 1859, he has given
us his latest views of its nature, founded on a valuable series of cases.
From that date, 1826, the disorder has been mentioned, at any rate, by
all who have written on insanity, and many have devoted much labor to
the investigation of the pathology thereof. Among the French, Bayle,
Delaye, Calmeil, Georget, Parchappe, Baillarger, J. Falret, Moreau of
Tours, Brierre de Boismont, Magnan, Voisin; among the Germans,
Duchek, Hoffman, Joffe, L. Meyer, Erlenmeyer, Rokitansky, Wedl, Meschede, Westphal; in England, Austin, Sankey, Lockhart Clarke, Wilks,

Bucknill, Batty Tuke, Crichton Browne, Major, Mickle, Clouston, and others, have contributed the result of their observations of the disease.

There have been many names proposed for it. Some of the best are démence paralytique, folie paralytique, paralysie générale incomplete, paralysie générale progressive; Geisteskrankheit mit Paralysie, allgemeine progressive Gehirnlähmung, paralytischer Blödsinn. In this country it has usually been called general paralysis, or general paralysis of the insane, or paralytic insanity. At one time some proposed to call it paresis, instead of paralysis. As by so doing we only substitute for the old name another equally vague and unscientific, it seems scarcely worth while to make the change. Various names have been given, based on the supposed pathology, some calling it a chronic meningitis, others an inflammation of the cortical structure—" périencéphalite chronique diffuse " of Calmeil. It is thoroughly recognized by all alienists under the term general paralysis, and under that I shall speak of it here.

This fatal malady destroys hundreds every year in our asylums, chiefly men in the prime of life. Its symptoms and progress, and the mode in which you are to recognize it, I shall now endeavor to describe. General paralysis has been said to present three or four stages. Like everything else, it has a beginning and an end; and there is, of course, an intermediate period of varying duration; but frequently a patient advances with gradual but certain progress from the beginning to the end, without our being able to fix the dates of any stages in the disorder. It may be convenient, however, to describe it at different periods, and for this purpose we may consider three:—

1st, The commencement, or period of incubation.

2nd, The acute maniacal period.

3rd, The period of chronic mania lapsing into dementia, with utter prostration both of mind and body.

In the beginning those who are familiar with the patient will notice in him an *alteration*. Like other lunatics, he will arrest attention by his altered manner and habits, before it is plain that he has become insane. I have already described to you the alteration that frequently takes place in a man who is drifting into melancholia or mania. We shall see a change preceding general paralysis similar in some respects, but having its own peculiarities. These are not very easy to describe or detect, and are not present in every case; where we do see them, they are valuable aids to diagnosis.

Paralytics, like others, show an *alteration* by extravagant acts, often by an extravagant expenditure of money, by making presents to those they know and those they don't know. Now, in all they do there is a silliness betokening a greater want and defect of mind than is evinced by melancholic or maniacal patients. The same act will be done in two different ways. Maniacal and melancholic patients often show considerable vigor and power of intellect albeit insane.   They will reason and argue sharply, and defend their delusions with much acumen.   A general paralytic asserts his delusions, or commits his outrageous acts, but he does not argue keenly in defense of them.   He does things which even a patient afflicted with what is termed moral insanity would not do; exposes his person, often apparently half unconscious of what he is about; commits assaults in a foolish manner upon women, without regard to opportunity, place, or consequences.   He is extremely restless, regardless of appointments, or the time of meals, bedtime, and the like; he comes and goes, scarcely noticing those about him, gives conflicting and absurd orders to his servants, and rages with passion if they are not executed on the instant.   There is a want of plan and method in his madness, which may be contrasted with that of other patients, who in the early stage are far more suspicious and careful, if they are in the state analogous to this.   One, and a most important symptom, is early to be observed, that is, forgetfulness.   There is a want of memory.   A man forgets what he has done, and what he has said, and this explains much that he does not do.   He fails to keep appointments and regular hours.   For the same reason he cannot sustain an argument.   His business and occupation are neglected, and he forgets, too, the consequences of indecent acts, dishonesty, or debauchery.

This loss of memory will be observable in many ways: especially is he likely to forget what he has done a day or two previously; and he will not only be forgetful, he will be careless, apathetic, and indifferent about that which formerly interested him, and when he takes up new schemes and projects, his attention soon flags, and his interest vanishes.   We see, in short, in his whole manner of life a weakening of mind, such as may be noticed at the commencement of senile dementia, but which, occurring in a fine and vigorous man of, it may be thirty-five, too surely indicates the ruin even now commencing.

In this stage patients are rarely seen by an alienist, but the family adviser will probably be consulted concerning them.   In it they may continue for a variable period, a few weeks, perhaps a month or two, never for a very lengthy time.   The disease is essentially progressive, and the second stage follows rapidly upon the first in the majority of cases.   The patient's mood in this early state is often dull and sulky, less frequently is he actually depressed and melancholic; but, careless of all save the idea of the moment, he wakes into violent rage when remonstrated with or thwarted.   In one of these fits of passion his real state is often much more recognizable than when he is quiet and reserved.   You will hear that he sleeps badly, that he eats and drinks irregularly, often voraciously, drinking to excess from inattention and forgetfulness of what he has taken.   He spills his food on his dress, eating in careless haste, and is neglectful of his person and appearance, often dressing in incongruous garb.

By degrees his dull and morose condition is converted into one of excitement.   Something occurs which necessitates opposition or interference, and the mental *alteration* becomes manifest insanity of a kind which requires immediate care and treatment.   We now see the patient in a maniacal condition, marked by several peculiarities to which I must call your earnest attention.   Almost all, certainly nineteen out of twenty, paralytic patients are full of ideas of their greatness, importance, and riches.   They are self-satisfied in no ordinary degree, and think themselves the most wonderful people that ever lived.   Here and there we may find one depressed and melancholic, but his melancholia is different from ordinary melancholia, as the man with grandiose notions differs from him who suffers from ordinary mania with delusions of greatness or wealth.   There is in the latter a certain probability and reasonableness, even in his wildest fancies; but the ideas of the paralytic are altogether absurd, impossible, and unintelligible, evincing his loss of mind as well as aberration.   An ordinary maniac may think himself a duke, or may purchase a carriage and horses which he cannot pay for; but a paralytic will tell us that he is a duke, a marquis, a king, and an emperor all at once, that he is going to marry the Queen and all the princesses, that he has a hundred million of horses, and is going to pull down all London to-day and rebuild it to-morrow.   Ordinary maniacs do not talk in this wild and absurd fashion.   They invent wonderful machines which will make their fortunes, and discover the method of squaring the circle, and so on, but do not ramble on to such a foolish extent.   Another difference is this: patients in ordinary mania generally hold to their delusions, at any rate for a time.   The inventor holds to his machine, the grandee to his title; but the paralytic to-day has forgotten his delusions of yesterday, and in his eager desire to be great, he increases his horses and carriages from thousands to millions, and invents half-a-dozen fresh fancies to add to what he has already announced.   There is scarcely such a thing as a fixed delusion in this stage.   It is all happiness, grandeur, and wealth in a rapid *crescendo*, and neither argument nor ridicule arrests it in the slightest degree.   Everything around is pressed into the cause, trumpery articles of dress or ornament become robes and orders, a cottage becomes a palace, the housemaid an empress.   Even an asylum, in which the unfortunate man complains that he is confined, is a regal abode, and the other patients courtiers and nobles.   And when strength is failing, and the patient can scarcely stand or lift his hand to his head, he tells us that he can write his name on the ceiling with a 500 lb. weight hung on his little finger.

Here and there we meet with one who, instead of the gayety and excited joyousness which characterize the majority, presents many of the symptoms of melancholia.   Two gentlemen have been under my care whose paralytic insanity was of this kind.   One had many of the commonest delusions of melancholia, thought he was going to be arrested, that people were about to injure him, that they were maligning and going to rob him. Yet he was not melancholic as other men are. He never refused his food, but, on the contrary, was very fond of it, and very particular as to what he ate. He had a very good opinion of himself, was very vain of his personal appearance, and, with all his melancholy ideas, was quite cheerful and chatty. His mind was dull, lethargic, and void of excitement during the whole illness. The other patient was very feeble when first admitted, but he insisted that he could play the violin at a concert, though his left hand was so paralyzed that he could hardly hold a book. The melancholia of general paralytics is often marked by hypochondriacal delusions, which may even cause refusal of food.

So strong is the feeling of *bien être* in these patients, that they will declare that they never were better and never stronger, when they cannot place their food in their mouths or rise from the chair. They remain to the last, throughout the gradual degeneration of mind down to the lowest depths of fatuity, not only contented, but proud of themselves, their position, health, and strength. Even when they are long past expressing fixed ideas or delusions, we may recognize in vacant dementia their intense happiness.

Along with the notions of greatness, the *délire ambitieux*, as it has been called, which specially marks this disease, we find many of the common delusions of non-paralytic insanity, and we may hence conjecture that the seat and origin are the same in both diseases. Such delusions as these —" Believes himself given over to the devil," " thinks poison is put in his food," " believes he has committed sins too enormous to be forgiven," " thinks he is going to be arrested,"—I heard from four paralytics.

You are, then, to recollect that general paralysis is to be suspected, if we hear a lunatic boasting of his grandeur, riches, or strength, especially if his delusions on this point are altogether wild, and far beyond the bounds of possibility. To confirm your diagnosis, you will next look for a physical symptom, which is as nearly constant as the exalted ideas. If you watch the man closely while giving utterance to his boastings, and recounting, in high excitement and exuberant spirits, his good fortune and exploits, you will notice a defect in his articulation more or less marked, a stopping or stutter in the enunciation of a word, or the various syllables of a word, which may recall to your mind the speech of a man somewhat in liquor, and from this circumstance such patients, when disorderly in public places, are frequently thought to be drunk. This is not the same as stammering, nor is it the defective articulation of ordinary hemiplegia. The patient is obliged to make an effort to get the word out, and possibly is compelled to shout it aloud, and then succeeds in saying it distinctly. It may be very slight. Dr. Conolly says that at the very commencement " there is in these patients not a stammer, no letter or syllable is repeated, but a slight delay, a lingering, a quivering in the formation of the successive words or syllables, apparently from a want of prompt nervous influence in the lips and tongue." Not merely in the sound of the articulated word will you detect this; the muscular action of the lips, particularly the upper, will aid the diagnosis. Dr. Bucknill draws attention to a tremulous motion of the lips like that seen in persons about to burst into passionate weeping. If you closely watch the lips while patients are speaking, you will note this tremulousness in some, whereas in others you will observe a stiffness and unnatural immobility of the lips, especially the upper. You will notice also a fibrillar tremor of the muscles of the tongue, which is jerked in or out in a convulsive manner, as if the patient had not full control over it. Not only are the organs of articulation affected, but the other facial muscles, those of the forehead, cheeks, and nostrils, are also subject to twitchings and contractions. These are the first indications of paralysis, occurring most frequently at about the time of the outbreak of decided insanity, sometimes earlier, sometimes later. For them you will look whenever you meet with a patient whose insanity is marked by exalted ideas and delusions that he is a very great man. You may not discover the defect of speech on every occasion. Throughout the illness you will find that it varies considerably on different days, but it is apparent in the majority of cases, and, placing our suspicions of the nature of the disorder beyond a doubt, enables us to give a certain, but most unfavorable prognosis. We may have paralytics without exalted notions, and others in whom we can detect no fault of articulation, and concerning these we may hesitate, and look further for other symptoms; but where we find the stutter and the characteristic delusions, we can have but little hesitation.

The defect of speech varies from the slight imperfection I have mentioned, which may be overlooked by all except the practiced eye and ear of one watching keenly, up to a degree which renders the patient absolutely unintelligible; but this occurs later in the progress of the disease.

The patient becomes more and more altered in habits, demeanor, and appearance. A grave parson dresses himself in a white hat and a sporting coat; a decorous father of a family walks about the house half-naked, or takes liberties with the maid servants—absents himself by night and day, buys quantities of useless or absurd articles, or writes letters to all manner of people, signing himself, King, Duke, or Commander-in-Chief. Something soon occurs which leaves no doubt in the minds of the friends, and obliges them to interfere, and then the patient bursts out into furious mania when subjected to control.

Now, you are to recollect that these patients are for the most part men in their fullest strength—the finest and most muscular that we can meet with. We have not now to deal with boys or the aged, very rarely with women—never, I may say, with ladies. Such men are reckless in their violence and resistance beyond any other class of patients. As I have told you, there is an incipient imbecility which prevents them from reflecting on what they are about. So, in a blind fury, they will attack all around, exerting their muscular power to the utmost, regardless of all consequences; and they are not paralyzed in their limbs at this stage to any extent that will interfere with their violence. Consequently they are not to be managed out of an asylum, and even in one they cause frequently much anxiety. These are the patients whose ribs are broken by attendants in efforts to overpower them, and who in various ways cause so much confusion in asylum wards.

There are by this time other symptoms which you will hear of or notice. The patient may have, or have had, a "fit." Sometimes this occurs quite early in the disease, before the mind is much affected, and you are disposed to think the mental symptoms are due to it. You may look upon it as an attack of apoplexy, or epilepsy, according to what you see or hear. Such attacks happen more or less frequently in the course of almost every case of this malady, which they often divide into stages, the patient never quite regaining what he lost by one of them. They are called "congestive," or "paralytic," or "epileptiform" attacks. They often resemble the *petit mal* of epilepsy; but sometimes reach the intensity of the *grand mal*, or are of a negative character, their presence being indicated, not by convulsion, but by a sudden collapse and paralysis slowly passing away again. Few go from first to last without some of them; but many, especially the younger patients, do not suffer in the early part of the disorder. There is not the definite fit of epilepsy; but we may see convulsions lasting for an hour or two, or, if slight, it may pass off without the patient falling to the ground. Pathologically, we connect these convulsive discharges with the twitchings and tremors noticed in the early stage. It is important to distinguish them from true epilepsy, as the latter is far more amenable to treatment, even if complicated with insanity. Patients in the fits of general paralysis seldom bite the tongue, the convulsions are not so violent, there is not the *aura*, nor cry, and the mental symptoms will, of course, be quite different. In some the disease runs its course with very few fits; in others they are numerous, and occasionally cause death even at an early stage. This has happened to patients under my observation much more frequently in the last six or seven years than it did fifteen or twenty years ago. A succession of convulsive attacks takes place, and the sufferer never rallies, but dies of exhaustion. It generally occurs in elderly patients, or in those who were not in robust health before the disease commenced.

You may also notice that the pupils of the eyes are unequal—not always, but very frequently; and when this symptom is present, it is important and pathognomonic. Dr. Nasse, of Siegburg,[1] tells us that of 108

---

[1] Allgemeine Zeitschrift für Psychiatrie, 1868.

cases of general paralysis examined by him, only in three was no inequality detected. Austin found only two exceptions in 100 cases of paralysis. You are not to forget, however, that inequality may exist in non-paralytic insanity as it may in sane persons, or in those suffering from other affections of the brain. Sometimes, instead of inequality, we find both pupils contracted to pin-points—a condition which may remain for a considerable period—and then be succeeded by inequality. Some think it always precedes the latter, but this is doubtful.

It has been said that the gait of these patients is peculiar—that they walk with a slow, cautious step, short and shuffling—that they walk as if about to run, jerking the legs forward. But it is not discoverable at an early stage. I have seen many whose mental condition was unmistakable, who could run, walk, or ride on horseback perfectly well. At an early stage we may be able to detect a difference in the handwriting, and the defect of mind and memory is shown by the frequent omission of words, the repetition of the same sentence, and the incoherent jumble of the whole, which differs altogether from the coherent though insane letter of an ordinary monomaniac. The writings of these, as of others, often afford us most valuable information.

Now, when a patient in this stage is brought to an asylum, what do we see? He is very angry, very violent, and very good-humored by turns, easily pacified and turned aside from his wrath, but dangerous if not judiciously managed. Together with his anger there is great silliness about him, and his countenance betokens vacancy of mind. He often looks stupid and blank while at rest; and when excited, we may notice twitchings and tremors of the various facial muscles, for not only the organs involved in speaking, but all the muscles of the face, tongue, and pharynx may be more or less affected by the incipient paralysis. His patellar tendon reflex will at this stage be exaggerated; later on it will become dull, and at last altogether absent. There is seldom any marked hemiplegic symptom, but occasionally I have seen ptosis of one eyelid. If this be the case, I should expect to hear that there have been epileptiform seizures. Atrophy of the optic disks was observed by Dr. Clifford Allbutt in 41 out of 53 cases examined by him at the West Riding Asylum; and Dr. Aldridge[1] examined 43 cases and found atrophy of the disk in many, and it was always most complete in the female patients. The atrophic changes were noticed to be most advanced in the left eye; and this again was most distinctly seen in the female patients. A later observer, Dr. Arbuckle,[2] tells us that, as regards general paralysis, so far from atrophy of the optic nerve existing in most cases, he has not met with it in one, nor has he seen a paralytic blind before death from amaurosis setting in during his

[1] West-Riding Asylum Reports, ii, 223.
[2] Ibid., 148

13

disease. My own experience agrees with the latter statement. Sometimes with an air of defiance, and sometimes with the greatest delight and self-satisfaction, the paralytic relates his accession of rank or fortune. By turns he likes or dislikes those about him, will make sudden attacks on the attendants in charge, and desperate attempts to escape. He may be in a state of what we may call acute mania—noisy, destructive, and dirty, breaking windows, tearing up bedding and clothes, and going about naked. He may sleep little, yet he will not go many days and nights without sleep, so as to cause fatal exhaustion; in fact, it is rare for a paralytic patient to die in this early stage, unless from an accident, or a series of epileptiform attacks. The less acutely maniacal patients often sleep well, and almost all eat well—nay, voraciously—bolting their food, swallowing often with some difficulty, for the paralysis which affects the tongue and lips may extend likewise to the pharynx. Frequently they are filthy in their habits, daubing themselves with fæces.

I consider the maniacal period to be the second stage of the disorder, in which that of incubation culminates. After it has lasted for a variable time, from a week to a month, or even longer, it generally yields to treatment, and one of two conditions follows: either the patient gets better, so that he may even leave the asylum and pass for a sane man, or the stage of imbecility comes on, and he progresses downward with more or less rapidity to extinction of mind and body.

I have not always found that patients improve in mind and body *pari passu*. Some of those who had the best right to be called "recovered" in mind, bore traces of bodily weakness or paralysis—a limp, or defect of speech, irregular pupils, or general feebleness, so that they were hardly equal to a walk of a mile. In some, however, I have seen a wonderful disappearance both of bodily and mental symptoms, the improvement lasting for some time. These are the cases which are said to be recoveries from the disease. I have seen some who certainly would not have been pronounced insane by any jury. They had either lost their delusions, or were competent to deny and conceal them. I have received letters from them detailing their travels or amusements, written without a mistake. They have spent their money without extravagance, and lived in their families as decent members of society. But those who have best recovered are long since dead, nor do I know one in whom the disease did not reappear in a longer or shorter time. Moreover—and this is the real test—I never knew one who was capable of work or business. Some lived for a time quietly and rationally in country houses, but the instant they returned to London and attempted to resume their former occupations, they broke down, and were obliged again to be placed in confinement. Yet these men, had they remained in enforced idleness, might have stayed among their friends during the decline of their failing strength. The mental defect is essentially mental weakness. They are incapable of effort or continued application, and deficient in memory, so that the attempt causes exhaustion, and lights up again the acute symptoms which had been allayed. Those who know them intimately in times past see a difference, a slowness or childishness, not apparent to a stranger. I mention this, because you may be asked if you consider a patient who has thus apparently recovered from general paralysis competent to manage his affairs. He may desire to supersede the commission of lunacy which has placed his property in the care of the Court of Chancery, and he may ask you to assist him by your affidavit of recovery. His improvement, however, is but the semblance of recovery—a remission, not the removal, of the disease. It may be quite safe to release him from an asylum, and to allow him a certain voice in the direction of his household; but if his affairs are of such a nature as to have necessitated a commission, they had better remain *in statu quo.*

In many cases the cessation of the acutely maniacal symptoms is not followed by the improvement I have spoken of, but the patient passes along through a period of chronic mania into ever-increasing dementia. Though the excitement and emotional display are less, the delusions remain. He is still a king, a duke, or general; he issues his orders and writes to tradesmen and others, giving commissions to the extent of thousands; and though these are never executed, and he is kept confined within asylum walls, he never recognizes the incongruity. He is always going away " to-morrow," and to-morrow finds him writing the same kind of letters, and doing and saying the same things. The present and the future he gilds with his exalted fancy, and of the past he takes no heed, frequently caring nothing about family or friends. Frequently, too, we find hallucinations—hallucinations of sight or hearing, or both, and there may be disordered sensations of taste, or the muscular sense, or the viscera. A man in this stage often gets very stout, and remains so for some time. His strength, mental and bodily, varies considerably. Memory is sometimes completely gone, sometimes he remembers a good deal, and the articulation and power of walking fluctuate in the same way. If he suffers from an attack of epileptiform convulsions, he loses much ground, and for some days may be quite lost and paralyzed, often one side more than the other. This chronic condition often lasts a long time, even years. Such patients are always more feeble in cold weather; in the heat of summer they regain strength, often to a surprising degree; with the first frosts they fall back, and, it may be, sink. Whether at home or in an asylum they are generally happy and easily amused; the annoyance of to-day, if any arise, is forgotten to-morrow, or can be turned aside without difficulty by calling up before them the glories they are expecting, in that brilliant future which is forever coming.

We now come to the last stage of all, hopeless dementia, utter fatuity. The patient can just walk round the garden, slowly and shuffling, an

attendant holding his arm.   His countenance is vacant and puffy, and he
takes little notice of what is said, or of the person speaking.   He begins
to get thin, losing the fat which has accumulated, and if he is confined to
bed for a day or two by an attack of convulsions, the skin of his back
rapidly gives way, and the bed-sores resulting are difficult to heal.   He
can with great difficulty hold anything in his hands, which tremble like
those of a person palsied by age.   A symptom frequently noticed in his
stage is loud grinding of the teeth.   For hours together a patient will sit
and grind his teeth, making a most horrible and discordant noise.   The
appetite is still good, and he looks forward to and enjoys his meals.   The
power of deglutition, however, is very feeble, and he will go on filling his
mouth without swallowing, till he has crammed it full of food; and the
consequence is that he either gets it impacted in the œsophagus so as to
compress the larynx, or else it gets into the larynx and trachea.   From
one or other of these accidents choking is a very frequent mode of death
in these cases, and the greatest care ought to be taken that a patient shall
never eat alone, or in fact without an attendant at his elbow, for instant
suffocation may be caused by a mass of food becoming impacted.   The
patient now requires to be nursed, like any other far advanced in paralysis.
There is complete annihilation both of bodily and mental activity, and yet
by careful nursing even this stage may be prolonged for a very indefinite
time.

And this brings me to another point—the duration of the disorder
—which often is of considerable importance, and on which authorities
differ widely.   When you pronounce an opinion that the insanity is incurable, nay, that life itself will soon be extinguished, it may be of the
utmost consequence to the friends to know the time likely to elapse before
the latter must take place.   If you turn to one of the chief authorities,
M. Calmeil, you read: "Some paralytic patients live eight months, a
year, eighteen months; others linger for two or three years, rarely beyond."   Dr. E. Salmon says: "The course of the disease may extend
from some months to three years: in rarer cases it may reach to five years,
but scarcely ever exceeds that time."   Griesinger says: "The duration
of general paralysis varies from several months to about three years."   In
my own experience I should say that the average duration was considerably longer.   The reason of this is, that my patients have been all of a
class able to command the best food and nursing.   Griesinger says:
"When nursed in their families these patients live longer than in asylums,
as they require the same attention in the latter stages as a young child."
Life may be prolonged for an indefinite time by dint of unstinted diet and
thorough nursing.   In the year 1858, a commission of lunacy was held on
a baronet of large fortune, who was at that time said to be suffering from
general paralysis, and who had shown symptoms of brain affection and
epileptiform attacks so far back as 1856, he having been married in 1855.

This gentleman lived till the year 1883. Dr. Clouston,[1] relates the case of a patient who was undoubtedly suffering from general paralysis on his admission in 1860, and was alive in 1882, in a state "facile, morbidly contented and exalted, walk uncertain and dragging, his articulation affected, just like a typical general paralytic in the end of the second stage of the disease." I should say that, with careful nursing, and with every appliance and means of taking care of a patient, we might put down the duration of life as from three to five years; but in crowded asylums, and with the diet which the poorer classes receive, a much shorter period must be assigned.

Who are the subjects of this malady? We shall find that it is unlike other forms of insanity, for it especially attacks men. Comparatively few women die of it, and these are almost all of the lower classes; it is the rarest thing to find a lady the subject of general paralysis. The ratio of liability, according to Dr. Sankey, runs thus: 1. Males of the lower classes; 2. Males of the upper classes; 3. Females of the lower classes; 4. Females of the upper classes. Whether the males of the upper or lower classes are more liable, is a moot point not easy to be solved. The proportion of paralytic patients admitted into a first-class private asylum in twelve years was 20 per cent. of all the males. M. Calmeil says the males are to the females as 50 to 15. There are also peculiarities with regard to the age of the persons attacked. General paralysis does not make its appearance in the very young or the very old, but chiefly attacks those in middle life. Although there are cases on record of its occurring in boys of 16[2] and even 12,[3] yet we do not expect it even at the age of 20; at 25 it is rare, at 60 it is rare, at 70 it is unknown; chiefly at 35 or 40 it commences, and the patients are not only in their greatest vigor, but often fine handsome powerful men—men who have enjoyed life and have lived hard. We do not find it amongst weak, nervous valetudinarians, the subjects of hypochondria and melancholia. The paralytic patient has rarely had to seek aid from doctors, and in the exuberant feeling of health and gayety he derides the notion of there being anything the matter with him, and refuses to have anything to do with medicine.

Those who consider general paralysis a special form of disease, point to the remarkable fact that it seldom or never attacks the chronic inhabitants of asylums, or supervenes upon other forms of insanity. If a young man is insane for a number of years, he does not, after a long period of mania or melancholia, develop symptoms of paralytic insanity. And if a patient recovers, and recovers perfectly, from an attack of mania or melancholia, he does not, if he has a second attack of insanity, show the

---

[1] Clinical Lects., p. 369.

[2] Dr. Clouston, Journ. Mental Science, xxiii., 419.

[3] Dr. Turnbull, Journ. Mental Science, xxvii., 279.

symptoms of general paralysis. There may be forms of the latter of which the diagnosis is difficult, and there may be apparent recoveries which I call remissions; but if the disease is really general paralysis it would seem to run a progressive course to dementia and death, being throughout a malady resembling in many respects ordinary insanity, yet differing altogether in its fatal character.

LECTURE XIII.

General Paralysis continued—Diagnosis—Illustrative Cases—Diseases simulating
General Paralysis — Varieties of General Paralysis — Prognosis — Causes —
Treatment—Post-mortem Appearances—Pathology.

I NOW pass to the diagnosis—a matter of the greatest consequence, seeing that the disease is fatal in such a vast majority of cases. You will have
to distinguish it on the one hand from varieties of insanity which are not
fatal to life, but in which the mental symptoms closely resemble those of
general paralysis; on the other, from affections which resemble it in the
bodily and paralytic symptoms, but which may yield to treatment, and
occasionally recover under it. We often see some patient presenting the
delusions of general paralysis, whose malady, nevertheless, is ordinary
mania with exaltation; while others may be paralytic, who, nevertheless,
lack the best-marked characteristics of the disease. I once saw a gentleman, aged forty-six, who had been in an asylum about a week, having
been brought over from Ireland by the medical man who accompanied
me. There was no stutter in his speech, no tremor or immobility of the
lips, no irregularity of pupil, no contraction nor dilatation. He had apparently full power and perfect co-ordination of both hands and feet. He
played both billiards and the piano in my presence, and did both well.
He walked with a long swinging stride, but whether this was habitual to
him or not I cannot say. Speaking generally, one might say that the
bodily signs of paralysis were wanting. The mental symptoms afforded
more information, though these were not very marked. He had no very
extravagant delusions, but he thought himself a wonderfully lucky individual, as he had bought five or six horses for small sums, by which he was
to realize some hundreds. He was gay and jocose, on the best of terms
with his friend, though he said that it was an infernal shame to have
brought him there. He showed loss of memory, for he said that he had
left Ireland three weeks, whereas it was only one. Although told that
another physician and myself were doctors come to examine him, he never
tried to persuade us to let him go, though he said he was quite well and
wanted no doctors. This gentleman was pronounced by us to be paralytic:
first, on account of the peculiar "larkiness" and hilarity exhibited under
the circumstances to two perfect strangers who had come to examine him;
secondly, his self-satisfaction and ideas of general good luck and success;
thirdly, his indifference with regard to being released; fourthly, the loss
of memory;—all of which went to prove that he was suffering, not merely

from aberration of intellect, but from incipient paralytic dementia. Another case remained for some time in doubt, and presented in its early stages symptoms by no means characteristic of general paralysis. In February, 1862, a gentleman was brought to an asylum, whose insanity was stated to be of only a few day's duration. He has been riding on the pavement and assaulting the police, was incoherent and rambling, said the sun was turned into the moon, and such things, but had no grandiose delusions, and was frequently taciturn, not speaking perhaps for a whole day. On alternate days his condition varied; on one he was dull and depressed, refused his food, and would not speak; on the other he was gay and excited, but though his conduct was manifestly insane, delusions were not a prominent feature, and he said little except that he " wanted to go." He was wet and dirty; there was no stutter, and the signs of general paralysis were mostly absent; there was, however, irregularity of the pupils, and when he improved somewhat, and talked more freely, it was evident that there was great defect of memory. He got so much better, however, that in July he went into the country with his wife, and was reported to be quite well. In the following February he was again admitted, and now the symptoms of general paralysis were well marked. His sons were dukes, he was worth millions, and so on. There were great doubts as to the nature of the disorder at the commencement, and only the irregularity of the pupils, the defect of memory, and general absence of mind, made the prognosis unfavorable. This gentleman complained constantly of pain in the head. When said to be recovered he remained in the country idle: the moment he resumed work the symptoms returned, and this time with unmistakable features of the disorder.

About the same time I saw a gentleman from whose extravagant delusions one might have imagined that he was the subject of general paralysis. He boasted of his extraordinary intellect and strength. He was going into parliament, and, as a preliminary step, was to assemble 10,000 people in his park, have them photographed, and sell the photographs at £5 apiece, thus paying off the mortgages on his estate, and making £100,000. He thought that people might live a thousand years if they bathed in beef-tea and beer, wanted to make a tunnel through the earth to the antipodes, and various things of this kind. He was dressed in a most extraordinary costume, and was insane beyond all question. Yet there was no stutter and no loss of memory, and I heard that he had had, and had recovered from, a similiar attack in India some years before. This was conclusive to my mind. Had the former attack been general paralysis, he could not have recovered. The mania which had passed away before, passed away again, again to return.

Another gentleman was pronounced paralytic, and certainly there were many symptoms of this disorder. He had been spending money in a most reckless manner, but defended all he had done. He thought himself a man of rank, and had a ducal coronet engraved for his paper and envelopes. He was gay and expansive, although he knew he was under legal restraint, quarreled with his friends, and yet joked with them, and at times there appeared to be a slight hesitation in his speech. I doubt if any one could have pronounced unhesitatingly on one inspection that this was or was not a case of general paralysis. Against it was his age. He was upward of sixty. He had long suffered from acute bodily disease —disease of bladder and kidney, consequent upon stricture, and in the midst of it mental symptoms showed themselves. But paralysis rarely attacks a man weakened by other disease, while it is common for ordinary mania to commence in such a subject. Time very soon made it plain that it was not paralysis. Though the bodily health got worse, the mind did not. Instead of paralytic delusions appearing and becoming more and more extravagant and absurd, all delusions disappeared, and there remained what we may term moral insanity, a weakness and degeneracy of mind shown in a desire to waste money and buy useless articles, to tell indecent stories, and generally to behave himself in a way the reverse of what he formerly had done. So matters went on till his death, no other symptoms of paralysis ever appearing. I believe the apparent stutter was due to nervous agitation, when obliged to discuss his conduct with strangers in his extreme state of bodily weakness. For the same cause he not unfrequently shed tears. Had this gentleman's bodily ailments been curable, I have no doubt that his mental aberration would have been removed. You will recollect that when a patient is over sixty the symptoms of general paralysis are to be examined with great care, and we are to doubt them unless they be of a most clear and unquestionable character. In this last case there was no irregularity of pupils, no difficulty of walking, no loss of memory; above all, the patient wrote an excellent formal business letter without leaving out a word or making mistakes of any kind. Such letters would by themselves almost decide the point.

The cases of non-paralytic insanity about which there may be doubt, are not chronic cases of mania or dementia. They are recent cases of mania, with exalted delusions, or what is termed moral insanity, with extravagance and indecent conduct. Here you will look for loss of memory, hesitation in speech, defects in letter-writing, especially words left out or repeated, inequality of pupils, tremor of lips, and other facial muscles, and possibly a slow or halting gait. You will consider the sex and age of the patient, the history, the occurrence or non-occurrence of former attacks, or of epileptiform seizures. If the patient some years ago has recovered, and thoroughly recovered, from some similar attack, and has since gone about his work like any one else, the disorder is not paralysis.

Although three out of four cases of general paralysis present no difficulties of diagnosis, yet we see a certain proportion every year in which the symptoms are obscure; the mental phenomena may point strongly to this form of disease, and yet physical symptoms may be wanting, or physical symptoms exist, which may depend on other varieties of brain disorder. It sometimes happens that the physical symptoms are not developed till long after the insanity. In a patient now under care, exalted mania followed an attack of depression. The case resembled one of *folie circulaire,* but instead of the melancholy returning, difficulties of articulation and epileptiform attacks supervened, and the patient is now in the last stage of the disease. In another case the mental symptoms were by no means conclusive, and the physical were absent; but all doubt was ended by a succession of epileptiform attacks and the speedy death of the patient. On the other hand, two patients who presented many mental symptoms, and were thought by the majority of those who saw them to be examples of the disease, unexpectedly and rapidly recovered. About one there is now little doubt, but it will be interesting to know whether the other will at a later date develop further signs of the malady with which he was at any rate sorely threatened. In doubtful cases it is impossible to come to a certain conclusion at an early stage, and the most experienced observer will wait for time to assist or confirm his diagnosis.

I now pass to the consideration of certain maladies characterized by paralytic as well as mental symptoms which may be mistaken for general paralysis of the insane. One is the paralysis of alcoholism, in which we sometimes find not only the impairment of muscular power, but the exalted delusions which are usually supposed to be pathognomonic of general paralysis. Dr. Batty Tuke[1] mentions several patients where he observed impairment of speech, walk, and memory, with marked optimism and exaltation. Yet two recovered, three remained in a stationary condition, and one died of Bright's disease without developing further symptoms. A fifth had been transferred to Dr. Tuke's care at the Fife Asylum eleven years previous, certified to be suffering from general paralysis by an eminent specialist. There was nothing which militated against such a diagnosis, except that the man gradually improved, was discharged, and for years after supported himself by his handicraft.

There was one symptom absent in these cases—the pupils were not affected. Dr. Tuke says:—"The indication which weighs most heavily with me is the condition of the pupil and retina. If after careful weighing of the history and symptoms of a suspected incipient case, we find contracted or irregular pupils, and on ophthalmoscopic examination, hyperæmia of the retina, the weight of evidence leans toward general paralysis; if the pupils are unaffected, and the fundus of the eye is anæmic, even although there may be some slight degree of atrophy of the disks, chronic alcoholism may be suspected, and diagnosis made with

---

[1] Edin. Med. Jour., April, 1877.

greater caution." The color test of Leber may, in Dr. Tuke's opinion, serve to confirm the diagnosis.

In my own experience I have not found the exalted delusions which characterize general paralysis in patients suffering from chronic alcoholism. There has also been less of the convulsive stutter and tremulousness of the lips and facial muscles, and greater tremor of hands. Their delusions have been mainly due to the almost complete obliteration of memory. But recovery has taken place after various periods of time. It is very important that we should arrive at a correct determination of the nature of such cases, and I think it quite possible that the instances of recovery from general paralysis of which one sometimes hears, may be in truth recoveries from this or some other form of insanity which simulates the usual symptoms of the graver malady. We meet with a number of cases which in a former edition I called instances of spurious or *pseudo* general paralysis. They do not present the ordinary characteristics or even the ordinary course of the disease, but advance slowly, yet surely, to death without much excitement, violence, or exaltation Dr. Crichton Browne has described[1] a certain variety as cases of chronic "brain-wasting," and his description fits those I have witnessed, to which I have applied the above term. The symptoms, as he tells us, are headache of a dull, heavy character, with pallor of countenance, cramp, or other anomalous sensations, or a sense of numbness or weight. Later a convulsive attack occurs affecting one or both sides, or, more commonly, paralysis is insidiously developed without convulsions. Muscular power is much diminished in one side or limb, or in all the limbs; articulation is affected, and the pupils are unequal. The temperature is rather depressed, and does not rise in the evening to such a degree as that of the general paralytic. The mental symptoms are confusion and failure of attention, with sluggishness and loss of memory. The emotional state is one of depression, rather than exaltation, with vague dread and apprehension. All this goes on to complete fatuity, which cannot be distinguished from that of general paralysis. From this condition, as from the disorders just mentioned, recovery sometimes, though rarely occurs.

There are cases of syphilitic brain disease which may, in some degree, simulate general paralysis. We may find the dementia and the paralytic symptoms with the feeling of *bien être*, so often seen in all kinds of paralytic dementia, yet here recovery may occur under treatment suitable to syphilis. Dr. Duckworth Williams reports a case of recovery from general paralysis, in which the symptoms were well marked, thickness of speech, tremulous tongue and lips, uncertain gait, and *délire ambitieux*. Recovery took place under the administration of bichloride of mercury, which raises the question whether the malady was syphilitic.[2]

---

[1] Brit. Med. Jour., 29th April, 1871.

[2] Med. Times and Gazette, 30th May, 1868. This has been treated at length by Dr. Mickle in British and Foreign Medico-Chirurg. Rev., Ap., 1877.

We sometimes find acute delirium breaking out in patients poisoned by lead, and with it there may be the paralysis produced by that substance. The delirium, however, is usually transient, the palsy does not affect the speech, the history and the blue line on the gums will assist our diagnosis, and the peculiar exalted delusions will be absent.

Senile dementia, which may commence at a comparatively early age, may be characterized by loss of memory, extravagant and indecent conduct, and delusions. There will, however, be an absence of the specific delusions and the maniacal condition; neither shall we find the inequality of pupils, the stutter, nor stumbling gait. In fact, the failing mind in senile dementia is usually manifested long before any symptoms of bodily paralysis, and you will recollect that general paralysis is rare at the age of sixty, senile dementia seldom beginning so soon.

But we may meet with dementia or mania following apoplexy or hemiplegia, and may notice a stutter in the speech, a defective walk, loss of memory, and dirty habits. The patient, however, presents none of the mental symptoms of general paralysis. If the condition is one of dementia, he will be dull, vacant, torpid, with none of the elation and expansiveness that characterize insane paralytics. If he be maniacal, the special delusions will be wanting, and when the mania passes off, the difference will be clearly seen. Moreover, we shall hear a history which will leave no doubt. The paralytic attack will have preceded the mental symptoms, whether these be maniacal, or only the imbecility of dementia, and the paralysis will be the result of a sudden attack, not of a slowly advancing progressive disease.

Mania and dementia, the result of epilepsy, may be confounded with the state of a paralytic patient who has lately had an epileptiform seizure, which you may be told was an epileptic fit, or series of fits. Here the history of the patient previous to the fit will be our best guide. Were there at that time any of the characteristic delusions of general paralysis? I was called to a gentleman who had had a seizure of this kind. He was lying in bed, not able to stand alone, or to lift his food to his mouth; he kept repeating one word, and was quite childish and lost. But for the previous history, it would have been difficult to say what was the origin of this paralytic dementia. Yet a few questions made it perfectly plain, and it was possible to say that in a few days he would be walking about again, but that he would not live beyond a year or two. He did not live a year. An attack of mania following epilepsy generally subsides in a week or less, and them the patient returns to the state he was in prior to the fit. His mental condition will at no time resemble that of a paralytic patient, though he may be very furious and dangerous. There will be no self-satisfied contentment or exaltation, but rather angry suspicion, rage, or panic, leading him to homicidal or suicidal violence. When this subsides, he may be apparently well and restored to reason, whereas the paralytic will show the effects of the seizure for a long time, and probably will never regain his former mental power.

Locomotor ataxy and other disorders affecting the muscular powers are not likely to be confounded with general paralysis of the insane, inasmuch as mental symptoms are wanting. I shall hereafter have to consider whether the paralytic symptoms of the latter disorder ever precede the mental. On this point there is considerable difference of opinion.

Cases of general paralysis vary in many respects, in the mental symptoms, the degree of paralysis, the mode of onset and course of the disease, and the morbid appearances seen after death. Dr. Mickle has laid down certain varieties of the disease,[1] which he has divided into five groups, according to the clinical features and the pathological lesions. The first group consists of cases of a common kind, marked by exalted delusions, maniacal excitement, and hallucinations. The duration is short, cerebral hy; cræmia and softening are observed, with adhesions and decortication. In the second group there is a protracted stage of dementia, the quiet self-satisfaction of the early stage is replaced by peevishness or apprehension, till the habits become foul and brutish. The duration is lengthy, and the brain is seen to be atrophied, with considerable increase of intracranial serum. The gyri of the upper surface and frontal region are wasted, adhesion and decortication are moderate, and the white substance is pale. In the third group dementia is early and predominant, and melancholic delusions frequent, the latter course being one of extreme dementia. Hemiplegia is marked and frequent, epileptiform attacks being very common. The duration is short, and after death the *left* cerebral hemisphere is found more diseased than the right, and more or less atrophied. In the fourth group the morbid lesions are much more marked in the *right* than in the left hemisphere. The outbreak begins with ambitious delirium and maniacal agitation, the symptoms of dementia and melancholia noticeable in the third group being wanting. The duration is somewhat lengthy. The fifth group is not well defined. There is much local induration of the cortex, and the interstitial changes tend to sclerosis; the mental symptoms are various: epileptiform fits, hemiplegia, and spasm are frequent, and the duration somewhat long. It will be interesting to see whether the observations of other pathologists will bear out these generalizations of Dr. Mickle, which at any rate are valuable and suggestive.

If the diagnosis leads us to the conclusion that the patient is suffering from true general paralysis, it follows that the *prognosis* must be extremely unfavorable. Practically, we look upon the disorder as fatal, and probably fatal in three or four years. Indeed, it is a question whether any

---

[1] General Paralysis of the Insane, p. 177.

have ever recovered from it.   Patients are dying of it in all our asylums by the hundred, yet our best authorities record no recoveries.   Here and there we may see a patient who is said to have recovered; but unless a long period has elapsed, we cannot be sure that the recovery is anything more than a remission.

The consideration of the pathology brings me to another point.   Can we in any way account for the disease? what is its cause in any one individual?   I have held for some years the opinion, based altogether on my observation of cases, that sexual excess has more to do with the causation of it than anything else.   It is difficult to get at the history of this excess.   If a man has led a very loose life, we may hear of it; but there may be great excess in married life, and of this we hear little.   I have known several cases where the patients were men not very young, who had married young wives, having in former days led very dissipated lives. One man had been a great masturbator, and when married had never had complete intercourse, but had broken down in constant attempts.   Some had led lives of great profligacy, which they had carefully concealed from everybody.   And when we speak of sexual excess, it is not to be forgotten that what is excess to one man may not be to another.   As one drinks with impunity an amount which kills another, so sexual indulgence which is harmless to this man may produce disorder in that.   These observations of my own are confirmed both by the opinion of others, and by various circumstances to which I would draw your attention.   General paralysis does not attack the young or the old, and is not in general found amongst the weakly, but rather invades the strong and vigorous, those most likely to be guilty of excess.   It rarely attacks women, especially of the higher classes.   I myself have no experience of female paralytics; but Dr. Sankey, who was the superintendent of the woman's side at Hanwell, and has made this disorder his special study, tell us,[1] "it is remarkable how many of them," that is, the women, "had led irregular lives, and especially had been guilty of sexual impropriety of some sort;" and he gives the particulars of seven cases.   He, moreover, tells us, "out of 34 cases, of which the history of the disease is complete, 11 are known to have led an habitually irregular life with respect to sexual indulgence, and of 14 only was the information satisfactory as to the contrary state of things; even of these 14, one had borne an illegitimate child in early life, but since, according to her mother, had lived correctly; and one other was a married woman who had left her husband on the day after marriage."

It is not to be supposed, however, that sexual excess is the cause in every case.   There are other excesses which may bring it about.   One is excessive drinking, which in some may produce not chronic alcoholism, but general paralysis.   Excess of any kind in certain constitutions may

---

[1] Lectures on Mental Diseases, p. 181.

light up this inflammatory cerebral mischief—excess of work, bodily or mental, excessive anxiety, which in one produces mania or melancholia, may in another light up the disorder I am speaking of.

Our inquiries into the causation of general paralysis ought to receive assistance from the fact that in certain countries it is unknown. Especially is it absent in Ireland, although alcohol is freely indulged in, syphilis and venereal excesses must be present in the large towns, and exciting mental causes must be as common as elsewhere. Yet Dr. Robert Stewart, for forty years superintendent of the Belfast Asylum, never saw a single case of the disease; and his successor, Dr. Merrick, wrote that in 1881 he had had no case. In this asylum the inmates average 440, drawn from an agricultural population and the large manufacturing towns of Belfast and Lisburn. In all the Irish asylums the cases are extremely rare, and not more frequent in those parts of Ireland which are the most Saxon than in the Celtic districts. Its frequent occurrence in Wales shows that the Celts have no special immunity; and though it is rare.among the Scotch Highlanders, they do not escape if they take up their residence in the large towns. Dr. Ashe,[1] seeking for the reason of this absence in Ireland, remarks that Irishmen eat the potato, and that this point is worthy of experimental investigation. The cereal diet of England and Scotland may account for their superior energy and cerebral activity, but the men of most energy and activity are those most exposed to the invasion of general paralysis. Whatever weight may be attached to these speculations of Dr. Ashe, the immunity of Ireland is a curious and interesting subject, and not to be explained like that of Asiatic and savage races who are free from the excesses of civilized life.

Next, I have to speak of the treatment of these patients. Although we do not cure them, although the disorder is still one of the *opprobria medicorum*, we must never give up the attempt. One cannot understand as yet why this progressive disease should not be arrested, for clearly the remissions and recoveries are at times so great that very little of the disease can be left, although the tendency to recur may be there, as in other forms of insanity. At any rate, our object in every case is to restore the patient to his friends, and enable him, if their means allow, to pass the close of his life amidst his family. These patients are rarely to be managed out of an asylum in the early stage. If they are wealthy, and a complete establishment can be provided, they may be so surrounded that they are virtually in an asylum for one; otherwise they are better off in a well-conducted asylum, and quite as happy. At the outset, when first restrained, they are subject to paroxysms of blind imbecile fury and violence, and are at this time very dangerous, requiring the appliances and skilled officers of an asylum. They also require exercise within secure grounds, and this

---

[1] Trans. of Internat. Med. Congres, 1881, vol. iii., p. 655.

they can rarely obtain elsewhere. And they are so elated by reason of their malady, that they do not feel restraint like other patients, who are more conscious of their position. They write myriads of letters, ordering horses, carriages, and diamonds, and never wonder why they are not sent, and why they get no answer. When by nature good-humored and pleasant people, no patients so enjoy themselves in an asylum, or are so easily pleased and humored. A promise that what they want shall come some day, turns aside their present ill-temper, and their failing memory has the next moment forgotten the desire. But they are childish and child-like in mind, and can be led like children by tact and kindness. They are, at the same time, dangerous and treacherous. Their weak minds regard not consequences, and they will set fire to the house, or secrete a stone, or some such article, to attack the object of a delusion. One can never trust them; many patients we may believe implicitly, these we cannot. Much may be done by medicine in the violent excitement which characterizes the early stages of the disorder; and the drug which above all others seems to act beneficially here is digitalis, which is largely given in our public and private asylums: administered in doses of ηxv. to ηxxx. of the tincture, repeated, if necessary, every three or four hours, it often produces a wonderful effect, soothing their noisy turbulence, and restoring them to a comparative state of rationality, so that they cease their destructive habits and filthy practices, wear clothes in decent fashion, and take food. In some cases opium and morphia are serviceable, and frequently we may give them in conjunction with digitalis with greater benefit than alone. In the latter stages of dementia, where we find often great restlessness and want of sleep, along with an advanced stage of paralysis and prostration, opium or morphia may be given without digitalis; and for the mere production of sleep, chloral is as valuable here as in other forms of insanity. A few years ago the Calabar bean, the seed of *Physostigma venenosum*, was highly recommended for the treatment of this disease. I cannot learn, however, that any recoveries have resulted from it, though it would appear that the duration of the disease has been prolonged under its use.

I know no other sedatives that are worth a trial in this disease, unless it be bromide of potassium. This I have not given in large doses to paralytics, as I look upon its effects as decidedly enfeebling. It may be given, however, in doses of 10 or 20 grains with benefit when epileptiform attacks are recurring frequently.

Some years ago it was the fashion to administer to these patients the bichloride of mercury. General paralysis was an inflammation: mercury arrested inflammation, therefore mercury was given to paralytics. But I never saw the least good done thereby in the many cases where it was tried. Tonics, on the contrary, are often of the greatest service when the great excitement has passed away; and, as in other head cases, I believe no tonic equals iron and its preparations.   Quinine and bark seem of secondary importance; but iron—the tinct. ferri perchloridi especially—often seems to infuse new vigor into the failing limbs of the paralytic.[1]   Next to iron I should place arsenic.

In the height of the excitement, beware how you give stimulants, especially brandy.   It renders patients furious, deprives them of sleep, and undoes the effect of other remedies.   As in acute delirium, give at first the more soporific stimulants, stout and ale, if you give any, and reserve your brandy and wine until a later period.   When the excitement has passed off and reason is returning, and when friends are beginning to think that, in spite of your gloomy predictions, the patient is recovering, he will require a generous diet, with a liberal supply of port wine.   And this must be continued during the whole period of his convalescence and subsequent decline into dementia; and the latter will, *cæteris paribus*, be prolonged according to the plenteousness of the food taken.   About this time of gradual decay there is little to say here.   Such patients must be nursed according to the ordinary rules of nursing.   They must not be allowed to lie in bed.   They must be taken out of bed, thoroughly washed —for in the advanced stage it is very difficult to keep them dry—and must sit by day in an easy chair, and, so long as they are able, taken for a walk in the garden, or a drive.   If allowed to lie in bed, they will very soon contract bed-sores, which, in their condition of depressed vitality, will be most difficult to cure.   A strong solution of sulphate of zinc forms a good lotion when the skin is threatening to give way.   When it has happened, I know no better application than the oxide of zinc, thickly strewn in powder on the sore, which, by repeated dredgings, may be coated over and preserved from the air, and thus will often quickly heal.

Considering the opportunities afforded for post-mortem examination of patients dying of this disease, it may seem strange that doubt should still exist as to its seat, and special pathological character.   Every portion of the brain has been thought to be the part affected.   The early discoverers looked upon it is as a chronic meningitis, and this view has been reproduced in our own times by L. Meyer.   Calmeil, who did so much for its accurate description, thought it an inflammation of the cortical portion; and of late many observers have held the same opinion.   The cortical substance by some is supposed to be affected by atrophy, and pigmentary or fatty degeneration of the cells, while by others the morbid change is thought to be an increase of the connective tissue, invading both the gray and white cerebral substance.   Changes in the blood-vessels of the cortical substance have been pointed out by various observers in cases of general paralysis, changes in the walls of the vessels, increase of

---

[1] Dr. Mickle (*op. cit.* page 173) speaks higly of the beneficial effect of this drug. He also finds veratrum viride useful.

14

the nuclei, twisting and aneurismal dilatations of the arteries and capillaries, and obliteration and amyloid degeneration of them.   Others point to an atrophy of the nerve-tubes, and degeneration of the white matter; while Dr. Lockhart Clarke describes holes or vacuoles, seen by him in various portions of the white matter of the brain as well as in the convolutions.   These he believes to be perivascular spaces or canals which originally contained blood-vessels, surrounded by their peculiar sheaths, and which subsequently became empty by the destruction and absorption of those vessels.   I have already spoken of these in my fifth lecture.[1]

The majority of observers have described only the morbid appearances found in the brain,' which they have considered the seat of the disorder. But others, as Drs. Joffe, Boyd, and Westphal, have called attention to the diseased condition of the spinal cord.   The last named has described minutely the morbid processes thereof: he has found inflammatory disease of the spinal dura mater (pachymeningitis), alteration, opacity and thickening of the pia mater, and three different forms of disease of the cord: (1) disease of the posterior columns alone throughout their length, from the cervical to the lumbar region, consisting of an atrophy of the nerve-tissues, and a growth of connective tissue which sometimes takes the place of the nerve-tubes, the morbid process being specially developed at the periphery of the posterior columns.   In addition to this he found (2) an affection of the posterior section of the lateral columns throughout their whole extent, and (3) a mixed form of affection of the posterior columns and of the posterior portion of the lateral columns.   In the two latter varieties Dr. Westphal looks on the disease as a *chronic myelitis.*   There are present granular cells, and a reticulated network of connective tissue surrounding the nerve-tubes; but the large plates of connective tissue are not found as in the first variety.   Dr. Westphal does not connect these appearances pathologically with the diseased conditions found in the brain.   He considers that the disorder does not spread from the brain downward, or from the cord upward, for he has not found disease in the mesocephalic parts.   " If we consider all the circumstances, we must for the present regard the cerebral and spinal diseases which simultaneously exist in general paralysis of the insane as, in so far, existing *per se,* and in certain respects independent of each other, as it is imposssible for us to define more minutely the nature of the cerebral malady, and to establish a connection between it on the one hand, and the processes of gray degeneration and chronic myelitis of the spinal cord or medulla oblongata, on the other."   I have known several patients in whom paraplegic symptoms existed for some years before the mind was affected, and who afterward displayed all the characteristic mental symptoms of general paralysis.   The question whether the disease is propagated from one part to the other,

---

[1] Page 81.

from the spinal cord to the brain, is one on which pathologists are by no means agreed.

After much reflection and consideration of the symptoms of this disorder, I have to come round to the opinion which I held concerning its nature in the year 1866. I then said [1] that general paralysis "seems to be the peculiar degenerative inflammation of the cortical part of the brain, ending in total annihilation of the life—that is, of the functional activity of the part." In the former edition of these lectures, I inclined to the theory of Drs. Poincaré, Bonnet, and Westphal, that the disease is primarily one of the sympathetic ganglia, the lesions of the cerebrum and spinal cord being secondary. My chief reason for advocating this theory was the fact that paraplegia in certain cases had existed for years before mental symptoms appeared. But I now think that Professor Ferrier's experiments warrant our belief that such paraplegic symptoms may be sometimes due to cerebral lesions and not to spinal, and that the cortical portion of the brain, and not the spinal cord, may be the seat of the disease throughout. Whether this be so or not, we may at any rate assume that the gray matter either of the cerebral or the spinal centres is the seat of the morbid lesion, the essential nature of which is a slowly-progressing degeneration of an inflammatory character, the latter being denoted by the postmortem appearances, which, although they have much in common with those found in other forms of insanity, yet present certain peculiarities which you must bear in mind. When we open the heads of general paralytics, we notice adhesions of the dura mater to the cranium, and on its inner surface are frequently seen layers of thin, hæmorrhagic, pseudomembranous exudation, the residua of a great afflux of blood. The arachnoid is milky and thickened, studded with Pacchionian granules, and so united to the dura mater. It is in the pia mater, however, that we find the most pathognomonic signs of the disorder. When we attempt to strip it off we find it thickened, coarse, and tough, and it adheres to the convolutions so intimately that we tear up the latter, which come away with the membrane. This is a very constant and peculiar appearance, and is characteristic of general paralysis, and at the same time points to an adhesive inflammation which has existed in this region. Such appearances are chiefly noticed in the frontal and parietal regions. "The points or areas of adhesions," says Dr. Crichton Browne,[2] "may be few or numerous, large or small, distinct or confluent, but they are invariably confined to the summits of the gyri, and do not spread down their sides nor occur at the bottom of the sulci. They have an irregular outline and jagged edges, and incline perhaps to an oval shape, having their long diameter parallel to that of the convolution on which they are situated. The sur-

---

[1] Journal of Mental Science, xii., 353.
[2] West Riding Asylum Reports, vi., p. 180.

face of the convolutions, where not implicated in the adhesions, is generally smooth and normal.   It is always paler than the eroded patches,
which are darker than the cortical substance usually is, and have often a
red and engorged aspect."

When we remove the pia mater and examine the cortical substance
itself, we discover, even with the naked eye, that in cases of average duration its structure is atrophied to a considerable extent.   "We have," says
Meschede, "hyperæmia and parenchymatous swelling of the inner layer
of the cortical substance on the one hand, and fatty pigmentous degeneration on the other, as the beginning and the end of the organic changes
in general paralysis.   Between these poles lie the destructive powers,
which by analogy we conclude to be a parenchymatous inflammation."[1]

Not only is there atrophy of the cortical substance.   Various appearances denote the afflux of blood which has taken place.   Stains of various
hues, from pale rose to violet, and small capillary apoplexies may be seen,
even without the aid of the microscope.   And if by chance we see the
brain of a patient who by some accident has died in an early stage of the
disease, we shall not find the adhesions which characterize the later stages,
but none the less shall we notice the evidences of the inflammatory condition which existed at the time of death.   In 1861 I saw the brain of
a man whose malady was only of some few weeks' duration.   There was
great thickening of the cranium, and slight opalescence of the arachnoid;
there were no adhesions of the pia mater, and beyond evidences of great
hyperæmia there were no morbid conditions discoverable by the naked eye.

The condition of the vessel also confirms this view.   They are tortuous,
looped, and varicose.   Their walls are thickened, and they present all
the appearances we should expect to find in a part where long-continued
hyperæmia had been raging.

The microscopic appearances bear out this view with even greater
force.   We find growths of lower tissues, patches of molecular degeneration, abnormal condition of the nerve-cells.   Meschede, in the article
already quoted, says that the alteration in the cells is found in different
degrees, from mere parenchymatous swelling down to their reduction to
molecular detritus.   Dr. Major's observations seem to confirm this.
"There exists," he says,[2] "also another condition, which has been described by some as 'inflated.'   The size of the cell is not much altered,
but it exhibits an irregular inflated appearance, this being not due to
excess of pigment or fatty degeneration.   It would seem probable that
this is the earlier stage, and is succeeded by the sinking and atrophy
before described."

Other chronic brain diseases, as chronic brain-wasting and senile atro-

---

[1] Journal of Mental Science, October, 1866.
[2] West Riding Reports, iii., p. 110

phy, present many appearances after death which closely resemble those seen in general paralysis, but, although there is no phenomenon peculiar to one disease and absent in the others, yet those most commonly found in general paralysis point to an inflammatory condition which is not present in the others. Whereas in the former, adhesions of the pia mater are constantly found, they are rarely seen in brain-wasting. The pia mater in this is easily stripped off, and the convolutions are found bathed in serous fluid, "waterlogged" in appearance, the sulci broad, the convolutions wasted. "Atrophy of the brain is here denoted, with compensatory effusion and thickening of the membranes without adhesion." The atrophy of chronic brain-wasting closely resembles the atrophy of old age, and is in fact a premature old age. Very different in all its characteristic symptoms is the inflammatory exaltation and violence of general paralysis. While mind and strength persist before the last prostration of dementia is reached, nothing can be more different than the optimism of the one, and the dull and feeble depression so generally witnessed in the other.

Another indication of an inflammatory disorder is the temperature, which is higher than the normal. "There is in general paralysis just that rise of temperature which we should expect in a chronic inflammatory process in the cortex, and that rise in temperature occurs just at those stages and epochs in the disease when the inflammatory action may be supposed to be most active." [1] A series of careful observations made by Dr. Mickle [2] shows that the evening temperature is usually higher than the morning temperature in general paralysis, and an absolutely high temperature occurs in cases rapidly progressing toward death. The latter kind of cases may show temperatures above the average both in the morning and evening, for a long time before any complication exists. A rise in temperature often accompanies a maniacal paroxysm, and precedes an approaching congestive or convulsive seizure, which it nearly always accompanies.

That the cerebral cortex is the seat of this disease can hardly be doubted, if we reflect that after all it is an insanity resembling in many respects the ordinary non-fatal insanity which we term mania. As I wrote in 1866, the "pars affecta" in general paralysis and non-paralytic insanity is one and the same. We may arrive at this conclusion without postmortem examination of the diseased brain. The line of demarcation between ordinary insanity and general paralysis is exceedingly fine, and the whole history of the latter points to a difference in degree rather than in kind. That general paralysis is intractable and progressive is the fact we are certain of. The hyperæmia which accompanies curable mania subsides, and the disease vanishes. The hyperæmia of general paralysis becomes a slowly-advancing degeneration which destroys the tissue.

[1] Dr. Crichton Browne, West Riding Asylum Reports, vi., p. 191.
[2] Journal of Mental Science, April, 1872.

LECTURE XIV.

Of Patients whose Insanity is doubtful—Insanity without Delusions—Are Delu-
sions the Test of Insanity?—On Moral Insanity, so called—Dr. Prichard—His
Illustrative Cases considered—Intellectual Defect in the Morally Insane—
Moral Insanity in connection with Epilepsy and Old Age—Emotional Insanity
—Prognosis and Treatment.

HITHERTO I have spoken of patients whose insanity is plain and unmistakable. The question for us is, how are we to cure them? When
the insanity is recent and acute, and the disease is not paralytic, we shall
be able to cure a considerable number. But there are many persons about
whom your opinion will be sought on other grounds. You will have to
say, either to the friends or in a court of law, whether a patient is or is
not legally of unsound mind—so unsound in mind as to be incapable of
taking care of himself and his affairs, and a fit and proper person to be
detained under legal restraint. Most difficult is it in many cases to come
to a decision upon such a question; still more difficult to give the grounds
of our opinion, and to give them publicly in the witness-box. Your opinion
will be required, speaking generally, for one of four purposes: 1. To place
a patient under legal restraint in an asylum or quasi-asylum for the purpose of treatment; 2. To deprive a person of the management of his
affairs by a commission *de lunatico inquirendo;* 3. To relieve a patient
from the responsibility of some crime committed or contract entered into;
4. To inquire into the state of mind of a testator at the time he executed
his will. The legal portion of the subject I shall leave till hereafter. I
wish now to bring before you certain classes of patients and certain varieties of insanity which most frequently give rise to forensic contests.
These are "moral insanity," "emotional," and "impulsive insanity."
Under these names you will find cases quoted in books, and may be questioned concerning them by counsel. Such patients may be more accurately described by other names, and I wish to indicate to you the method
of examining and testing them.

Certain forms are discussed in courts of law, and hotly contested, because they are said by some to lack the symptoms necessary to be demon·
strated before we can pronounce any one legally insane. Chiefly, the absence or presence of *delusions* is the point at issue. Can a person be found
lunatic who has no delusions? Most lawyers deny this, even now. Our
own profession affirms it, and points to many persons of undoubtedly unsound mind, in whom no delusions are to be found.

In reading the history of the great cases of disputed insanity, you may be astonished to find that what is considered unsoundness of mind in a civil cause is by no means looked upon as legal insanity in a criminal trial. That which is deemed proof of a man's being unable to manage his affairs, does not necessarily absolve him from responsibility if he has committed murder. When giving evidence upon a commission in lunacy, you are asked, in plain terms, if you think the patient is fit to take care of himself and manage his affairs—a question which the facts of the case generally render easy to answer. But, in a crown case, you are asked if the prisoner knew what he was doing when he committed the crime, or if he knew right from wrong, which are questions which cannot be decided by medical science. I shall have to refer to this again, but I mention it here inasmuch as it is held that the proof of insanity in criminal cases must be stronger than in others, and this must be borne in mind by all of us who may be called upon to give evidence.

Lawyers and judges vary indefinitely according to their humanity and idiosyncrasy in their opinions as to what constitutes irresponsibility. Some say that a patient is responsible unless he is so insane that he cannot know right from wrong; others hold that he is not insane unless he has delusions. This latter theory is constantly propounded in both civil and criminal courts. It was held and believed at one time by both lawyers and doctors. If we look back at the definitions and doctrines of the great medical luminaries of the seventeenth and eighteenth centuries, we shall find that they almost invariably divided insanity into *melancholy* and *mania*, which latter they also called frenzy or fury. *Melancholy* they defined to be "a permanent delirium, without fury or fever, in which the mind is dejected and timorous, and usually employed about one object." "*Mania* is a permanent delirium, with fury and audacity, but without fever."[1]

Now delirium in Arnold's time, 1782, did not mean what it does nowadays. You know very well what the delirium of fever is, or delirium tremens. We should not say that a monomaniac laboring under the delusion that he was the rightful heir to the throne, but in all other respects rational, was suffering from delirium, yet this name would have been given to his malady by the writers of the last century. *Delirium sine febre* was our "delusion," *delirium cum febre* was our "delirium." "Phrenitis," says Hoffman, "est insania cum febre, a stasi sanguinis inflammatoria in vasis cerebri orta."

Melancholy by the older writers, as far back as the time of Burton, is used to signify monomania or partial insanity in contradistinction to *mania*, which meant general insanity, fury, or frenzy. This distinction between general and partial insanity we hear maintained by lawyers in our

[1] *Vide* Arnold on Insanity, vol. i., p. 30.

own day, and we can trace it back at any rate as far as the seventeenth century, for Lord Hale says [1]—" There is a partial insanity and a total insanity. The former is either in respect to things, *quoad hoc, vel illud insanire.* Some persons that have a competent use of reason in respect of some subjects, are yet under a particular *dementia* in respect of some particular discourses, subjects, or applications, or else it is partial in respect of degrees; and this is the condition of very many, *especially melancholy persons,* who for the most part discover their defect in excessive fears and griefs, and yet are not wholly destitute of the use of reason; and this partial insanity seems not to excuse them in the committing of any offense for its matter capital, for, doubtless, most persons that are felons themselves and others are under a degree of partial insanity when they commit these offenses. It is very difficult to define the invisible line that divides perfect and partial insanity; but it must rest upon circumstances duly to be weighed and considered, both by judge and jury, lest on the one side there be a kind of inhumanity toward the defects of human nature, or on the other side too great an indulgence given to great crimes."

Here Lord Hale lays down the doctrine that partial insanity, melancholy, or monomania, does not absolve from responsibility in criminal cases. Those lawyers who do not go so far as to say this, but would allow that insanity of any kind absolves a man from punishment, nevertheless almost all assert that to prove insanity we must prove delusion. This opinion is maintained by both civil and criminal lawyers, and I have heard it enunciated frequently within the last few years. Two legal opinions I may quote, given by men of great eminence on this side. Mr., afterward Lord, Erskine, at the trial of Hadfield, who shot at the King in Drury Lane Theatre in the year 1800, said—" Delusion, when there is no frenzy or raving madness, is the true character of insanity; and when it cannot be predicated of a man standing for life or death for a crime, he ought not, in my opinion, to be acquitted." The second is that of Sir John Nicholl, whose judgments are highly esteemed and honored by all lawyers. In a very celebrated judgment pronounced by him in the Court of Probate, in Dew *v.* Clark, he said—" The true criterion, the true test of the absence or presence of insanity, I take to be the absence or presence of what, used in a certain sense of it, is comprisable in a single term, namely, *delusion.* In short, I look upon delusion, in this sense of it, and insanity, to be almost, if not altogether, convertible terms. On the contrary, in the absence of any such delusion, with whatever extravagances a supposed lunatic may be justly chargeable, and how like soever to a real madman he may think or act on some one or all subjects; still, in the absence, I repeat, of anything in the nature of delusion, so understood as above, the supposed lunatic is in my judgment not properly or essentially insane."

---

[1] Pleas of the Crown, 30.

This is the opinion of many, I may say most, lawyers. But lawyers, I need not tell you, have no practical acquaintance with the insane. Their doctrines are based on the traditions and judgments of preceding dignitaries, and in no way depend on the advance of scientific research or more accurate observation of patients. I shall endeavor to show that many patients, undoubtedly of unsound mind, have no delusions, that delusions are not the one "test" of unsoundness of mind, nor even of insanity so-called; and, further, that there are many who are beyond all question of unsound mind, who cannot properly be called insane. Counsel will try to trip you on this point, and ask if you consider the patient insane. He may not be insane, strictly speaking, and you may have to admit it; they will then argue that he is not *legally* of unsound mind. But a great lawyer, Lord Coke, in his commentary upon Littleton, forestalled this objection. He, too, uses the expression "unsound of mind." "*Non compos mentis*," saith he, "explaineth the true sense, and calleth him not *amens, demens, furiosus, lunaticus, fatuus, stultus*, or the like, for *non compos mentis* is most sure and legal." And the Lunacy Act of 1845, 8 and 9 Vict. c. 100, in the interpretation clause, § cxiv., says—"'Lunatic' shall mean every insane person, and every person being an idiot, or lunatic, or of unsound mind."

I proceed, therefore, to describe to you various patients who are legally unsound in mind, yet cannot be included in any of the classes already mentioned. And, first, let us take those whose malady is called by some *moral*, by others *emotional* insanity, *affective* insanity, and so forth. I do not myself employ this nomenclature, but it is necessary that you should be familiar with it. The essential feature, whatever name we give it, is, that there are no delusions.

Pinel was, I believe, the first to lay down distinctly the doctrine that insanity may exist without delusion. In 1802, he described this as *manie sans délire*, mania without delusion, and he gives a brief history of three patients thus affected. One of them was liberated by the revolutionary mob which broke open the Bicêtre, but being soon roused to fury by the excitement around, was quickly led back to his cell by his new-found friends. After Pinel came Esquirol: to the partial insanity which hitherto had been called *melancholia*, he gave the name of *monomania*, and described two varieties as existing *sans délire*, without delusion, *monomanie instinctive* and *monomanie affective* or *raisonnante*. From his time to our own, all the most illustrious authorities of our profession have recognized this fact, and under one name or other have described patients whose insanity was free from delusions. I may cite, amongst others, the names of Hoffbauer, Georget, Gall, Marc, Combe, Prichard, Ray, Reil, Rush, Bucknill, Hack Tuke, and Maudsley.

Various names have been bestowed on this insanity, various divisions and classifications of it and its varieties have been constructed, according to the theories held as to the mind and its component parts.

The mind is commonly said to be divisible into the intellect, emotions, and will. In accordance with this, Dr. Bucknill speaks of *intellectual, emotional,* and *volitional* insanity, and Dr. D. Hack Tuke points out that some such division as this has been adopted by many authors. Reid, in his analysis of the mind, divides it into the *understanding* and the *will;* Dr. Thomas Browne speaks of the *intellectual states of the mind* and the *emotions;* Morel gives a triple division, *the animal passions, the moral feelings,* and *the intellect;* Bain's division is *intellect, emotion,* and *volition,* while the emotional or affective part of the mind is by many subdivided into *propensities* and *sentiments,* or *moral sentiments.*

Besides the division of insanity into *intellectual, emotional,* and *volitional,* it has been divided into *intellectual* or ideational, and *emotional* or affective, with a subdivision of the latter into *moral alienation,* or insanity of the moral sentiments, and *insanity affecting the propensities,* or impulsive insanity. Here we come to the *moral insanity,* concerning which so much contention has arisen, and which is so often said to have been invented by doctors as an excuse for crime.

The great teacher of the doctrine of *moral insanity* was Dr. Prichard, who, in his well-known Treatise on Insanity, published in 1835, insists strongly on this division, and on the fact that "insanity exists sometimes with an apparently unimpaired state of the intellectual faculties." Moral insanity he defines to be "madness consisting in a morbid perversion of the natural feelings, affections, inclinations, temper, habits, moral dispositions, and natural impulses, without any remarkable disorder or defect of the intellect, or knowing and reasoning faculties, and particularly without any insane illusion or hallucination."

Now, I deny that the absence of the moral sense proves or constitutes insanity, any more than its presence proves sanity. It is perfectly true that it is absent in many lunatics, all notions of duty, propriety, and decency being destroyed in the general overthrow of the mind; but it is also true that we can find perfectly sane people who, either from early education and habit—the habit of continual vice—and also hereditary transmission, are devoid of moral sense to an equal or greater degree. Probably greater wickedness is daily perpetrated by sane than ever was committed by insane men and women; so that when immorality makes us question a man's state of mind, it must be remembered that insanity, if it exists, is to be demonstrated by other mental symptoms and concomitant facts and circumstances, and not by the act of wickedness alone. Writers who, like M. Despine, think that the committal of great crimes without concern or remorse indicates an absence of the "moral sense" amounting to irresponsible defect, overlook the fact that the habit of wrong-doing may be acquired to such an extent that the thing done excites no feeling whatever. An habitual murderer, as a Thug or a brigand, thinks no more of taking life than does a veteran soldier. It is his ἕξις, his everyday habit. As there are, according to Aristotle, spurious forms of courage, one of which is ἐμπειρια, experience, or, in other words, habit; so criminal and bad acts may be so habitual as to be unaccompanied by any feeling of having done wrong. The gradual effacing of the moral sense, and gradual hardening in vice, have been portrayed by many a moralist; but something else is needed to prove the disease or deficiency of mind we look for in the inhabitants of an asylum.

I cannot help thinking that the authors who have most strongly upheld the doctrine of a moral insanity and morbid perversion of the moral sentiments, have often underrated or neglected the intellectual defect or alteration observable in the patients. Because no delusion has been found, it has been assumed that the intellect is not impaired, intellectual insanity and insanity with delusions being spoken of as synonymous. But many patients of defective intelligence have no delusions, as all kinds of idiots and imbeciles. Many of altered intelligence have not as yet reached the stage of delusion, and many recover from the latter, or from the stage at which delusions are present, yet do not recover their full intellectual powers, but remain semi-cured and semi-insane. This class is a very large one. Dr. Prichard gives as illustrations of moral insanity seven cases in his own practice, one communicated by Dr. Symonds, and nine by Dr. Hitch. These are entitled, "Cases of Moral Insanity and Monomania;" and, as they are quoted frequently in courts of law, I think it right to say something about them here. They are descriptions, and very valuable ones, of insane patients, not of varieties of insanity, and there is all the difference between the two. In strictness I ought to give these cases in full, but time prevents. Dr. Prichard's work, however, is to be found in most medical libraries, and I must refer you to it. The first patient was a gentleman of high intellectual attainments, who married a lady no less endowed, and well known in the literary world. He was greatly attached to her, but extremely fearful lest it should be supposed that she dictated what he wrote, and he never let her know what he was thinking or writing about. He then acquired strange habits, placing everything in a certain order, whether in his own or other people's rooms. He would run up and down the garden a certain number of times, rinsing his mouth with water, and spitting alternately on one side and the other in regular succession. He employed a good deal of time in rolling up little pieces of writing-paper, which he used for cleaning his nose. It did not surprise those who were best acquainted with his peculiarities to hear that in a short time he became notoriously insane. He committed several acts of violence, argued vehemently in favor of suicide, and was shortly afterward found drowned in a canal. Such is the epitome of the history of one who gradually drifted into unmistakable insanity through the stages of alteration and eccentricity. Even if his ideas concerning his wife, and concerning his arrangements of the articles of furniture, could not strictly be called delusions, it is impossible to say that in him was no defect of intellect. He would appear at dinner in his dressing-gown, apologizing for not having had time to dress, when he had been dressing all the morning; and he would go out for a walk in a winter's evening, with a lantern, "because he had not been able to get ready earlier in the day."

The second case is that of a patient who was at first melancholic, timid and irresolute, suspicious of all about him, always changing both his studies and residence. Soon he fancied himself the object of dislike to all in the house, and when questioned confessed that he heard whispers of malevolence and abhorrence. Here we have another plain case of a patient gradually drifting into delusions and hallucinations.

The third patient, a maiden lady of forty-eight, after an attack of pneumonia, evinced alternations of depression and exaltation for eight years, without showing any delusions or marked symptoms of insanity, and then broke down, had many delusions, "saw people looking at her," "thought it was the devil who made her act so." Yet when Dr. Prichard saw her, he could find no insane delusion. She herself confessed, however, "that she was formerly very different."

The fourth, a farmer, displayed no hallucination or disturbance of the intellectual faculties. Yet we hear that from being a man of sober and domestic habits, frugal and steady in his conduct, he became wild, excitable, thoughtless, full of schemes and absurd projects. He bought cattle and farming-stock of which he had no means of disposing; bought a number of carriages; called on the steward of a gentleman in the middle of the night to survey an estate; and yet we are told that "his intellectual faculties were not disturbed." He was, however, placed in an asylum, and recovered.

The fifth was a gentleman who was for many years in an asylum, but was released by a jury, in opposition to the unanimous opinion of several physicians, who all thought him insane. He scarcely performed any action in the same manner as other men, and some of his habits were filthy and disgusting. If a physician came near him, he recoiled with horror, exclaiming, "If you were to feel my pulse, you would be lord paramount over me for the rest of my life." This individual, however, a chronic monomaniac, "has lived for many years on his estate, where his conduct, though eccentric and not that of a sane man, has been without injury to himself and others."

The sixth was a working tradesman, who, after the death of his wife, denied himself the necessaries of life, and lived in a state of starvation and filth, till removed to an asylum. Here he so far recovered that he was released, and married a servant of the asylum. She soon, however, brought him back to the asylum, where he remained, his derangement being afterward marked by various delusions.

The seventh, and last, was a gentleman who had had epileptic fits.

He became totally changed, remained always in bed, was dirty, irascible, and violent. He was in a state of constant despondency. His friends would not place him in an asylum, and we are told that "the event was calamitous," which, I presume, means that he committed suicide.

To say that "the reasoning faculty remains unaffected" in patients who commit every variety of outrageous and filthy act, and justify it, seems to me to involve a total misapprehension of what the "reasoning faculty" is.

But I will pass on to the cases reported to Dr. Prichard by others. Dr. Symonds's is that of a man who had never fully recovered from an attack of acute mania. He was what is commonly designated flighty or cracked. He advertised for sale property which he knew to be entailed, left his family without cause and without means of subsistence, while his letters were confused and incoherent, the expressions being ridiculously dispro-tionate to the subjects. He printed a pamphlet concerning his domestic history: "A more insane document than this I scarcely ever perused. The sentences were so involved and undistinguished that, although the ideas were not absolutely incongruous with each other, it was impossible to collect more than the general tenor of a long passage." He was filthy in habits, constantly passing his evacuations in his bed.

Dr. Symonds signed a certificate, though, as he says, "he was unable to trace any positive intellectual error,"—that is, I take it, he could find no delusion.

In all these cases, and in cases 1, 2, 8, and 9, reported by Dr. Hitch, it appears to me that there was plain and palpable intellectual alteration and defect, to such an extent as to make it evident that the patients were insane, whether their actions were moral or immoral; and in many of them delusions appeared after a time, though in some instances the patients took a considerable period to go through the stages of alteration and alienation, which so many pass through in a few days or weeks.

Dr. Hitch's third is an excellent illustration of intermittent "dipso-mania." "At times the gentleman is in habits most abstemious; he never drinks anything stronger than beer, and frequently tastes only water for weeks together. Then comes on a thirst for ardent spirits, and a fond-ness for low society. He drinks in a pot-house till he can drink no more, or get no more to drink, falls asleep for from twenty to thirty hours, awakes to the horrors of his situation, and is the humblest of the meek for several weeks. In about three months the same thing occurs." This form deserves the name of recurrent moral insanity more than any of the foregoing, and must be studied in connection with the propensity to drink.

No. 4. This patient serves as a good example of what may be called moral insanity, if the term is to be used at all. "He had been the inmate of several asylums, but his early history is not given. No delusions were ascertainable; but he enjoyed in a high degree the art of lying and the pleasure of boasting. The former was applied to the production of mischief and disturbance. He was an adept at stealing, and hoarded and secreted in his clothes and bedding articles of all kinds; yet he possessed many good qualities, would be kind and useful in the gallery, and corrected obscene or impious language in others.

"His judgment was quick and correct, he had quick perceptions, strong memory, and great discretion in matters of business. His madness appeared to me to consist in part *in a morbid love of being noticed*. He is now at large, and has been in the management of his affairs for three years, in which time he has sold an estate advantageously, and conducts his business with profit."

The next patient also deserves to be called morally insane. "Always of a bad temper, she gradually gave way to paroxysms of passion, followed by a morose and unyielding sullenness. A change came over her: she neglected her children, and abused her husband; she smashed all the windows in her own house and the workhouse, and then was sent to an asylum, where she would remain constantly in bed if allowed, or suddenly roll on the ground and scream if questioned, or cry and sigh as if in the greatest distress. As a disagreeable and unmanageable patient, without actual violence, she exceeds most with whom I have met. Her mind appears totally unaffected as to its understanding portion, but in the moral part completely perverted." This case is a very good instance of insanity without delusions, shown, as in the last patient, by outrageous conduct wholly irreconcilable with reason.

The same may be said of No. 6, a man who by many might be called bad rather than mad. "I found him one of the most mischievous of beings; his constant delight was in creating disorder to effect what he called 'fun;' but he had no *motive*, no *impression* on his mind, which induced him to this conduct; he was merely impelled by his immediate feelings. In his state of health I found nothing wrong, except that he did not sleep."

No. 7 is the case of a gentleman who suffered from cerebral symptoms, and eventually died of apoplexy. He entertained deep feeling of hostility to certain persons who had affronted him; but he was never insane, at least according to this report.

No. 8 was a young lady distinguished in the *beau monde*, whose father's embarrassments caused her withdrawal from society. She became coarse and abusive, neglectful of her person, altogether changed in habit and feelings, hated her family, and said "she was dead." After three months in an asylum she was dismissed "much relieved." Intellectual change must have been observable here, as in so many of Dr. Prichard's cases.

Lastly, we have the particulars of a young man whose defect was a morbid want of self-confidence, a fear that he was unequal to his business, and that he should ruin it. He tried to run away, and when stopped, took poison. He was placed in an asylum and recovered, but periodically absented himself from home. In time his affections were altered toward his father and family, and he became suspicious of them all. An elder brother, without any symptoms of madness, had destroyed himself.

Here we have a well-marked instance of intermitting melancholia, with self-depreciation amounting to delusion.

Thus have I briefly commented on all the cases of moral insanity related by Dr. Prichard. That in some of them insanity existed without insane delusion for some time, I admit; but that it existed without alteration or defect of the intelligence, I deny. Whether we choose to call it *emotional* insanity or *affective* or *moral*, I hold that the intellect is also involved, that we cannot divorce the emotional and intellectual functions of the mind, and that this view may be upheld both on *a priori* and *a posteriori* grounds; for *a priori* we should say that the ideational portion of the mind is so intimately joined in operation to the emotional—the stored ideas of the brain are so influenced by the feelings of the moment, whether these arise from within or without—that the two must be sound together, or unsound together; so *a posteriori*, we see the insanity displayed in absurd and extravagant acts, groundless and foolish likes and dislikes, or reckless squandering of health and property. We say of such a person, either he is mad or he is a fool, thereby attributing to him at once intellectual defect. And we shall see by-and-by that these acts are characteristic of persons notably deficient in intelligence, congenitally or from disease.

It is quite certain that various patients are undoubtedly insane, who present none of the ordinary delusions of insanity. They may not have reached the stage of delusions, and they may go on to recovery without ever reaching it; or they may recover from the stage of delusions, and yet not perfectly recover, remaining in a chronic state of what Dr. Prichard calls moral insanity.

It happens, however, in many cases, that the moral side of a patient's mind shows alteration or defect at an earlier date, and more plainly and markedly than the intellectual. For in the daily relations of life—the domestic circle and business connection—in all the surroundings, or, as it is now the fashion to call it, the environment, changed feelings, undue depression, increased irascibility or hilarity, with corresponding conduct, will be noticed long before the stage of delusions, or even intellectual defect is reached. This alteration in conduct may be very slight, so as only to cause apprehension and distress to those most intimate with the patient, or it may increase and be notorious to all, and at last may call for legal interference and restraint. But throughout the moral alteration may be more marked than the intellectual; the insanity will be characterized by conduct rather than delusions, and it will in many cases be most difficult for those who do not live constantly with the individual, and have not known him in former times, to say whether his acts amount to legal insanity or not.

As some people present this stage of moral insanity before they reach that wherein delusions exist, so, as they recover from the latter, they may regain their mental health to the extent of losing, or being able to deny and conceal the delusions, still remaining eccentric in conduct and behavior, hostile to their dearest friends, vindictive, litigious, impossible to live with, leading solitary or peculiar lives, or drifting into odd and unclean habits. Here again we may find great difficulty in dealing with them legally.

We may find at any age this emotion or moral insanity without delusions. It is the insanity of the young. The second of the cases communicated to Dr. Prichard by Dr. Hitch is that of a little girl aged seven, who, from being a quick, lively, and affectionate child, became rude, passionate, vulgar, and unmanageable. She was morbid in appetite, and would sleep on the ground rather than in bed. She was cruel to her sister, could not apply to anything, and her health was much disordered. She would eat her fæces and drink her urine—in short, was in a complete state of mania, but had no fixed ideas or delusion. She recovered in two months.

Besides the young who have an attack of insanity like the one just described from which they recover in due time, there is a class of boys and girls who from an early age present peculiarities which they retain throughout life. I shall speak of them in my next lecture under the head of weak-minded patients. Many have a fair amount of intellectual development, but all are peculiar in their conduct, and many present excellent examples of moral insanity. Of course such defect is sure to be inherited. We see also odd and eccentric people who figure in the newspapers, and are notorious in various ways. They are the outcome of an inherited neurosis, and are morally, if not legally, insane.

Moral insanity may be brought about by various causes which have "reduced" the higher brain centres, but have not caused an ordinary attack of insanity with delusions, such as we call mania or melancholia.

After an accident, as a blow on the head, or after an attack of epilepsy, we may perceive a change come over an individual, insidiously gaining ground, till it is plain that he is insane, and his insanity may for a long time be of this merely moral description. I have myself seen two gentlemen who might have been called morally insane, whose disorder could be traced to an epileptic attack. I have already recorded these in the "Journal of Mental Science," April, 1869. One gentleman had originally received a concussion of the brain in a railway accident, and subsequently had an epileptic fit, and after a long interval a second. He was noisy, ostentatious and boastful, irritable, occasionally low and hysterical. There was an extreme want of consecutiveness in his conversation, and this, with his extravagant notions and bombastic language—all quite foreign to him—plainly revealed his insanity; but he had no positive delusions. He was in an asylum six times, yet he died at home, weak in mind, but sane. Twice he recovered after an attack of gout; and although there were no delusions, his conversation on one occasion became perfectly incoherent, and his excitement arose into a state of sub-acute mania. It was very difficult to sign a certificate in his case—at any rate for one not previously acquainted with him—and yet no one who was with him for twenty-four hours could doubt his insanity.

The other patient's malady also commenced with an epileptic fit, from which time he gradually became an altered man, and had periodical fits of drinking, though he never pushed this to extremes. Three years elapsed before anything occurred which justified the interference of his friends. He then rode on horseback to the end of Brighton chain-pier, afterward assaulted the police, and when bailed did not appear. He was then sent to an asylum. When there, he justified his acts in a manner markedly insane, and wrote hundreds of absurd letters to the same effect. He rambled both in conversation and ideas. Certain officious people, who thought him unjustly detained, sent a solicitor to see him. He betrayed no delusion whatever, but treated the whole matter in so frivolous a way, and wandered away from the point to such a degree, that the lawyer pronounced him insane. He was subsequently discharged, and died within a year, I was told, of abscesses of the liver.

Moral insanity may be the result of drink, and add to the difficulties we almost always encounter, when we have to deal with drinking patients. A man or woman may drink till the condition is one of moral insanity without delusions. The drink may be even a symptom of the insanity, but still it is there. If we shut them up and deprive them of drink, they get better, and we are forced to release them. If not legally restrained, they appear again and again in police courts, and are fined and punished. Such have generally inherited a defective nerve constitution, and there is little hope of reclamation.

This form of insanity may make its appearance in old age, and constitute what Drs. Prichard and Burrows call "senile insanity." An old gentleman whom I knew well for years as the pattern of strict propriety, honor, and paternal affection, took in his latter days to going with loose women of a low class, on whom he squandered money to a considerable extent. Fortunately for his family his health gave way, and he was obliged to stay at home and be nursed, his death preventing the necessity of any legal restraint. It would have been impossible, I think, for any medical man to have signed a certificate in this case at the time that he was running after these women, though his friends had no doubt that his mind was deranged. Although he had never shown signs of insanity till his old age, it is worthy of note that two, at least, of his children have

15

evinced symptoms of mental derangement in middle life. In such patients this moral insanity is the forerunner of senile fatuity and dementia, but some time may elapse before the mind is so decayed as to warrant our giving it this name, and the term senile insanity would appear to be more fitting.

When a man or woman becomes insane, the character of the emotion displayed depends on the general physical condition of the individual, and varies at different times, and the delusions will follow the feeling of the moment. So, when there are no delusions, and the insanity is manifested only in the appearance, habits, and acts, we shall still find that the emotion displayed will vary from the extreme of melancholy to the highest hilarity, according to the general condition of the patient. In melancholia there is almost always a premonitory stage in which no delusion is perceptible, and the depression may, and often does, last for a long time, and then pass off without any definite delusion. I say without any definite delusion; but on close cross-examination we find an all-pervading feeling of gloom, or a feeling that everything is wrong, everything miserable and sad, and life itself a burden,—which is, in fact, one great and universal delusion. This is the state of many who commit suicide—a state of undoubted insanity, yet with so little derangement of ideas and intelligence that the friends shrink from interfering, and think that cheering up and change of scene are all that is wanted. When you hear that an individual has become low spirited, even to an apparently slight extent, set it down that he or she is suicidal—even if the friends tell you to the contrary—even if the patient himself expresses a horror of such an act. This he will do, and an hour afterward will be impelled to throw himself out of the window.

Abnormal irascibility, or abnormal hilarity, may, no less than depression, characterize insanity without delusions. This we see illustrated by various cases narrated by Dr. Prichard and Dr. Hitch. Sullenness, causeless anger, with violent outburst of temper and corresponding acts, accompany the insanity of some, and hold a place midway between the extreme depression called melancholia, and the gayety and hilarity which others display. In fact, when a man is entirely changed, no matter how, you may reasonably suspect his sanity, having first thoroughly ascertained the truth of this change. By dint of patient and constant examination, you will probably come to the bottom of the whole matter, to the *fons et origo* of the change, and will find that nothing but insanity can account for it. Men do not suddenly change their nature without a cause, and when such a change takes place, the cause must be sought. And if you suspect insanity, you must examine carefully the individual and his history, the history *quoad* insanity of his parents, brothers, and sisters, his history as regards fits, blows on the head, or other exciting or predisposing causes of mental disorder. Then you must well weigh the changes alleged to have taken place in his temper, habits, ideas, and acts, and must hear what he himself has to say about his feelings and doings. You will probably find such an amount of irrationality, incompetence to argue, rambling from the point, or justification of palpably insane acts, as to warrant your deciding that he is of unsound intellect, whether the ordinary delusions of insanity appear or not. And although your opinion may be asked chiefly upon the question of sanity or insanity, yet you are not to forget that the question of treatment is involved therein. Many are curable, especially those whose insanity has not yet reached the stage of delusions, and is of recent origin. Change, rest, and restraint, in or out of an asylum, according to circumstances, may effect a cure, and this you are bound to promote. Insanity following epilepsy or blows on the head cannot be favorably looked at as regards prognosis; but even this may be intermittent, and the remissions may be treated till there is at any rate temporary recovery, and such patients may be restored for a time to their homes or friends. If you meet with this form in patients who have recovered from severe attacks up to a certain point, and there have remained stationary, not being thoroughly restored to their right and sound mind, it may be possible, by judicious care and supervision, to regulate the lives of such in a way which will enable them to live out of an asylum, in the family of a friend or stranger; and you may witness their perfect recovery even after a long period of alienation. These semi-insane people, if free from mischievous dangerous habits or impulses, have a better chance of improvement out of than in an asylum. Their self-respect is more encouraged; they feel that they have more to live for, and the dread of going back often operates as a wholesome check upon their insane propensities.

Patients affected with "moral insanity" are fruitful sources of legal disputes if they come before courts of law; and are accused of serious crimes, as homicide. I shall have to speak in my next lecture of the legal test of insanity and the mode of dealing with the question of responsibility adopted at our criminal trials; but before leaving the subject, I may mention the case of Christiana Edmunds. No better illustration of homicide committed by a person in a state of moral insanity can be adduced. She was tried in 1872, at the Central Criminal Court, for the murder of a little boy, poisoned by some chocolate drops bought by his uncle at a confectioner's at Brighton. These poisoned drops had been substituted by the prisoner in place of others bought at the shop through the instrumentality of another boy, in order to fix upon the confectioner the responsibility of selling noxious sweets; and this was proved by anonymous letters addressed by her to the father of the child, and various other circumstances. Several medical men gave it as their opinion that she was insane; but the judge summed up very strongly for a conviction, chiefly on the point that she knew right from wrong. She was convicted

and sentenced.   An inquiry was afterwards instituted by the Home
Secretary, and two physicians pronounced her insane, whereupon she was
sent to Broadmoor Asylum, where she now is.   The chief feature in her
case was the fact that her family was saturated with insanity.   Her father
died in an asylum; her brother died in Earlswood Asylum; a sister was
"hysterical," and on one occasion suicidal; her mother's father died
childish and paralyzed at the age of forty-three; a first cousin on the
mother's side was imbecile, and another brother had died of brain disease
since the trial.

Christiana Edmunds is an example of a person utterly devoid of moral
sense or moral feeling in matters relating to herself, though theoretically
she doubtless knew that murder was a wrong thing.   "'That she knew'
perfectly well what she was doing in purchasing poison and disseminating
it broadcast through the town by means of poisoned chocolate creams and
that she knew she was therein doing wrong, were equally beyond dispute.
Her whole conduct before the crime, and her perfectly rational conversation in jail, clearly proved she could have taught a school-room of
children the Ten Commandments, and explain to them clearly that it was
a wicked act to break any one of them, and a most wicked act to break
the Sixth Commandment.   But no one could have talked with her in
jail without being convinced that in her own case she had no real feeling
of the wicked nature of her acts, and that she would have poisoned a
whole city full of people without hesitation, compunction, or remorse.
Indeed, it may be doubted whether in her later experiments she was
really so much influenced by the inadequate motive which no doubt insti-
gated them at the beginning, as by a morbid pleasure in poisoning for its
own sake and in the sensation which her secret crimes excited.   The
terrible story of insanity in her family furnished the real explanation
of her state of mind; she had the heritage of the insane temperament." [1]

Dr. Orange, the Superintendent of Broadmoor, under whose care
Christiana Edmunds has been ever since her trial, says "that she has
been regarded as of unsound mind by all the medical officers of the
asylum who have been here during the last five years.   She is at present
fairly tranquil, and her conduct is much better than formerly; but I do
not regard her as being sane, or fit to be trusted to keep out of mischief.
She formerly had periods of depression alternating with periods of sub-
acute mania, but latterly her condition has been more equable.   There
is not one really sound member of all the brothers and sisters, although
some are worse than the others."

---

[1] Journal of Mental Science, xviii., 112.

LECTURE XV.

Impulsive Insanity—Characterized by Criminal Acts—Explanation of the Impulse —Rules for Diagnosis—Other Symptoms of Insanity usually discoverable— The Legal Test of Insanity, the Knowledge of Right and Wrong—Weak-minded or Imbecile Patients—Characteristics—Cases—Demented Patients— Chief Defects—Cases.

THERE is another variety of insanity, or rather another class of insane patients, in whom, as in the last mentioned, no delusions are to be discovered, and whose insanity is manifested rather by what they do than what they say. This is *impulsive* or *instinctive* insanity, the victims of which under an *impulse* or *instinct* do something, commit some crime or act of violence, for which, being insane, they are not to be held responsible, but for which, were they sane, they would be punished. This form of insanity is considered by many to be closely allied to the last, the "moral insanity" of Prichard. By some it is considered as a variety of "emotional insanity." Dr. Maudsley describes it as one division of "affective insanity," "moral insanity" being the other.

As we saw that a great many patients in various states and stages of insanity may be, and have been, grouped together by one feature common to all, viz., the absence of delusions, and are called "morally insane," because, there being no delusions, their intellect is supposed to be sound, so we shall find that a number of different lunatics have been said to be suffering from *impulsive insanity*, having this feature in common, that they committed, tried, or desired to commit, some act of violence, yet appeared to be free from everything like delusion.

The crime itself is held to be evidence of the insanity, and is accounted for upon the theory that the person is suddenly and insanely impelled to commit it. We cannot be surprised that in courts of law "impulsive insanity without delusion" is regarded with great suspicion, and together with "moral insanity," is looked upon by lawyers as a chimera of philanthropic doctors, who consider all grave crimes to be acts of madness. I have no doubt, however, that if we could sufficiently examine all the cases of so-called "impulsive insanity," and really ascertain the entire history and phenomena previous and subsequent to the commission of the act, we should find many other signs and symptoms of insanity, and should be able to assign to the majority of the patients places in the ranks of the ordinary varieties and classes of the insane. And this you must endeavor to do, considering the individual before you rather than any particular variety or theory of insanity, and bringing together every fact in his life and history—his bodily ailments, demeanor, words, and acts—which will prove to your own satisfaction and that of others that he is unsound of mind.

I believe that the mistake made by those who describe so many cases as *impulsive insanity* is this; they have not proved too much, but too little. Instead of saying there is a form of insanity which impels patients to commit violent acts, they might have gone much further, and said that all insanity is impulsive, all insane patients are liable to commit such acts. A. is an insane patient, therefore it is likely that he will do such an act. We shall then have to prove, not that A. is afflicted with a particular kind of insanity, but that being insane in some degree or other he committed the act. We have already seen, and laid it down as a truth, that patients may be insane, and yet have no delusions. If one of these commits a crime, we need go no further than to say that he, being insane, has committed an impulsive act of violence, his insanity being manifest from such and such symptoms. And these symptoms we must very carefully analyze and set forth, so that they may carry to others the conviction they bring to us. It may be that the act itself is the chief or only symptom. I think close observation will generally disclose others, possibly not always. Yet, as we pronounce patients who have no delusions to be insane because of their outrageous and absurd doings, totally inconsistent with their ordinary character and habits, so we may recognize insanity in an act of violence equally inconsistent with the known character of the individual, or in one which none but a madman would commit. There are such cases, but I believe them to be few. Where these crimes are committed, I think close and skilled observation will generally link the act with other symptoms or causes of mental aberration, and as science and the study of brain-disorder advance, we shall recognize more and more the alliance of the different neuroses, the affinity—nay, close relationship—of epilepsy and insanity, chorea and insanity, hypochondria and insanity, and even drunkenness and insanity. The extent to which drunkenness can produce irresponsibility is a problem which jurists have never settled, and probably never will settle. Dipsomania may proceed from insanity caused by hereditary taint, or a variety of circumstances; or habitual drinking may itself produce dipsomania, or other forms of unsoundness of mind, to say nothing of *delirium tremens*. But all this is to be observed in view of each particular case. My present purpose is to consider the impulsive acts of the insane, especially of those whose insanity, not being marked by delusions, is chiefly indicated by the act itself. The act, however, is plainly the outcome of some idea present for the moment in the mind, but present, possibly, only for the moment, and then so obliterated that the individual afterward has lost all trace of it. As Dr. Maudsley says:[1] " It is no longer an idea the relations of which the mind can contemplate,

---

[1] Op. cit., p. 310.

but a violent impulse into which the mind is absorbed, and which irresistibly utters itself in action." This being done, the feeling and idea, having expended themselves in action, may cease for a time, till the morbid process is enacted over again in the brain. These speculations tally with what we observe in so many instances. The patient having committed the act he desired to commit, whether one of homicide or merely of violence and mischief, wakes up, so to speak; and whether horrified at what he has done, or satisfied at having given vent in action to the craving experienced, he at any rate feels the latter no longer. This irresistble desire to do something—to commit suicide or homicide, to smash windows, or merely to strike a blow at something or somebody —is quite a different thing from acting under a delusion, under a fear of coming harm, a fancied command from on high, or a causeless enmity. Yet that such feelings are felt is admitted, certainly by all who make insanity their study, and by others also who are most removed from any predilections for the theories of lunacy doctors. A writer in the "Saturday Review"[1] says: "*The law must recognize facts*, and many cases (of homicidal impulse) have occurred which can hardly be described by any other name." And Casper, the great Prussian medico-jurist, whom none will accuse of undue leniency toward alleged lunatics, says: "There are still other cases whose actual existence I am all the less inclined to deny, as I myself have had occasion to make similar observations. These pure cases, that is, those in which, without the individual having labored under any form whatever of insanity, or having been, from any bodily cause, suddenly and transitorily affected by mental disturbance—those cases, therefore, in which there coexisted with otherwise mental integrity an 'inexplicable something,' an instinctive desire to kill (Esquirol, Marc, Georget, etc.), are extremely rare, or rather there are extremely few of these cases published; for I am convinced that such pure cases actually occur far more frequently than their literary history would seem to show."[2] That such cases exist, and are not merely invented by doctors to excuse crime, is sufficiently proved, first by the observation of patients actually secluded or treated for this one form of insanity; secondly, by the confession of those who have suffered from an impulse, and have either controlled it, or have come voluntarily and begged to be restrained, feeling unable to control themselves longer. In fact, there can be no doubt about the existence of insanity marked by such impulses. The disputed question is, whether the insanity is not always recognizable by symptoms other than the impulsive act.

The impulse and craving may occur in any insane person, in those whose insanity is patent in many ways, and in those where it is hid from

---

[1] April 25, 1863.

[2] Casper's Forensic Medicine, iv., 334, Sydn. Soc. Trans.

the eyes of all who have hitherto seen the individual.   Yet I believe we shall generally be able to find evidence of mental disease if we only have full opportunities of observation, and a full history of the life and ante-cedents of the man and his ancestors.   I say, if we have full and sufficient opportunity of observation.   This may fail us, as it has failed many.   In that case we had better decline to give an opinion, as we should do in any ordinary medical or surgical case where opportunity of examining the patient thoroughly was denied.   The scandal which has come upon evidence given in doubtful cases of insanity has arisen from medical men giving their opinions after an amount of knowledge and examination of the patient which in no degree warranted any opinion at all.   Half an hour's conversation with a patient may tell us very little about him; it may be necessary to see him again and again; to see a woman at various periods of the month; above all, to observe a patient without his knowledge —in the night, at meals, in various occupations, and to see what he writes. if he can be got to write.   We shall have to consider his motive for the crime—if it be a crime—the method of its performance, the preparations for it, and his present feelings with regard to it; to ascertain, so far as we can, the presence or absence of hereditary taint, any illnesses or peculiari-ties observable, his history as regards former attacks of insanity, epilepsy, blows on the head, or drunkenness; to learn, either from personal inspec-tion or reliable evidence, his conduct and demeanor after the committal up to the time of our examination; and to compare all that we see with what we hear.

When patients are in asylums, and there are ample opportunities of watching them, there is seldom any difficulty in recognizing insanity. Dr. Gray, of Utica, gives the particulars of no less than fifty-two homicidal cases.[1]   In no one of these was there any doubt about the insanity; and in Bethlehem formerly, and now at Broadmoor, the medical officers can distin-guish the insane from those who are sane, though acquitted on the ground of insanity.   I am not, however, disposed to think that it is an easy thing to diagnose insanity by merely visiting a prisoner awaiting trial in one of our prisons, and seeing him perhaps on one occasion only.   If the ex-amination were conducted with the care and consideration displayed in French cases, there would be less violent writing and dissatisfaction expressed in our journals when a murderer is acquitted on the ground of insanity, or when the sentence of the jury is reversed in the office of the Home Secretary.

I think you will find, if you go to the root of the matter, that the act which is supposed to be committed under the influence of insane impulse is rarely, if ever, the first symptom of insanity or brain affection shown by the alleged lunatic.   You may be told by friends that they have never

---

[1] American Journal of Insanity, October, 1857

seen any insanity in him; but some people cannot see it in five out of six of the patients in an asylum. If you can get sufficient information, you will probably discover that he has had former attacks, from which he may or may not have been considered as recovered. Some may have thought him well, while others may have always looked upon him as "odd." At any rate, he will have had previous symptoms; or, possibly, he may have been noticed as being changed and peculiar for some time, short or long, prior to the committal of the act. The latter being the outcome and culmination of a morbid process which the mind has undergone, it would be extraordinary if it occurred quite suddenly in a moment of time. It is foreign to what we know of the pathology of disease generally, to suppose that such sudden disorder can arise. It is quite possible that symptoms may be disregarded; but careful inquiry will often lead to the discovery of a connected history of premonitory indications, even if the individual has never before been under restraint. It often happens that after a man is condemned to death this kind of inquiry is instituted; a history of insanity is revealed, the sentence is reversed, and scandal caused by the whole proceeding.

If a criminal has had at some time or other an attack, the present act may arise from a recurrence of insanity, or it may have been committed under the influence of a long-hidden delusion, a relic of the former attack, never lost, though kept under and concealed for years. I believe delusion to be common in these cases, more common than is suspected, and that many so-called impulsive acts are really those where delusions have been hidden, whose promptings the patient has obeyed. And no class of patients is so liable to act upon sudden impulse as those who have hallucinations of hearing. A man hears himself called some insulting name, or accused of some filthy act, and he turns round, and, deeming it to come from some person near him, violently assaults the nearest he sees. Here, however, he will justify the act and we get a clue to the real state of the case; but it may be thus discovered for the first time that he has had such hallucinations for years.

Besides the history we receive of the individual, of the occurrence of former insanity, epilepsy, strangeness, or alteration of character and habits, and what we learn by our own observation and interrogation, we shall have to take into consideration the character of the act, the mode of committal, and the absence or presence of motive. This is a test of a more uncertain character, but one which cannot be entirely overlooked. The act may be so motiveless, that no one can doubt it must have been that of a madman. When a man murders one known to have been most dear to him, we may suspect insanity, and more than suspect it. It is not the amount of wickedness displayed in the act, but the senselessness of it that we are to regard. As I said, speaking of moral insanity, that there was always evidence of intellectual defect and alteration as well as of mere wrong-doing, so in the impulsive acts of the insane there is not only a wickedness, but an eccentricity, a want of motive, or a motive palpably insane, which points to intellectual and ideational defect or alteration, and not merely to crime such as we recognize in the acts of a Greenacre or Courvoisier.

On examining the recorded examples of homicidal impulse—and these are the cases to which the theory of impulsive insanity is chiefly applied—we shall find that in almost all that are reported in such detail as to be worthy of notice, and many are not, there was, or had been, general mental derangement. Of the fifty-two cases reported by Dr. Gray, there was manifest insanity in all. I quote from Casper:[1] "Marc has collected eight cases of so-called homicidal mania. There is, however, not one among them in which general mental disease did not indubitably exist. Cazauvielh[2] has collected as many as four-and-twenty French cases, among which there are several cases of newly-delivered women who felt an impulse to murder their children. This, of course, was no permanent monomania, but rapidly passed off, and only one of these falls to be considered here as coming under this category. All the others, without exception, refer to lunatics." A celebrated case, reported by Dr. Lockhart Robertson in the "Journal of Mental Science,"[3] was that of a man who, when first admitted into an asylum, had been insane nine months, and heard voices. Such a patient I should consider an incurable lunatic, and if afterward he committed homicidal acts, I should set them down to the voices, or at any rate to his general condition. There would be no need to have recourse to a theory of impulsive insanity. And many cases so called are those of patients suffering from melancholia.

A good instance of what I have been saying is brought forward by Dr. Orange in his Presidential Address.[4] He mentions the case of a woman who murdered her child, and was tried by Lord Blackburn, who took on himself to tell the jury to acquit her on the ground of insanity, though it was clear that she knew right from wrong, and knew the character of the act. This murder was looked upon as one of sudden and uncontrollable impulse; but Dr. Orange, under whose care she has since been, tells us that the act was committed not from impulse, but delusion. She did not want to live, and she remembers that she thought it would be a right thing to kill the children before she killed herself. She lay awake all night thinking it over; and when her husband in the morning asked how she was, she said "Better," in order to induce him to go to his work and leave her, and then she committed the murder. So little of sudden impulse was there about her.

---

[1] Op. cit., p. 333, note.
[2] Annales d'Hygiène Publ., t. xvi., p. 121.
[3] April, 1861.
[4] Journ. Mental Science, xxix., 351.

A striking case of impulsive insanity is recorded by Dr. Maclaren.[1] The patient, a lady aged forty-three, was admitted into the Royal Edin-. burgh Asylum, laboring under great excitement, and bleeding from wounds in the mouth caused by attempts to swallow pieces of the broken glass of the cab window. She was not epileptic, had no delusions or hallucinations, and had only occasional and often no consciousness of the irresistible impulse to violence. In her the paroxysm was periodic, accompanied by always partial, frequently total, unconsciousness, and consequently followed by a similar state of forgetfulness of her acts; it was preceded by a sharp pain in the head, dizziness, and confusion of ideas. When free from excitement, she was devout and her memory unimpaired. Her state was, however, one of depression on account of her calamity, and separation from husband and children. While reading the Bible, or talking quietly to her attendant, she would without a moment's warning rush at the fire or attack the attendant and try to strangle her. There was no noise or shouting, but the eyes were fixed and suffused, the face flushed, the teeth clenched, and every muscle on the strain. She had had an insane ear, and after an attack there was a slight stutter and thickness of speech. The right pupil was more dilated than the left, and the tongue pointed to the right side, yet she was not epileptic, as might have been expected. This lady became by degrees less dangerous, and the impulsive attacks less intense. But her mind became more enfeebled, and she is now, after seven years, almost demented and quite incurable.

Dr. Claye Shaw has also contributed an interesting case[2] of a male patient, aged forty, admitted into the Leavesden Asylum. He could read and get his living, was active, had good memory, was not epileptic, and had no delusions or hallucinations of any kind. He confessed to an unconquerable feeling at times to " do something." He would then smash windows, being always very pale before the act, like a person in the first stage of an epileptic fit; after the act he became quiet, confessing his sorrow, and protesting his inability to prevent it. He frequently came on the access of an impulse and asked for a shower-bath, which for a time checked it. He would try and commit violence on himself, and this necessitated his constant incarceration.

Dr. Yellowlees has narrated[3] the case of a murder committed in a police-cell by a soldier, who was locked up when drunk with another drunken man. At 11.10 P.M. the latter was found dead on the floor covered with blood, the soldier sitting quietly by the fire with his arms folded. When questioned, he said he knew nothing about it; but his boots and clothes showed that he must have kicked the other man to death. He seemed not drunk, but dazed; but afterward appeared all right. He

---

[1] Med. Times and Gazette, Jan., 1876.
[2] St. Bartholomew's Hospital Reports, vol. xi.
[3] Journ. Mental Science, xxix. 382.

stated that the previous day he remembers drinking with some comrades, but nothing more till he found himself in a cell and his hands stained with blood. He did not remember coming from Hamilton to Glasgow, or being taken to the police-office. He had previously had attacks of "dizziness," and fits of violence; and on one occasion had flung himself over a bridge in Glasgow, of which he remembered nothing. This was probably a case of alcoholic epilepsy, followed by an attack of violence, arising from some hallucination or delusion.

This is followed [1] by another case, contributed by Dr. Richard Greene, which seems to deserve the name of homicidal impulse more than most of those to which it is usually applied. A lad of nineteen years, after some odd behavior, arose one night, not being able to sleep, went to the coal cellar and laid hold of the coal pick, returned with it to his mother's bedroom, and struck her three violent blows in the neck. He then "felt as though his brain was on fire," wandered away and slept in fields and barns for several days, and then came back and gave himself up. Yet this youth had conducted himself in a very odd manner for six months previously, so that his fellow-clerks had said that "something ought to be done" about him. All that he could tell of the occurrence was, that he tried his utmost to resist the impulse, but found it uncontrollable. He said that for a year or more he had suffered from almost constant headache, and referred the pain to the parts corresponding to the longitudinal fissure. He was acquitted on the ground of insanity.

The subject of moral and impulsive insanity brings me to the medico-legal question of the relation of insanity to crime, of the responsibility or irresponsibility of the insane, and the so-called legal test of insanity. The interpretation of the law as enunciated to juries by the judges of the present day is based on certain dicta of the bench of judges given after the trial of Macnaghten in 1843. He had murdered Mr. Drummond under a delusion, was tried, and acquitted on the ground of insanity. The matter was made a subject of debate in the House of Lords, and Lord Chancellor Lyndhurst stated that the law, as laid down by various judges, was, that if a man when he commits an offense is capable of distinguishing right from wrong, and is not under the influence of such a delusion as disables him from distinguishing that he is doing a wrong act, he is answerable to the law. Lord Brougham, with his clear perception, saw the difficulty of this interpretation. "One judge," he said, "lays down the law that a man is responsible if he is 'capable of knowing right from wrong;' another says, 'if he is capable of distinguishing good from evil;' another, 'capable of knowing what was proper;' another, 'what was wicked.' He was not sure that the public at large 'knew right from wrong,' though their Lordships knew that 'distinguishing right from wrong' meant a knowledge that the act a person was about to commit was punishable by law." The question was referred to the entire bench of fifteen judges, and they returned certain answers, of which this is the most important and the most debated:— " To establish a defense on the ground of insanity, it must be clearly proved that at the time of committing the act the accused was laboring under such a defect of reason from disease of the mind, as not to know the nature and quality of the act he was doing, or if he did know it, that he did not know he was doing what was wrong." Also, a delusion does not always excuse a crime:—" The judges are unanimous in opinion that if the delusion were only partial, the party accused was equally liable with a person of sane mind. If the accused killed another in self-defense, he would be entitled to an acquittal; but if committed for any supposed injury, he would then be liable to the punishment awarded by the laws to his crime." With this interpretation of the law some judges are content, but many, and among them some of great eminence, are not. Between the legal and the medical professions there has always been a conflict, and always will be, so long as this test remains. For what we say is, that an insane person may know right from wrong, may know that the act is unlawful and a wicked act, but may, through insanity, be totally unable to control himself, and may, either on account of a delusion or an insane impulse, commit a crime. Our test is not a knowledge of the nature or quality of an act, but the capability or incapability of abstaining from it.[1] If we examine the writings and charges of some of the judges of our own times, we shall see that this opinion is shared by a considerable number. And first I may mention Mr. Justice Stephen, who in the last edition of his work, " The History of the Criminal Law of England," writes thus:—" The proposition which I have to maintain and explain is that, if it is not, it ought to be the law of England, that no act is a crime if the person who does it is, at the time when it is done, prevented either by defective mental power, or by any disease affecting his mind, from controlling his own conduct, unless the absence of the power of control has been produced by his own default." The late Lord Chief Justice Cockburn, in a written communication made to the select committee appointed in 1874 to consider the amendment of the law of homicide, said, " I concur most cordially in the proposed alteration of the law, having been always strongly of opinion that, as the pathology of insanity abundantly establishes, there are forms of mental disease in which, though the patient is quite aware he is about to do wrong, the will becomes overpowered by the force of irresistible impulse; the power of self-control, when destroyed or suspended by mental disease, becomes, I think, an essential element of responsibility." Lord Blackburn, also, in his evidence given before the same committee, admits that there are exceptional cases in which an ac-

---

[1] Dr. Bucknill, Brit. Med. Journ., 15th March, 1884.

cused person ought to be acquitted on the ground of insanity, even if it is clear that he or she knew that the act was wrong, and knew the nature and quality thereof. Another eminent judge, Lord Chief Justice Bovill, at the trial of an American surgeon in 1872, is reported [1] to have told the jury that "if the evidence satisfied them that the prisoner at the time he committed the act, was not in a state to distinguish right from wrong, *and was not capable of controlling his actions*, then he would not be responsible for the act he committed." On the other hand we find judges in our own day insisting on the judges' test as strongly as ever. In the case of Gouldstone, who in 1883 was tried for the murder of his five children, the judge told the jury that "if the accused at the time he killed them knew the nature and quality of the act he was committing, and knew that he was doing wrong, he was guilty of murder;" and at the trial of Cole in the same year, another judge in passing sentence of death said, "although it was established in evidence that you had been suffering from delusions, I cannot entertain a doubt that on the occasion on which you violently caused the death of your child you knew you were doing wrong, and knew that you acted contrary to the law of this country, and that you did it under the influence of passion, which had got possession of your mind from want of sufficient control." Here, as Dr. Bucknill remarks, the judge infers not the innocence but the guilt of a man suffering from delusions, from not having exercised sufficient control over his passion. The natural result of charges like these is, that a second trial is necessary in every case where insanity exists, or is alleged to exist. This takes place not in public, but in private, being conducted by the Home Secretary, who orders certain experts to examine the prisoner. The sentences of these two men, Gouldstone and Cole, were promptly reversed by the doctors sent by the Home Office, and the public trial in court before a judge was absolutely nullified. That this system is full of evil no one can doubt. The Home Secretary does not order an inquiry, unless strongly moved thereunto by the press and public opinion. It is a matter very much of accident whether a poor man, unknown and unfriended, of whose antecedents nothing perhaps is known, can so reach the ear of the Secretary of State as to cause an investigation to be made. There is another evil. Inasmuch as the former trial before a jury is now looked on as only one stage of the proceedings, the plea of insanity is not put forward at all in some cases till the trial is over. This was notoriously the course pursued after the trials of Lefroy and Lamson. The plea of insanity was not put forward in court, but after conviction urgent attempts were made to obtain commutation of the sentence on this ground. And nothing more unscientific was ever brought forward than the evidence adduced. Lefroy was said to be suffering from homicidal mania, it being proved that the

---

murder was committed for the purpose of robbery, was accompanied by fraud, and followed by cunning hiding to escape detection. Lamson showed still greater plan and ingenuity to compass a death by which he was to reap pecuniary benefit. I saw a long rambling statement written by him, describing a morbid mental state produced, it was said, by morphia. If such had ever existed, it might have resulted in odd or even violent acts, done perhaps suddenly and on impulse, but how such a state of mind could lead to carefully planned and crafty scientific poisoning, I failed to see. More may be said on behalf of Guiteau, whose trial will always remain a *cause célèbre,* and as to whose insanity opinions will probably always differ. Those who thought him insane rested their belief, as did he himself, on a supposed delusion, that he thought the killing the President was an inspired act. I myself think Guiteau was responsible, and that his behavior at the trial was a simulation of insanity, but the arguments on either side are too long to reproduce here. The French authorities and those of other continental countries examine a criminal of whom insanity is alleged before trial, we examine him after. It is inconsistent with our maxims of law to examine a prisoner, or ask him any questions which may make him criminate himself. At the present time we must agree that substantial justice is done. It is long since an undoubted lunatic was hanged, and probably the practice of after-trial examination will prevail till something occurs to necessitate a change. The examination being conducted by medical experts, who examine into the insanity of the prisoner, and not into metaphysical questions of knowledge of right and wrong and the like, it matters little whether these views of the judges of 1843 are enunciated to juries or not.[1]

Although most of the homicides mentioned were committed under the influence of delusions, yet I must remind you once more before leaving the subject, that homicidal impulse does exist, as well as suicidal, or the impulse to merely smash or do some act of violence. Let us, however, not as witnesses in courts of law, but as physicians, carefully study the impulsive and unreasoning acts of people of unsound mind. So common are they amongst the insane that they attract no special attention when they occur in everyday asylum life. We do not call it impulsive insanity when a lunatic all day long tries to smash the windows, or tears his clothes and bedding to shreds, or incessantly endeavors to set himself and the house on fire. Yet, perchance, he can give no reason for any of these things. He has no delusion in connection therewith. He has very few delusions. He is in that state of partial insanity which would be unrecognizable by many; and yet he is in every sense of the word a lunatic, and his impulsive acts are the result of his general condition. Were every

---

[1] *Cf.* Dr. H. Kornfield, Handbuch der Gerichtlichen Médicin, Stuttgart, 1884; and Dr. E. Blanche, Des Homicides commis par les Aliénés, Paris, 1878.

impulsive act carefully recorded, we should see how numerous they are, and also see that the lunacy of those committing them is plain and undoubted. We should look upon them as peculiarities of lunatics, like dirty habits, shameless masturbation, or hallucinations and delusions.

Besides the various patients who are in one shape or other *insane*, who are changed and altered from their former sane condition, or who commit acts inconsistent with reason and healthy mind, your opinion will be sought in case of others, who though not insane in the ordinary sense of the word, are, nevertheless, of unsound mind, and incapable of taking care of themselves and their affairs. When an attempt is made to bring the lunacy laws to bear upon them, a quibble is always raised as to their not being *insane;* but the lunacy laws have for their subject all " persons of unsound mind," whatever the form of unsoundness may be, and this has been decided by the Court of Appeal. I shall mention two varieties of unsoundness, concerning which you may be consulted,—one a weakness or imbecility of mind, congenital, or the result of disease in early life, whence the individual is through life deficient, below the standard of other sound people, and incapable of taking care of himself, his condition being a destitution of powers that never were possessed; the other being an enfeeblement and decay of a once healthy mind coming on after insanity, epilepsy, and brain disease of all kinds, or being the dotage and decline of sheer old age. The latter is much easier to recognize and to deal with legally than the former, the subjects of which have given rise to some of our most celebrated forensic contests. They may come under your notice at any period of their lives, but, as a rule, it will be at the time when they are ceasing to be boys and girls, and beginning to be men and women. It is found that though men in years, they are still children in mind; if men, men only in wickedness and vice, children still in intellect and in the sense of duty and responsibility. These weak-minded youths are not to be called idiots, though they are but one grade higher. They are weak-minded imbeciles, and the imbecility may be congenital, or have been brought about by convulsions and fits in infancy or early life, arresting the due development of the brain. In a humble station these boys and girls swell the ranks of criminals, and become the constant inmates of a prison, unless they are fortunate enough to be carried off to the more permanent haven of an asylum. In higher society parents are horrified at finding them indulging in vices and propensities tending to the same end, and seek our advice and assistance. But it is not always easy to give them the latter, for it is often very difficult to deal with such patients legally. Testing them for the various symptoms of insanity, we shall find that there is very little to warrant our signing a certificate for their care and detention. Possibly they may be approaching the age of twenty-one, and the question will arise whether they are or are not fit to take care of property? Or they may be uncontrollable by any government, home or tutorial, and even at an early age are addicted to practices which entail the interference and punishment of the law. These children have no delusions; none need be looked for, for none will be found. They are not changed in habits and demeanor. They are now what they always have been—stubborn, eccentric, spiteful, mischievous, often horribly cruel, vain, perfectly devoid of truth, incapable of being taught, but picking up in a desultory way many scraps of information, and holding these with a most tenacious memory; given, perhaps, to some one amusement or hobby, and doing this for a time fairly well, but irregular and restless, fond of change and novelty, and wholly unable to settle down assiduously and constantly to one pursuit. The parents, if gentlefolks, seek the physician's assistance; but, as I have just said, this is often hard to afford. How are we to test these patients? Impressed with the symptoms of dementia, we try their memory, and find it excellent. We watch them at dinner and in society, and we find they conduct themselves with perfect propriety. We examine them as to the value of money, and they evince a keen appreciation of the amount of amusement to be got out of half-a-crown or half-a-sovereign. On the common topics of life they will converse readily and accurately, read the miscellaneous news of the journals, and recollect what they have read. And they may excuse their ignorance of other matters on the ground of their education having been neglected, or, if we ask them concerning their property and affairs, may say that these have always been managed for them, and that they have had no occasion to attend to them. In short, a formal examination of such people may tell us nothing. They are on their guard and good behavior, and if we tax them with their sins they confess them, allow it was wrong, and promise amendment. Only those who live with, and have opportunities of observing them at all times with and without their knowledge, can give a just and complete account of their mental state. If we ourselves have no such opportunity, we must receive the statement of those who have, and test it, so far as we can, by our own observation. Possibly we may not be able to pronounce an opinion. We must at any rate be careful how we give a negative opinion, based on our imperfect information, in opposition to that of others who have had ampler opportunities of coming to a sound conclusion.

The chief characteristics of these patients are a childishness in mind, showing itself in an inability to learn, think, or reason like others of the same age and social standing; frequently, but not always, a tendency to low and depraved habits, to vice of a kind not to be looked for at such an age, and an unnatural hatred and malice exhibited to parents, brothers, or sisters. Great stress has been laid on this moral depravity, and theories of moral insanity have been founded upon it. Without discussing the question of an innate moral sense, I would say that, in conjunction with the depravity, we shall, I feel confident, find in such patients a low and

16

imperfectly-developed intellect, incapable, because of its feebleness and childishness, of finding pleasure in anything but the brutish and sensual enjoyments of the body.   It may be able to lay up the facts of everyday life and experience—may know how much pleasure a shilling may buy, but of knowledge to be derived from reasoning, judgment, and reflection, it possesses none.   But it is most difficult to say this man is imbecile, and that one is not—to set up a standard, not of insanity, but of sufficiency of mind.   It was said by Sir Hugh Cairns in the Windham case that— "In a case of imbecility, where there is either no mind at all, or next to none, the task of coming to a right and just decision is comparatively easy. It is impossible for a man who is said to have only a limited amount of mind, or none at all, to assume at any moment or for any purpose a great- er amount of mind than he really possesses.   If the mind is not there, or only there in a certain small and limited quantity, no desire on the part of the individual to show a greater amount of mind, or to assume the ap- pearance of a greater amount of mind, can supply him with that which nature has denied him.   Hence, when a man is charged with imbecility, if it can be shown that for a considerable time, and in various situations he has acted like a natural being, any acts of folly which might be alleged against him should be carefully, deliberately, and keenly investigated, because at first sight it is next to impossible that a man can at certain times assume a mind and intelligence which are wholly absent."   These remarks show the difficulty we have to pronounce upon the absence of mind—in fact, to prove a negative—and not the entire absence, but the absence of a particular amount and measure of mind.   As Sir Hugh Cairns suggested, a man's acts of sanity must be weighed against his acts of folly. In all these cases acts are of far more importance than words; for there being no delusions, and the ideas of these weak and uneducated persons being but scanty, we are not likely to detect much that is erroneous or extraordinary in what they say.   By what they do, however, or would like to do, they betray their imbecility and incapacity for taking care of themselves.   They are not uncommon, and you will not be long in prac- tice without meeting with some examples.   The particulars of one or two I have already recorded in the "Journal of Mental Science,"[1] and will briefly give here.   One was a youth, well born, with every advantage of education which wealth could give.   When I first knew him he was be- tween fifteen and sixteen years of age.   As a child he was looked upon as weak-minded, and though he had been at various schools and tutors, edu- cation had stood still, and his handwriting, spelling, and letter-writing, would have been bad for a boy of eight.   When I first saw him he was living with a man who was to him virtually an attendant, whom he hated no less than feared.   Thence he was sent to a farmer's to learn farming,

---

[1] April, 1869.

but one evening he assaulted the maid, took out a horse from the stable, rode off to the nearest town, and took up his quarters at a small public-house. Thence brought back to his first residence, he escaped to Brighton, pawning all he could carry off. He was placed in lodgings with various attendants, with each of whom he got on very well for the time during which he was on his good behavior; then he had an outburst of passion because he was not allowed to do as he liked, and would do nothing right till the attendant was changed. His tastes were low, his pleasures either depraved or childish; yet he was not utterly bad; he valued the good opinion of his father and mother and my own, but this he was constantly forfeiting, for which he was sorry, but, as he said, "he could not help it." Next he went to a medical man's house, and behaved for a month or two so well that the character he took with him was thought to be unjustly exaggerated. But he broke down and behaved outrageously, and then an attempt was made to place him in an asylum. However, the certifying medical gentlemen stated nothing in the certificates but acts of depravity, and the Commissioners in Lunacy refused to receive them, so he was released. Since then he twice enlisted in the army, but was bought out again by his friends. On the second occasion, however, he was in the regiment for some months, kept clear of scrapes, and had a good character from his sergeant, but he was looked upon as "not right." He then took to race-courses and set up as a betting man, and after this threatened his father's life, was brought before a magistrate and locked up. When last I heard of him, twenty years later, he was said to be driving a cab.

Now, of those who by constant intercourse with this youth had the best opportunity of rightly judging of his mental state, no one thought him of sound mind. And yet his unsoundness was not at once manifest, nor was it easy to reduce it to a short verbal description. Probably, on no one given day could any medical man have seen sufficient to enable him to sign a valid certificate, yet he was imbecile and childish. His attainments and mental calibre were those of a child of eight or nine; and although in certain strata of society education marks little, yet in the highest ranks an incapacity to receive even the elements of education is significant. He would repeat the same question over and over again like a demented patient. This, again, though not much in itself, is a common symptom of the loss of memory or attention which characterizes the feeble-minded. He displayed that love of change, that periodicity of outbreak and restlessness, so often met with. He could go on quietly and well for a certain time, but then he found a vent, either in passionate quarrelling, drinking, or riotous behavior of one kind or other. And this I verily believe he could not control or help. He had a good memory and a certain sharpness about details, which are not uncommon even among idiots. He knew the times of all the omnibuses on the road, and could give the

times of trains according to " Bradshaw " by the column.   He was sharp
enough in calculations of pence and shillings, but he would have been
perfectly incapable of taking care of an estate, or any large property.
His head was very small, his whole development of brain and mind in
defect.   He had no powers of reasoning, and he lived, in fact, the life of
an animal, only caring to gratify his appetites.

Such another was a youth who was possessed of some few hundred
pounds—or would have been in a few months, as he was approaching the
age of twenty-one.   This his friends wished to protect, and applied to the
Lords Justices for an order.[1]   He had run the same course as the former,
but instead of being sent to learn farming, he had been sent to sea.   He
knew, however, nothing about a ship, though he had twice sailed to Australia.   He had run away while out there, and ran away, in fact, wherever
he was.   He could tell me nothing about what he had seen, neither could
he tell the name of the street in London in which he was living.   He was
defective both in attention and in judgment.   He had no idea of doing
anything with his money when he got it, but thought he should set up a
dog-cart.   He seemed to be entirely ignorant of everything connected
with property, securities, and investments.   He was plainly unfit to have
the care of property, and the Lords Justices made the order accordingly.

One of these doubtful cases was that of a girl whose father's family
was saturated with insanity.   She had quarreled with her sisters, her
mother, and innumerable governesses.   She had fits of obstinacy, during
which she refused to do anything required of her.   She was peculiar,
would dress fantastically, would cut off the toes of her stockings, and do
other odd things which were not mere child's mischief.   Vain and con-
ceited, she spoke of her mother and sisters as poor unenlightened creatures, affected much knowledge, and pretended to read deep books; yet
she was intellectually deficient and backward for her age.   She spelt badly,
and could not be taught.   When walking with her mother she would sig-
nal to strange men in the street, and talk of her wish to get married.
She was such an extraordinary liar that it was a work of time to realize the
fact that all her stories were lies.   She went to Scotland and I lost sight
of her, but I have no doubt that her unsoundness of mind will become
more and more patent.

Your opinion will be asked with regard to the training of these imbecile children, as well as the restraint by legal means.   When all tutors and
governesses have been exhausted, the friends will seek medical aid, finding their own efforts more productive of harm than good.   Parents, as a
rule, have but little influence, neither are they in general judicious in
their conduct toward them.   You must recollect that such unsoundness
is for the most part inherited, and you will detect peculiarities in the

father or mother—great irritability, intemperance, or weakness of character. Fathers are often very harsh and severe to these children; and mothers, on the other hand, screen their faults, and so encourage vicious propensities. Those of you who see them from childhood may do much by counsel and advice to promote their welfare and improvement. You will rescue them from blows, imprisonment, and undue punishment—from the irritation of angry parents, and the indulgence of foolish ones: above all, from being handed over entirely to the mercy of servants and attendants. Sooner or later these boys and girls are found such a pest at home, that they are sent away, first of all, to school. But few schools can keep them; and then the boys go to a tutor, the girls to a family. Here everything depends on the character of the individual who controls them. It requires a high order of mind, together with unwearied assiduity and vigilance, to train with success these blighted waifs of humanity. Yet it may sometimes be done. The great problem is to find out something, some walk or occupation, for which the child is fitted. Many are capable of doing something. We are dealing with a defective mind, a mind incapable of following the pursuits of those in the same sphere of life—incapable of commanding a regiment or a ship—incapable of studying for a learned profession; but capable, it may be, of executing the mechanical work of doing such things as we see done at Earlswood; or pleased at being occupied about animals—horses, or dogs—and, under judicious and kindly surveillance, capable of a habit of self-control and regularity. Parents may shrink from having their son put to a trade; many would far rather shut up in an asylum sons and daughters who are likely to disgrace the family. But I need not use argument to prove that it is better to bring up boys or girls in a humble occupation, in which they may cultivate self-restraint and self-respect, than to apply restraint by force of law.

I now come to the second class, patients whose minds have fallen into decay from disease of the brain of some kind, or from old age. There is little difficulty in recognizing this condition. I only mention it because you are to recollect that they come under the provisions of the Lunacy Acts, just as maniacs or any other insane persons; and if sent away from home, they must be placed under certificates of lunacy, if they cannot be pronounced able to take care of themselves and their affairs. I have known the mental condition of these individuals disputed in a court of law under such circumstances as I have alluded to; but, as a rule, there is not much contention about it during their lives. Their defect is palpable, irremediable, and ever increases till death; but after that, a contest often arises over their wills, and we hear evidence to prove, on the one hand, that the testator was a drivelling dotard, on the other that he was like other people—so differently do witnesses regard the sayings and doings of other men.

The most constant defect met with amongst these patients is loss of memory, varying at different times and in degree. Chiefly, the individual forgets recent occurrences, retaining a vivid remembrance of the days of his youth. He may forget the names of those most near to him, the name of the place in which he is residing, and may be unable to give any accurate information respecting his business and affairs. Now, to be of disposing mind—to be capable of making a will—a man must be, as the lawyers say, "of sound mind, memory, and understanding." It is clear that memory is essential to sanity, so clear that I need not dwell upon it. But the degree of failure of memory may vary much. Some people have naturally bad memories; some have great difficulty in recollecting names; others forget dates; and you will have to consider, in view of each individual case, whether the failure of memory is to such an extent as to separate the patient from all who can be called sound and sane in mind, and render him palpably unable to take care of himself and his affairs. You will ask, could such a one shift for himself, take a lodging or house, come and go unattended, and pay his accounts? If he could not, he must be allowed to be unsound of mind. It is more difficult, however, to come to a conclusion in respect of a lady. For ladies frequently do not take care of themselves or their affairs at any time of their lives, nor do they pretend or claim to do so. Many could not who are yet of sound mind, and able to make a will. You must take this into account when you are examining ladies with a view to testing their mental strength or weakness. I lately examined a lady who had been in former years an inmate of an asylum, and who since then had lived as a boarder in a family. She met all questions as to her affairs by saying that she left all that to her man of business, and on common topics she talked well enough. There was no loss of memory, and though I do not suppose she could have lived entirely by herself, yet there was nothing to make me certify that she was legally of unsound mind. It was just a case for the guardianship of trustees, who already existed. Failure of memory you will look for in these cases of dementia, and will estimate it according to its gravity; and you will often find yourself assisted by other symptoms, such as dirty habits and tricks, wetting or fouling clothes and bedding, or conversation devoid of delicacy or decency.

Two gentlemen I saw in this condition, who up to the time of inquiries made by the Commissioners in Lunacy, had been living under care and guardianship away from their friends, who, nevertheless, did not consider them insane, because they had no delusions

One of them was in an advanced stage of dementia. His memory was gone. He did not know the name of the proprietor of the house, nor that of his daughter, nor, in fact, any name but his own. He did not recollect how long he had been there, whether months or years. By night he had forgotten that he had seen his daughter in the morning, though her visits were of rare occurrence. He kept repeating over and over again the same sentence, without reason and without being addressed. His habits and person were filthy beyond description, as was the miserable room where I found him lying on bedding that was literally a dung-heap, yet in which he remained voluntarily and contentedly. This gentleman's condition was that of dementia following on hemiplegic attacks. He was not hemiplegic, however, when I saw him, but, with slow and shuffling step, could walk some miles in a day. The state of his person sufficiently indicated that of his mind, which was altogether deficient and gone; yet lawyers were found to argue that, because he had no delusions, he was not "insane."

The other had been insane, and frequently an inmate of asylums. His insanity was, as I understood, the result of drink, and it had terminated in dementia. His memory was gone: he did not recollect that he had placed all his affairs in the hands of trustees; but told me that he had a balance at his banker's, and that he drew cheques, which his servant got cashed, when he had done nothing of the kind for years. He, too, would repeat the same sentence and ask the same question over and over again; and was dirty in habits, though not neglected like the former patient.

Now, you will have no difficulty in appreciating the condition of such people. They cannot conceal their defects, especially this great loss of memory; but the opinion you are to give concerning them may have reference to one of several things. You may be consulted as to the power of such a patient to make a will or execute a legal instrument, as the sale of an estate, and you may be requested to act as one of the witnesses in such a matter. You may be asked to sign a certificate of lunacy, or give an affidavit and evidence for a commission in lunacy; or you may have to advise as to the chances of amelioration or recovery. And to take the last of these, we may look as a rule upon chronic dementia as hopeless and incurable. But we must inquire into the history; for it occasionally happens that what at first sight appears very like incurable dementia passes off in course of time. I am not now alluding to what I have already described as "stupor with dementia," which is a variety of acute insanity; but we sometimes find that after an apoplectic or epileptic attack, or even after an acute disease such as fever, there is for weeks or months' great weakness of intellect, with loss of memory and complete inability to transact business, and yet the patient may perfectly recover and be himself again. Time, therefore, is our great guide in prognosis here, as in other mental affections: where the dementia is beyond question chronic, it is not likely that the individual will again be able to take care of himself and his affairs. There may be also great improvement in the mental and bodily condition of a demented patient, even when recovery is out of the question. Those who are much neglected, and are to be found occasionally in an abject state of filth and destitution in private houses, are susceptible of much amelioration if properly tended and fed. You will recollect that certificates will be required if they are not under the care of their relations, whether they are in an asylum or in a private house.   As regards their competency to make a will or transact business, you will, of course, carefully weigh the extent of the imbecility, and the importance of that which they are about to do.   You may allow a patient to sign a receipt for money, whom you might think unequal to transact any involved or lengthy business.   Let it be your rule, generally, not to sanction with your presence, or attesting signature, the execution of any document by a person whose mind is in any way affected, for you may find yourselves involved in troublesome legal contests.

LECTURE XVI.

Terminations of Insanity—Liability of Recurrence—Recovery often Imperfect—
How recognizable—Release of Dangerous Patients to be refused—Concealed
Insanity—A Trial to be advised—Recurring Insanity—Lucid Intervals—Re-
coveries numerous—Chance of Life—Causes of Death—Diagnosis of Bodily
Disease—Care of the Chronic Insane.

I PROPOSE in the present lecture to consider briefly the terminations
of an attack' of insanity; for you will be called upon in practice to pronounce an opinion upon various points in the after-history of one who
has at any time so suffered, whether he has recovered or not.   He may
recover from an acute attack, only to continue in a state of chronic insanity.   In this condition is his life likely to be of long or short duration?
If he recover altogether, is he liable to a recurrence, and what is his
chance in a second or third attack?   If he were to insure his life after
having recovered from one attack, what is the value of his life?   To answer
these questions is no easy matter.   We must have recourse to those
asylums whose numbers are large, otherwise our deductions must be
formed upon very insufficient data; but it is very difficult to follow the
fortunes of all those who are discharged from a large asylum, and to speak
with accuracy of the subsequent history of their life and death.   The late
Dr. Thurnam, however, while at the York Retreat—the asylum belonging to the Society of Friends—had singular facilities for tracing the subsequent history of the patients discharged thence; and, among many most
interesting tables of statistics, he gives one, which I will quote:—

" Table showing the history of two hundred and forty-four persons who
died at, or after discharge from, the York Retreat, from 1796 to 1840,
with the number who died during, and after recovery from, the first or
subsequent attack of mental disorder."   .

Cases followed through life.		Died Insane during the First Attack.	Recovered from the First Attack.				
			Total.	Recovery Permanent. Died Sane.	Had subsequent Attacks.		
					Died Sane.	Died Insane.	Total.
Males,	113	55	58	21	6	31	37
Females,	131	58	73	24	14	35	49
Total,	244	113	131	45	20	66	86

Now although, as Dr. Thurnam says, certain deductions must be made from the picture which this table exhibits, it must still be allowed to be a melancholy one. 244 persons of the middle ranks of life, not poor and destitute, but well-to-do people, as the Friends generally are, become insane, and of these only 131, or 53.6 per cent., recover from the first attack; the rest never recover, and die insane. But looking at the after-history of the 131, we find that only 45, or 18.4 per cent. of the whole, remain permanently sane. The rest are again insane, once or oftener, and of these only 20 die sane. " In round numbers, then, of ten persons attacked by insanity, five recover, and five die, sooner or later, during the attack. Of the five who recover, not more than two remain well during the rest of their lives, the other three sustain subsequent attacks, during which at least two of them die. But although the picture is thus an unfavorable one, it is very far from justifying the popular prejudice that insanity is virtually an incurable disease, and the view which it presents is much modified by the long intervals which often occur between the attacks, during which intervals of mental health (in many cases of from ten to twenty years' duration) the individual has lived in all the enjoyments of social life." [1] Although the statistics derived from my own experience would be too scanty to be worth anything, they would, I believe, fully bear out the assertion of Dr. Thurnam. If we could carefully watch all such cases of insanity from their commencement, I fear we should see that a less number than 53 per cent. recover from the first attack, so great is the proportion of those who are incurable from the first, or who, from the prejudices of friends, are not subjected to treatment till the chance of cure is gone; and if by dint of proper treatment the above percentage recover, they only recover again to become insane in a large proportion. Although it would not only be uncharitable, but unscientific and at variance with facts, to look on all who have once been insane as lunatics for the rest of their lives, it must yet be confessed that popular prejudice receives considerable support from these statistics, and men may look with reasonable suspicion on former inmates of an asylum, when they hear that of those said to have recovered, only two out of five remain permanently well. We may fairly say, that when a man or woman has once been insane, no one can tell when he or she may not again become so. The changes and chances of life are not to be guarded against. With the utmost caution a former patient may be suddenly exposed to the shock of some horrible sight or accident, to the loss of one most dear, or to reverses of fortune. And, therefore, if you are consulted about the propriety of such a person contracting marriage, or entering into a partnership or any

---

[1] It must be borne in mind that the insanity in these cases was such as required asylum treatment; and thus a number of cases of the most curable kinds of insanity are omitted from the calculation.

engagement whatever, recollect that he is exposed to extra risk on account of what he has already gone through, and that his previous recovery does not ensure his subsequent immunity or subsequent recovery from future attacks. I have said already that a woman who has at any time been insane ought to be preserved from the peril of childbirth for her own sake, to say nothing of the danger incurred by her children. A man is, of course, exposed to less personal risk by marriage; probably to him the married state is rather an advantage than otherwise. He is thereby induced to lead a regular life, and has at hand a constant companion and nurse, who is aware if his nights are sleepless, or if he has peculiar habits or ideas, such as often escape for a length of time the observation of friends or more distant relatives. But we are not to forget that the man may become insane again, nay, will most likely be so, that his wife will have all the anxiety, and be exposed to the dangers consequent upon such an event, and that children may inherit allied maladies, or the disorder itself in its many forms.

When we closely examine the state of those said to have recovered, we may find that the recovery is sufficient perhaps to warrant their being discharged from the asylum, and to be called cured by their friends, but that ever afterward they are odd and eccentric, or easily upset by the merest trifles, or they periodically break out into violence or an acute state, which ought rather to be called an exacerbation of their habitual condition than a fresh attack of insanity. I had formerly under my care on various occasions an old farmer whose first attack was, I believe, at the age of seventeen, his last was when he was upwards of eighty, and in the interval he had been in an asylum nearly thirty times. He used to stay some months, his excitement then passed away, and he returned home to be his own master. He was discharged in the books as "recovered," but he had not really regained sound mind—he had only recovered from an acute attack of excitement; he did not even lose his delusions, for he had, if I am not mistaken, a persistent delusion throughout his whole life that he was married to a noble lady. Dr. Sankey, in his Lectures, lays great stress on these half-recovered cases. He believes that recovery from the first attack is not so common as might be thought from statistics; "that, therefore, what appears to be, and is usually called, a second attack, is no such thing. I believe that there is a remnant of the old disease, a smouldering of the morbid processes still left in these cases, though often very difficult of detection. . . . . This under-current of disease is, as I have said, more marked or less marked. In those cases in which it is obvious, and constantly so, the patient would be simply called a chronic lunatic, or he would have perhaps that form of chronic insanity to which the title *folie circulaire* has been given; but in the class in which the mental symptoms are exaggerated at distant periods, and a great degree of intellectual integrity remains in the interval, the disease would be called by a host of names, according to the different views of different authors. For my own part, I would include all these cases under the one term of *recurrent mania, or recurrent insanity.*" [1]

Dr. Sankey goes on further to say, that he has examined the reports of a great many persons accused of acts of violence, and he found in every case that the violent deed was not the *first insane act* of the lunatic. When we hear of cases of impulsive insanity, and it is stated that no insane symptoms had ever been observed before the commission of the act, he is of opinion that such statements emanate from those who are not capable of making a correct diagnosis, or who ignore the fact that there has been a former attack of insanity, or suppose perfect recovery to have taken place, and the subsequent attack to be altogether a fresh and distinct event.

How are you to know when a patient is recovered? We have the same difficulty in deciding this as is so often experienced in determining, in the first instance, whether a man is or is not insane. We find patients' friends, lawyers, and others not versed in the study of mental diseases, contending that a man is cured when the chief symptoms of acute insanity have abated, and he can talk rationally on some points—when, in short, they are unable to see insanity plainly depicted in his words and actions. And then in this semi-recovered condition they demand his release. He may have only got rid of half his delusions, or have learned to conceal them, or lost them, and yet be in a weak and unstable nervous state, requiring repose and a considerable period of convalescence; yet the demand for his release may be loud and persistent, and in withstanding it, you will meet with many difficulties. More especially will this be the case with private patients. The friends of paupers care less for their release. In public asylums they are carefully kept at the expense of the county, but the friends of private patients, thinking that those who have charge of them have an interest in their detention, set up their own opinion concerning the question of recovery in opposition to the interested, as they suppose, advice of the medical attendant. If patients are ever discharged too early from public asylums, it is probably due to their overcrowded state. There can be no question that they are frequently released too soon from private establishments on account of the importunities of friends, and the unwillingness of proprietors to submit to the insinuations and misrepresentations of the latter. When there is brought against a medical man the accusation that he is detaining a sane man for the pecuniary advantage to be gained thereby, it requires considerable moral courage to withstand such pressure. Yet in many cases it is our duty to do so, and by dint of temperate arguments, and the assistance of collateral friends and advisers, to prevent the disastrous result which may follow the release of a half-cured patient.

[1] Lectures on Mental Diseases. p. 94.

Such a state of things is contemplated by the Legislature, and provided for. If a lunatic's friends determine to release him from an asylum, the medical attendant may, if he considers him dangerous, refuse to liberate him. In the Lunacy Act, 1845, 8 & 9 Vict. c. 100, § 75, we read:—" Be it enacted, that no patient shall be removed under any of the powers hereinbefore contained, from any licensed house or any hospital, if the physician, surgeon, or apothecary, by whom the same shall be kept, or who shall be the regular medical attendant thereof, shall by writing under his hand certify that in his opinion such patient is dangerous and unfit to be at large, together with the grounds on which such opinion is founded, unless the commissioners visiting such house, or the visitors of such house, shall, after such certificate shall have been produced to them, give their consent in writing that such patient shall be discharged or removed; provided that nothing herein contained shall prevent any patient from being transferred from any licensed house, or any hospital, to any other licensed house or any other hospital, or to any asylum; but in such case every such patient shall be placed under the control of an attendant belonging to the licensed house, hospital, or asylum to or from which he shall be about to be removed for the purpose of such removal, and shall remain under such control until such time as such removal shall be duly effected."

Here, then, you see that power is given to prevent the release of dangerous patients. We cannot, however, prevent their being transferred to another asylum; but when the friends apply to the commissioners for an " order of transfer," the latter always write to the medical attendant, and require from him a certificate that the patient is capable of being removed with safety; and without such certificate no " order of transfer " is granted. It often happens that removal to another asylum is of great service to a patient whom we cannot release: great soreness may have arisen between him and those who have had the control of him, and removal and change of scene and attendants may effect a cure which would not otherwise have come to pass.

But, to revert to our question, how are you to know when a patient is recovered, and may fitly be trusted with the management of himself and his affairs? As an alteration in the general bearing, demeanor, and habits of a man is the surest sign of mental disorder, so an alteration from the state in which we receive him as a patient is an indication of amelioration or recovery. But our difficulty in pronouncing an opinion as to perfect recovery is often great when we have never known the patient in his previous sane condition. You, who will become not asylum doctors, but family advisers, will first see a man sane, then insane; and if, at the termination of an attack of insanity, you are called to examine him, you may be able to say at once that he is, or is not, himself, and give most valuable assistance to the asylum doctor, who may erroneously suppose that he is not, or is, cured. Friends are so apt to be biassed, and near relatives are so often themselves crotchety and peculiar, that we hail with satisfaction the information to be gained from others. Friends are frequently over-eager to release the lunatic because of his displeasure, or over-fearful of setting him at liberty lest he relapse. Alteration in character and manner will be a test of recovery when the patient is greatly improved, and has got rid of delusions, when, in, fact, it might be very difficult to sign a certificate for him. With all this amendment, his manner may not be natural. He may be unduly depressed, excitable, or irritable. His friends will ascribe this to the detention, if they wish his release, and will tell you they are quite sure that it is thus produced, and that it will pass away when the cause is removed. But it is a fact within my experience, that we do not see this depression or excitement in those who are perfectly cured, and know that their stay in the asylum is only a question of weeks or days. Friends imagine that a patient cannot be aware that he is in an asylum without its having a prejudicial effect; but this is not so, for he will go on to perfect recovery in it in spite of their fears and remonstrances. Sometimes it happens that, when a patient is progressing favorably to recovery, and is not half but wholly cured, his friends are much more anxious for his release than he is himself, and in this case his wishes and opinion ought to be consulted rather than theirs.

We shall have to base our diagnosis of recovery upon what we hear from the patient himself concerning his illness, its cause, and symptoms, such as acts, delusions, or hallucinations; upon what we see of him, his forsaking or continuing eccentric habits, peculiarities of dress or demeanor; on what we hear of him from attendants and others, when he is out of our observation; and on what we are told by relatives, friends, or medical attendants. And then we shall have to decide whether a condition of apparent recovery is a genuine and perfect recovery, or merely an interval between attacks of recurrent insanity.

It is a bad sign when a patient will not allow that anything has been the matter with him, or insists that his condition has been caused by his friends shutting him up, ignoring all that occurred before he was shut up, and attributing evil motives to all concerned. A patient may assert wrongly that he has recovered, or that he never was ill at all; that his delusions, so called, were not delusions; and that his acts were justifiable. Of course, if his delusions are absurdities, and he holds them now, his state is not a matter of doubt; but a man who has recovered from delusions may be unwilling to allow that he has held any, and may explain them away, singling out the grain of truth that may be at the foundation of them, and justifying the whole by this. Now, it may be thought that a patient cannot be recovered who justifies previous delusions, even if he does not hold them; but much allowance must be made for individual temperament and character. Some men and women cannot bear to think that they have been insane, or have entertained insane fancies, or done insane acts, and they satisfy their consciences and salve their wounded pride by explaining away as much as they can. This we must often overlook, and must not too rigidly compel confession from patients, or too closely cross-examine them as to all the details of the past. As I have told you, delusions spring to such an extent out of the feeling of the moment, that a patient a month or two afterward, in an altered physical condition, cannot go back to the ideas he held in his former state, and may deny that he held them, or may justify them, because he is unable now to enter into a contemplation of another state of things. You must consider the whole manner in which he speaks of the past: if he is ashamed of himself, and would rather let the subject alone, and talks rather of the future, and of returning to work or home, and if all his talk of the present and future is healthy and hopeful, we must not be too particular in judging of the manner in which he speaks of the past. But if he is perpetually harping on the past, reviewing and discussing every detail, always complaining and threatening retaliation, law-suits, and the like—if he craves for liberty in order to set about such proceedings rather than to return to his usual avocations—we must look with suspicion on his condition, and advise further detention and surveillance, though possibly in a modified form. In coming to a decision on such cases there is no general rule to be laid down or observed. Experience, and the intuitive appreciation of insanity which experience gives, are the only guides to a right judgment.

Another patient may not deny that he has had delusions, and may not seek to explain away or justify them, but he will assure us that he holds them no longer. There may have been much discussion as to some one or two special delusions which remain after all acute symptoms have subsided, and our patient, grown cunning by experience, and gathering that so long as he holds these opinions he will be restrained and looked on as insane, suddenly gives them up, professes that he holds them no longer, and perhaps expresses an unnecessary degree of astonishment at his ever having held them at all. Yet he may hold them all the time, only denying them to regain his liberty. Here you must take into consideration the whole history of the case. A patient who, during a somewhat acute state of insanity, entertains various delusions, will probably lose them as he passes into a quiet convalescent state. There will be marked improvement in his whole condition; sleep will return, regular habits of eating, attention to cleanliness, fondness for ordinary occupations and amusements; and, in accordance with all this, we also expect that the delusions and fancies of the insane mind will pass away. But if a man tell you that he has lost his delusions, and yet you observe no change for the better in his habits and appearance—if he still dresses in an extraordinary way, and behaves outrageously—we cannot believe his assertions to be true. With the inconsistency of a lunatic, he may act a delusion at the very time that he denies that he entertains it. We must endeavor to discover whether he has really lost them, or whether he is merely making the assertion to deceive us. Possibly we may find that although he denies them to us, he will confess them to his relations and friends, to other patients, or the attendants. He may betray them in his letters. He will deny that he hears voices, yet we may overhear him talking when alone to imaginary people, and answering imaginary questions. We may notice ornaments about him illustrating delusions concerning imaginary rank and titles, or unfounded hostility toward wife or friends shows that he still entertains the former delusions concerning them.

Where a patient's insanity is displayed not so much in delusions as in acts, and partakes more of the nature of so-called moral insanity, it is not easy to say whether he is or is not recovered. For he is restrained and kept from acts of extravagance or vice, and we cannot therefore, be sure whether he would return to these or not. But from his general behavior, and by comparing his present with what we hear of his former mode of life—by observing whether there is anything absurd or bizarre in his ways or acts—we may arrive at a tolerably accurate diagnosis. We shall also take into consideration the way in which he justifies his former acts, for this he may do in a manner highly indicative of insanity. I had rather hear a patient deny than justify a very insane act. In the confusion of his brain he may have almost forgotten it; or he may prefer to deny it altogether, pleading "not guilty;" but I have known a lunatic justify acts that none but a lunatic would have perpetrated, and none but a lunatic would defend.

We may have our doubts as to a patient's recovery, may disbelieve his statements, and think his friends too sanguine, but may hesitate about detaining him longer in an asylum. There comes a period in the history of almost every convalescing patient at which change is necessary, when, if he be further restrained in the same place and in the same fashion, he is likely to go back rather than forward. Although we do not consider him fit to be restored at once to full and unrestricted control over himself and his affairs, we wish to test his recovery, to put him on his trial, and to give him change. The law provides for this emergency. The Commissioners in Lunacy, upon the receipt of a certificate of the patient's fitness, will grant "leave of absence" from the asylum, hospital, or house, for any reasonable length of time, provided the patient is removed "under proper control." The control may be that of relatives or friends, or an attendant, or medical man. This plan I advise you to adopt in every case in which you are not quite certain how the patient will go on, when the restraint is first removed. It is often an advantage that the individual should know that he is only away on trial and probation, and that he can be brought back at a moment's notice. It enables us to judge whether he is cured or not: many improve in a remarkable manner when thus sent away, and our forebodings are not realized; others show that their seeming recovery was not real, and may be brought back without trouble and delay; and their friends, seeing that the trial has been a failure, are more satisfied than they would have been had it not been granted. In many cases we can never, so long as a patient is subjected to the restraint of an asylum, ascertain his actual mental condition in the way that it is revealed by his being left comparatively to his own devices for two or three months.

I have already spoken [1] of a class of cases where the insanity is remittent or recurring, a period intervening in which the patient appears either quite or nearly recovered, or, at any rate, altogether different from what he is during the time of the attack. This recurring insanity is not uncommon, but is very unfavorable as regards prognosis, and very difficult to deal with when the periods of apparent sanity are of any duration; for the patient then demands his release, and may threaten us with the consequences of his detention. We may see a man apparently recover from an acute attack, and just as we think him well and able to go out into the world, without any reason or warning he breaks down, and the whole of the symptoms recur, and this may happen again and again during many years. I do not know that we have anything to warn us that a patient's insanity will be recurring, except that the recovery is usually very rapid. Rapid recoveries must always be looked on with suspicion. The slowest recoveries that I have ever seen in patients suffering from acute mania have been in those who have remained well ever since. If a patient recovers very rapidly, probably he will not remain well long; but we may not be able to detain him, and he will go out, break down, and be again admitted. Some are never well long enough to gain their release. They alternate, month about, between comparative sanity and most evident insanity, mania, or melancholia; and as time goes on their sanity will be less apparent, and the violent stage will alternate with a state of quiet and harmless imbecility, the mind wearing out, but the recurring disease being as potent as ever.

Now, it is important that we should endeavor to break through the habit of periodical attacks, and destroy the periodicity of disease. We cannot do it by medicine; but it may be done sometimes by change of scene. In most cases change of scene and surroundings will have some effect, will lengthen the period of sanity, or render less severe the attack; but instances have come under my notice where, by a judicious change or series of changes, the periodical attack was finally averted and the patient cured. This should always be tried where means are forthcoming, and where the attack is of a nature to allow of its being treated out of an asylum under proper control. It sometimes happens that patients will brook

---

[1] Page 176.

17

no control, and will have their entire liberty or nothing. If they are subject to paroxysms of sudden homicidal mania, it may be impossible to
allow of their leaving an asylum; but in many of these recurring cases of
mania or melancholia it is quite possible to try the effect of change of scene;
and I hold that no patient has been fairly tested till some such plan has
been tried.

A person who suffers from recurrent insanity is, above all others, such
as is described by Lord Coke as " a *lunatic* that hath sometimes his understanding and sometimes not, *aliquando gaudet lucidis intervallis*, and
therefore he is called *non compos mentis*, so long as he hath not understanding." The older lawyers contemplated the existence of what they
called lucid intervals in all lunatics, who during such lucid intervals were
held to be capable of entering into marriage, or contracts, or of making a
will. There was no legal difference between one lunatic or another as
regarded the probability of a lucid interval occurring; in fact, all that was
known in those days concerning lunacy was derived from the lawyers, the
medical profession being very little consulted in the matter. In the present day the doctrine of lucid intervals and of partial insanity has been
much upset by decisions, at any rate in civil courts. The existence of insanity, however slight, has been held to invalidate any civil act, and the
existence of a recurrent insanity, if thoroughly proved, might vitiate anything done in the lucid interval. At the same time, it is to be remembered that lunatics have been admitted as competent witnesses in courts
of law, and many lawyers would sanction the signature of a lunatic to a
deed, if it could be proved that he was at the time in a lucid interval,
and understood the nature of what he was doing. Signatures are constantly obtained, and the validity of them must depend upon the circumstances of the case—not upon any general principle—for unless a man is
pronounced insane by a commission *de lunatico inquirendo*, he is, *prima
facie*, supposed to be sane.

Although I have spoken in a gloomy strain of the subsequent history
and fortunes of recovered lunatics, it is not the less certain that recoveries do take place in great numbers, and that modern science tends to increase the number. If we take the records of an asylum, as Dr. Thurnam did, examining all the cases admitted, curable and incurable, the
percentage of cures will probably be about that which he gives. But if
we take curable cases only, it will be much higher,—nay, I venture to say
that if we were to examine curable cases only, and of these such as were
submitted to skilled treatment so soon as symptoms of mental derangement
were discovered, we should find that three-fourths, or even more, had recovered. Numbers of such patients never go into an asylum: their malady
is slight, and passes off, or a cure is effected without the necessity of removal from home, and so they do not swell the statistics of recorded recoveries. It stands to reason that the worst cases are sent to asylums.

It is a fact of experience, that many are not sent till they have reached the stage of incurability. And when we consider how many are sent thither afflicted with general paralysis, epilepsy, or congenital defect of mind, it is clear that any percentage of cures must be greatly affected thereby.

You must labor, then, in your position as medical advisers, to bring under treatment at as early a period as possible all who show any symptoms of mental disorder. Where you know of, or have reason to suspect, the presence of hereditary taint, it behoves you to watch narrowly for the earliest indications of evil, to ward them off by judicious treatment, medical and moral, and if this cannot be carried out without legal interference, to insist on its being at once had recourse to. The arrest of insanity in its very beginning is that which, above everything, should be studied by all medical men. The abolition of the restraint of chronic lunatics has brought undying fame to the name of Conolly, but asylums full of chronic lunatics are an *opprobrium medicorum*. Those who pass their lives in the management of them, in the invention of amusements, the planting of fields and gardens, and the feeding, tending, and cleansing of the patients, are apt to look upon all this as the end of their labors, and a favorable report from the visiting commissioners concerning the state of the house as the summit of their ambition. But he who could advance the cure of lunatics in an equal degree to that in which Conolly promoted their comfort and happiness, would win fame no less brilliant, and the gratitude of mankind throughout the ages.

Even now, in spite of relapses, recoveries are sufficiently numerous to repay us richly, and to form a satisfactory basis for scientific observation; and if a patient breaks down a second time, we may hope again to cure him. The old farmer of whom I spoke, after having been in an asylum some thirty times, died at last in his own house, among his own people; and on all these various occasions he had gone away so much better that he was called "recovered," though his mind was not in all respects sane and sound. And others I have known, who came again and again, though not so often, and finally died at home of general decay or ordinary disease. The ultimate fate of any one who has ever once been insane, is very grave, from one point of view—so grave as to make us dissuade others from intermarriage and such contracts; yet as regards the individual himself there is enough of hope to allow of our cheering him at the termination of an attack, and trusting to cure him should he have another. If, however, you are consulted as to his chance of life, you will undoubtedly give, as your opinion, that it is inferior to that of a sane person. A man, we will suppose, has recovered from an attack of insanity—his first attack. He is liable to a second—liable, therefore, to the various accidents which so often befall lunatics before they are placed under proper care and control. And the second attack, or the third, may be of a very acute nature, in which he may die. Then, being subject to attacks of insanity, he may squander his property, lose his business, and come to the condition of a pauper; and if recovery again takes place, he may recover only to undergo great privations, which may materially shorten life. If you are asked as to the probable duration of life of a chronic lunatic confined in an asylum, you may speak with great certainty in view of the particular case. Such patients are under constant medical care: they have everything that money can bring, and all that medical skill can do to promote health and ward off evil. Their diet and drink are regulated, as are the hours of sleep and exercise. If they are in good health, and the malady does not tend to wear them out by great excitement or depression, their lives may be in no way inferior to those of persons exposed to the accidents of every-day life. But each case must be judged apart. One could not say that the lives of all chronic cases were good. But in every asylum you will be shown some octogenarian inmates, who by their long sojourn prove that insanity of itself does not shorten life.

What is the mode and immediate cause of the death of the insane? Many die in the acute stage, and we often feel a difficulty in stating in our certificate the exact cause, whether after a post-mortem examination or not. We see a patient in an attack of acute mania which runs a rapid course. Sleep is absent, and in ten days or a fortnight he dies. He gets weaker and weaker, and at last collapse sets in, and the heart fails; profuse perspiration breaks out, and he gradually sinks from exhaustion of the heart's energy. On performing a post-mortem examination we find merely signs of great hyperæmia of the brain, or increased action, but do not perceive any actual lesion, or any trace of that which has caused the stoppage of the heart. The patient's strength and nerve-force have in fact been exhausted, and have never had the chance of being renewed. If we examine another case, of much longer duration, the same appearances may meet us. That which happens to one in a fortnight, may, in another, come about in two or three months. Sleep is not so completely wanting, the violence is not so great or so incessant, yet the waste is greater than the repair, and death follows; and on examination we are equally at a loss to give a definite reason for the termination of life. I have seen medical men who were unaccustomed to make post-mortem examinations of such patients, greatly surprised at finding so little after such severe disorder and rapid death; but the process by which the metamorphosis that goes on in our daily lives is arrested or terminated, sometimes leaves no marked signs for us to scrutinize after death. Acute mania, like the poison of the serpent or prussic acid, may kill and leave no trace. We know, by watching the patient's strength slowly ebb and fade, that exhaustion is the mode of death; but it is the disease which kills, as do typhus and cholera, and therefore it is vain to talk about there being no such disorder as acute amnia, or to say that lesions or marks of inflammation are always to be found, whether by the microscope or naked eye   Two patients are attacked with acute mania; in ten days one is well, the other is dead. The same thing may happen in less time in the case of two suffering from delirium tremens. Is it likely that anything like an appreciable organic lesion has existed in either of these cases which can have been perfectly removed by one long sleep? That there have been disturbances in the molecular constitution of the nerve-centres, we know; but we do not believe that these would be discernible, even if we could apply the microscope during life.

It may often be, then, that we shall have to describe patients as dying of acute mania or acute melancholia producing fatal exhaustion; and to this we may have to add that such exhaustion was accelerated by the impossibility of giving sufficient food. Many are brought to us who have been allowed to go so long without being fed, that all hope of sustaining life is past.

When, however, we survey the non-acute forms of insanity, or the patients who live for years in a chronic state, it appears that many, I might say most of them, die not of the insanity, but of diseases to which sane people are liable. According to the insanity, however, and the condition to which they are reduced by it, they are more or less liable to the attacks of other diseases. Patients suffering from chronic mania or monomania, who have a considerable amount of nervous energy and of intellect, albeit deranged, will live much longer, and withstand disease much better, than demented persons whose vital powers are at the lowest point. The demented are very prone to get fat; taking but little exercise, their whole system is feeble, and the heart and muscular tissue undergo retrograde metamorphosis. In this state their great foe is acute bronchitis; and this, in my experience, carries off the majority of them, and, indeed, of all chronic lunatics. These very fat patients seem especially its victims: the circulation becomes impeded, the heart cannot force the blood onward, they are choked with mucus, and die rapidly. I warn you to watch very closely the approach of this malady, if such patients are under your charge. The ordinary cough-medicines are of little use; but I have found the greatest benefit from the use of a steam-kettle in the room; it should be kept boiling night and day, and a good large jet of steam should constantly moisten the air. In the case of these patients, as with young children, and, in fact, in bronchitis generally, I believe this remedy to be of incalculable value. In my own experience—which has been only of the upper classes—I have not found phthisis at all prevalent amongst chronic lunatics. One gentleman died of acute mania after an attack of hæmoptysis, and in the lungs of two who died of general paralysis tubercles existed; but amongst 73 deaths in the last few years, only one was returned as caused by phthisis. In the public asylums the proportion of those who die of this disorder is large.

If you ever see much of the insane, you may have to form a diagnosis of bodily disorders occurring in patients suffering from a recent or chronic form of mental affection. Most difficult is it at times to ascertain if anything be the matter with such people, and, if anything, what the seat and nature of the ailment are. One class will simulate every kind of disorder, will complain of agonies, of obstruction in the bowels or urethra, inability to swallow, headache, or sickness—to say nothing of matters more palpably fanciful, as eruptions, broken bones, or paralyzed limbs. Another class will tell us nothing—nay, will strenuously deny serious illness —partly from a fear of medical interference and physic, partly because they are too demented or deluded to realize their true state, which possibly they attribute to supernatural or inevitable causes. In dealing with these we have need of an accurate knowledge of disease, and of patient and painstaking, investigation of every fact and every organ. We are prone to think that an individual, melancholic and hypochondriacal, may be narrating to us sufferings existing only in his or her hypochondriac fancy, as a reason for refusing food, and avoiding all exertion or occupation. But they may be real, and we should commit a grave error if we ignored their existence. Then a chronic case, a demented man, who tells us nothing, suddenly appears out of sorts, does not eat as he is wont, sits listless and dejected. We can extract no information, no complaint. Like an animal or child, he attracts attention only by his appearance and the alteration observable. We examine him, his pulse and tongue, the state of the urine, if it is possible to obtain some, the motions and temperature. If he is ordinarily a hearty feeder, we inquire as to his eating; has he lost appetite? If not, we do not think him very ill. There is no better test. But if he will, contrary to his usual custom, take no food; if he appears thirsty and will drink copiously, or if he rejects both food and drink, we try to discover what is amiss. Is he sick, has he diarrhœa or constipation, has he lung mischief? Very insidious is the latter. Great ravages may have been made in the lung without any cough or other symptom to draw attention to the uncomplaining sufferer. Loss of appetite in old-standing cases is perhaps the most valuable warning; but in those more recent, refusal of food is so common, that very close inquiry is necessary, and we may have to insist on its being taken in spite of the alleged indisposition. But here the latter will not be concealed, but put forward and dwelt upon, and it will generally be represented as most serious, causing great suffering and sense of illness. If we find that the tongue is perfectly clean, the temperature normal, the pulse quiet, and urine healthy, we shall with reason doubt the statement, and look upon it as having a purpose. Nevertheless, cases will often puzzle and cause us to hesitate, and we must never be content with anything short of a thorough examination of the patient. If there is frequent sickness, we must be sure there is no hernia. If there be little water passed, or a constant dribbling, we may find a distended bladder. And whenever we hear of an unexpected and unaccountable death, nothing but a post-mortem examination should satisfy us as to the cause thereof.

It may often fall to your lot to have to treat chronic cases of mania or dementia, patients who are in a state of unsoundness of mind, but are harmless, not requiring the restraint of an asylum; they demand, never-theless, a careful watching and nursing, being frequently in a state of second childhood, like children or even infants in uncleanliness, and utterly unfit to take care of themselves. Such patients are often found by the commissioners in private houses in a state of great neglect. They are perfectly manageable without the appliances of an asylum, but re-quire constant and watchful care. One difficulty you will have to encoun-ter is the keeping them clean. Their tendency is to sink into an apathetic state, in which they disharge their evacuations regardless of place or time, like wild animals or very young infants. In or out of an asylum they may cause trouble in this respect; but in or out of an asylum dirty habits may be much eradicated by careful attendants. Some attendants who have had no experience of such cases, are altogether amazed when told that they are to blame for wet or dirty beds or clothes. They think that it is a concomitant of the imbecile state, no more to be altered than the failing memory or the shuffling and feeble walk. You will find, however, that a patient must be very far gone indeed who cannot be taught the habit of relieving him-self at regular intervals, and in a proper receptacle. Even when a patient is unconscious, and sunk in the last stage of paralytic dementia, accidents, though they cannot be altogether avoided, may be made the exception in-stead of the rule. And I believe that any chronic demented patient, whose mind remains the same from year's end to year's end, may be taught to be cleanly. Some, who have a good deal of mind, whether recent or chronic cases, will be dirty willfully, to give trouble or annoyance. They must be dealt with very firmly, and forced to go to the closet or to get out of bed. Imbecile patients, who are dirty from sheer want of atten-tion, are at least as capable of being taught to be clean as a child of a twelve-month, or a dog or cat. In short, patients dirty by night or day imply careless or inefficient attendants. Do not listen to the excuse that it can-not be helped: change the attendant, or threaten to do so, and you will probably find that the habit is eradicated. Another circumstance, which is equally a test of the care of the attendant, is the presence or absence of bed-sores. No chronic lunatic should be kept in bed by day and night simply for infirmity, unless he is actually ill. He should be washed and dressed, and seated in an easy-chair, even if he is unable to walk about. By this method, by thorough cleansing, and by thickly powdering with oxide of zinc powder any part of the back which is likely to give way, bed-sores may be avoided in patients who linger on in an extreme stage of paralysis even for years. In such cases a wet bed cannot always be avoided; but proper precautions, and the establishment of systematic and regular times for micturition and defæcation, will reduce "accidents" to a mini-mum. More is to be done by these measures than by the use of urinals or other apparatus. Many will not suffer them to remain properly ad-justed, or cannot bear the pressure occasioned. Such appliances are costly, often out of order, and soon become very offensive.

LECTURE XVII.

General Remarks on Treatment—Importance of Early Treatment to be urged by Family Practitioner—Restraint to be advised when necessary—Objections of Friends to be met—Use of an Asylum—Attendants—Delusions, how to be met—Asylums' not necessary for all the Insane—On the Choice of an Asylum —Refusal of Food—Forcible Feeding by various Methods.

I HAVE a few remarks to make upon the general treatment and management of insane persons, which will occupy the present lecture.   In all probability but few of you will have to treat insanity as a specialty; the majority will meet and have to deal with it as it occurs in the course of the practice of a physician, surgeon, or general practitioner, and in this way you will see patients and their friends at an earlier period than those who practice more specially as lunacy doctors.   The friends shrink from calling for the latter's advice or assistance till every other means has been tried; upon you will devolve the responsibility of taking the earliest, often the most important, steps for the security and cure of the individual. And I assure you the friends will prove to you as great a source of difficulty as the patient himself.   They will refuse to believe that his mind is affected, and shut their ears and eyes to all they hear or see, insomuch that they will say that the disorder commenced quite suddenly, without any warning, on a particular day, when every one else has noticed its approach for months.   Now, in the earliest stages, insanity is a very curable disorder; but through the obstinacy of friends it happens over and over again that the curable stage is past and gone long before any remedial measures have been taken, and the patient is brought to us a confirmed and hopeless lunatic, requiring care not cure, to be shut up in restraint for the term of his natural life.   And often a patient in this stage is put in an asylum for the sake of avoiding trouble and expense, who might very well live outside, mixing under some sort of surveillance with a family, and with the world at large.   At a time when an asylum would have effected a cure, the friends would not hear of sending him thither; but when all hope of recovery is over, he is placed there because it is cheap and saves trouble.

Now, I hold that at the present day our method of dealing with the insane should be this: First, we should endeavor to ward off an attack of impending insanity, and this, I believe, may be done very frequently. If it is not done, if the storm breaks, and breaks with violence, so that the patient, together with those about him, is in danger, to an asylum he ought to be sent, where everything surrounding him is specially adapted to his wants. If he does not recover, but quiets down into a "partially insane" man, tranquil and orderly, yet requiring supervision, and unfit to be in his own home, he ought not to remain in an asylum, if he is capable of enjoying himself in a greater degree beyond its walls.

At the very commencement of symptoms threatening mental disorder, you will have the least difficulty in getting your advice followed. At this stage you may be able to advise and consult with the patient and with his friends at the same time, and you must, in forcible terms, lay down the necessity of change of scene, cessation of work, and attention to diet and medicines. At this time you have not to inculcate the necessity of resorting to legal measures. Either in his own house, or in a friend's, or on a tour with some member of his family, with or without an attendant, the patient may pass a period of rest and treatment, and you may reasonably hope that your advice as to all this will be followed, and if followed strictly, will be attended by recovery.

But if the patient gets worse instead of better—if he will take no advice, and submit to no treatment except on compulsion—if delusions show themselves, and become more and more formidable, it will be your duty to represent, not to the patient—for this is useless—but to his friends, the urgent necessity for legal restraint; that he may, in the first place, he kept in safety; in the second, be subjected to treatment with a view to cure. Here you will be met with every conceivable objection. Wives are afraid to take any step of the kind without the co-operation of the husband's relatives—husbands without those of the wife. They are afraid that the patient, even if he recovers, will never forgive the step you are urging. They would sooner wait a little longer, till something occurs that more loudly calls for legal interference. They are sure that if he is placed in an asylum it will drive him quite "mad," when he knows where he is, and sees the other patients. Now, you may assure such people that it is absolutely requisite that the individual shall be placed somewhere under legal restraint; and that if he is insane enough to require this, he himself will care little whether the place of restraint is an asylum or a private house. If he is indignant at being restrained, and clamorous for liberty, he will clamor as loudly in a private house as in an asylum, and probably make more determined attempts to get out, owing to facilities for escape being more numerous. If he is wildly maniacal, he will care little where he is; if profoundly melancholic, all places will be to him alike. And as regards the other patients, we find that each one is so wrapped in himself, in his own delusions and projects, in his own misery or his own greatness, that he little heeds the rest; and in the acute state, at any rate, their presence does him no harm, often the contrary. Later, perhaps, he may shrink from them, when mental health is returning, and then it may be advisable that he should be removed from such a scene

and placed among sane people.   It may be worth while to consider for a
moment the advantages gained by a patient who is placed in an asylum,
and the mode in which a cure is there effected.   There can be no question
that the perfection of treatment would be to place a patient in an asylum
where the other inmates were not insane, but sane people.   We should
then have all the advantages we have at present without any of the draw-
backs.   The patient would be surrounded by a fresh scene and fresh faces.
New subjects would be presented to his mind by ways of occupation and
amusement, to take the place of his morbid ideas; but this proceeding
cannot be carried out: we therefore fall back on asylums as they are.
Here the patient finds, above everything, rest and safety.   He is kept
from accident and suicide.   He is cut off from his friends and all with
whom his delusions are so often connected.   And I would urge you to
impress upon the friends the necessity of leaving a patient alone and
unvisited when he is first placed in an asylum.   To sever home-ties and
ideas is one of the main objects you have in placing him there; and if
friends, in mistaken kindness, visit him from day to day, he might as well
be at home.   All letters must, in the majority of cases, be interdicted,
at any rate at first, and the patient be told plainly and openly that he is
not well enough to carry on a correspondence, and that his letters, if
written, will not be sent.   Nothing irritates a man more than to be told
and his letters are sent, while he finds, by the absence of all replies, that
they are not; or concludes that if sent, they are disregarded by those to
whom they are addressed.   Patients are not to be treated entirely as
children, nor can they be satisfied with trifling excuses and evasions,
though they resemble children in that we are obliged to act for them, and
cannot consult with them.   In quiet, then, and forced inactivity, many a
man recovers in an asylum by rest alone without any very special treatment, so far as medicine is concerned.   His health may be tolerably good;
possibly he refuses to take medicine, and may not be in a condition in
which we care to force him to swallow it.   The struggle and ill-feeling
thence arising would more than counterbalance the probable good to be
gained from the physic, yet he may recover, and that in no long time.
He recovers simply because he has been kept in an asylum, or, as some
would say, because he has been subjected to moral treatment.   Doubtless,
you have all heard of the moral treatment of insanity.   But shutting a
man up in an asylum can hardly be called moral treatment.   It is simply
restraint, which may be highly beneficial, and even remedial, as it is a
means whereby the patient obtains rest and seclusion from all that is
harassing and vexing, but it is not what I understand by moral treatment.   For in old days men were placed in asylums, and then and there
confined in a restraint-chair or strait-waistcoat, by leg-locks and hand-
cuffs, and fed, washed, and dressed; and this, together with some purging
and blistering, constituted the treatment.   But we should hardly call this

moral treatment. By the latter, I mean that personal contact and influence of man over man, which the sane can exercise over the insane, and which we see so largely and beneficially exercised by those having the gift, whether superintendents, matrons, or attendants. There can be no proper treatment of an insane person without it, and, beyond all question, the recovery of many has been delayed or prevented by its absence. There are patients, however, who are not within its reach. A man or woman in a state of acute delirious mania is beyond moral treatment, and needs only that which is physical or medicinal. That is why it is of little importance whether we treat such in or out of an asylum, provided we can place them in a suitable apartment. But we may see another who will never get well out of an asylum. What do we notice here? A morbid and intense *philautia*, an extreme concentration of the whole thoughts and ideas on self and all that concerns self: whether the individual's feelings are those of self-satisfaction and elation or of depression, whether he thinks himself the greatest man in the world or the most miserable, he is constantly absorbed in the contemplation of self, and thinks the whole world has its attention directed to him. Now, when such a being is at home, he generally contrives to make himself the centre and focus of every one's regard; and if away from home; in a lodging or family, he may be able to do the same thing—nay, in the majority of cases, this cannot fail to be the case, for the arrangements of the household must more or less depend on the presence of such an inmate; but place him in an asylum of fifty patients, and he occupies at once merely the fiftieth part of the attention of those about him. He is given to understand that the establishment goes on just the same whether he is there or not, but that being there, he must conform to the rules, his going away depending to a considerable extent on his own efforts, and his observance of the precepts and advice which he receives. He is encouraged to follow the latter by the approval of those about him, whose approval he ought to value; he is dissuaded or even prevented from doing that which he ought not. He is indulged with a certain amount of liberty, according as he shows that he is fitted to enjoy it, with liberty to go beyond the premises, to visit places of amusement, to have money at his command, to choose his own recreation and occupation; and this liberty he forfeits if he abuses it, and strict surveillance and watching are exercised until he shows that he can control himself.

Now, in all this it is necessary that we have the co-operation of attendants. In an asylum such as I have mentioned attendants must be numerous, and for the purpose of judicious treatment an asylum should not be too small. A patient may keep one containing only half-a-dozen inmates in a continual turmoil, and his self-importance is only increased thereby; but merged among forty or fifty, he becomes at once a much smaller fraction of the whole. On the other hand, no asylum should be so large as to preclude that personal attention which constitutes real moral treatment. The day may come when ladies will devote themselves to the nursing of lunatics, as they now labor in general hospitals. But as matters stand at present, we have to control insane ladies and gentlemen by means of servants, and great difficulties thus arise. A good attendant is a treasure beyond price, but it is not in the power of every one, whatever the desire, to be a good attendant: *nascitur, non fit*. It requires a combination of patience, tact, and judgment, of boldness, firmness, and unvarying good temper, possessed by the few rather than the many. Yet we cannot cure patients without attendants. Male patients cannot, for obvious reasons, be attended in all cases by females, whether servants or ladies; and gentlemen cannot be procured to act the part of attendants, nor would they on many occasions be more fitting. It is incumbent on us, therefore, to select with the utmost care those to whose charge we are forced to commit the insane, to watch them with unceasing vigilance, to remove those who, by constitution and infirmity of temper, or weakness of health, are unfit for the arduous task, and to retain by ample pay and reward, by relaxation and indulgence, those we feel to be faithful servants. Were I writing a book about asylums, I might say more, but other topics demand attention.

By the moral control exercised personally by man over man, the patient's thoughts and feelings are to be directed from his morbid self-contemplation to that care and concern for others which is his normal state. Those about him will endeavor to supplant his delusions and insane thoughts by other ideas, subjects, and occupations. As the latter gain a foothold and predominance, the former fade and disappear. Direct controversy on the truth or falsehood of delusions does little. Toward the close of an attack of insanity, in the period of convalescence, a patient may now and then be convinced of the falsehood of one of his fancies by direct demonstration; but at the height of the disorder this cannot be done, and it is often unwise to attempt it. Controversy perpetually renewed only tends to fix and confirm the fancy, which often departs quickly if never alluded to.

Under this head of moral treatment must be considered the question of occupation, exercise, and amusement; for nothing is of greater importance, not only to the welfare of the chronic, but to the cure of recent cases. All three are in turn requisite and indispensable, though not all are equally required by the same individual. To one bodily exercise is a necessity. In sub-acute, restless, sleepless mania, protracted muscular work will bring sleep, and act as a sedative more efficacious than drugs. Hard exercise will distract another whose thoughts are fixed unceasingly on melancholy subjects. I have known a man dig all day in the garden —dig a pit and fill it up again if other occupation for his spade was not to be had—and profit thereby. In public asylums there are far more opportunities for giving the inmates hard bodily work than exist among private patients. It is very difficult to subject the latter, particularly ladies, to sufficient exercise. Many a lady would be the better, could she be made to do the hard day's work done by many in our public asylums; but beyond walking, it is next to impossible to provide any exercise for her. Gentlemen fare somewhat better: they can ride, play cricket, billiards, skittles, football; but play is not the same thing as regular work, and regular work and long-continued exercise are of more value than the short but severe labor of games. So with regard to mental exercise and occupation. There are many brains which require to lie fallow and do nothing; if they must be amused, we recommend a course of Pickwick, or such like fare, or backgammon, or bagatelle; but some patients require harder mental work. To distract their thoughts they need to fix their minds on a subject deep enough to engross attention, and employ them day after day, and week after week. Such are generally intellectual people, and their occupation must be intellectual. For them I have found no work so suitable as the study of new languages; it is intellectual without being emotional, and does not require a great number of books or much assistance. I have known ladies study Greek and Hebrew, to say nothing of German, Italian, and Spanish. There is no end of this occupation, and to a busy mind it is often very fascinating. But people differ; another may prefer some new fashion of embroidery or lace-work, and drawing and water-color painting should be encouraged in all who have the very slightest artistic leaning.

You will see recorded in books how, by various devices, delusions have been dispelled. A woman thought that frogs were in her inside. Her physician introduced some frogs into the nightstool: she believed that she had passed them, and was cured of her delusion. But by such a scheme you admit the truth of the delusion, and, by inference, of all other extraordinary fancies which may be alleged. A patient may say, "True, I have got rid of six frogs, but others are still left behind." You cannot then say the whole thing is an impossibility and an insane delusion. A patient of mine who hears voices and noises, once heard at night a knocking or ringing at the front door. Her nurse treated this at first as one of her delusions, but on its repetition discovered that a policeman had rung, owing to a window having been left open. The patient has ever since triumphantly quoted this as a proof that her so-called delusions are realities. Never be tempted by any present chance of success into admitting the truth of a delusion, or doing anything which by inference admits the same. Sooner or later, in the present or some subsequent attack of insanity, the patient will turn round and place you in a position of difficulty owing to your having made such a concession. And be not too anxious to prove the falsity of a delusion. Frequently the patient starts from some premises which cannot be absolutely disproved, and in logical argument may seem to have the best of it. Rather try to oust the idea by the substitution of others. It is astonishing how patients ignore proof and demonstration. The people they say are dead stand before them alive and well, yet they declare it is some one else. The partaking of food does not make them think the less that their throat is closed, or their inside completely gone.

The more acute the insanity, and the more variable and numerous the delusions, the more favorable is the prognosis. When there is considerable disorder of the bodily health, sleeplessness, disinclination to eat, emaciation, or constipation, we may hope that delusions will vanish as the health improves. If it improves, and there is not *pari passu* a corresponding improvement in the mental symptoms, the prognosis is favorable. If the health is completely restored, if the patient sleeps well, eats and drinks well, regains flesh and looks, and still delusions remain, our augury as to the final result will not be encouraging. Perhaps this is why patients get well after many years of melancholia. During all the time they remain in a depressed state, both of body and mind, and generally look thin and miserable, refusing food as much as possible, and being altogether out of health. The chronic cases in asylums who best preserve their health, and look fat and ruddy, and live the longest, are those whose delusions are not of deep depression, but of what we term mania or monomania.

I have already said that at a time when an asylum is necessary and offers the only chance of cure, the friends of a patient will often do anything rather than send him there, will go to any expense to avoid the stigma of the asylum, and run great risk of violating the law. But when the case has become chronic, and the patient is a harmless monomaniac or dement, they cast about to discover how he may most cheaply be kept for his natural life. An asylum offers great advantages in this respect—for there are asylums of all grades—and to an asylum he goes. The question of restraining chronic lunatics, whether private or pauper, in asylums, is one which is attracting, and will attract, attention more and more. The notion that all insane persons must dwell in them has arisen from various causes. For generations such people were looked upon, not as sick, but as a class apart from all others. They were handed over to be kept in houses, the proprietors of which were not medical men, but laymen, ignorant and uneducated. No one in those days thought the insane capable of mixing with sane members of society. In asylums they dwelt from year to year, a few walking beyond the premises, but none sleeping beyond, or going to any place of amusement like ordinary men. Now, from all asylums, patients are sent to the seaside, to the theatre, the picture galleries; each proprietor vies with his fellows in providing recreation and entertainment for his patients—in proving, in fact, how little they need the restraint of an asylum. There will always be a certain number who cannot be allowed so much liberty, who cannot be taken to the seaside, who cannot even walk beyond the bounds of the asylum grounds, whose life is one incessant struggle to escape by fraud or force, or execute, perchance, some insane project fraught with danger to themselves or others. Some there will be whose limited means procure for them greater luxury and enjoyment amongst the numerous boarders of an asylum than could be afforded were they placed alone in a private family. But there are many with ample means, patients who make the fortunes of asylum proprietors, whose lives would be infinitely happier did they live beyond asylum walls. I would refer you to what Dr. Maudsley has eloquently written on this subject. After mentioning various objections urged against their release by the advocates of the present state of things, he says: " Another objection to the liberation will be that the insane in private houses will not be so well cared for as they are, nor have any more comfort than they now have in well-conducted asylums. The quarter from which this objection is urged taints it with suspicion: I never heard it put forward but by those who are interested in the continuance of the present state of things. Those who make it, appear to fail entirely to appreciate the strength of the passion for liberty which there is in the human breast; and as I feel most earnestly that I should infinitely prefer a garret or a cellar for lodgings, with bread and water only for food, than to be clothed in purple and fine linen, and to fare sumptuously every day as a prisoner, I can well believe that all the comforts which an insane person has in his captivity are but a miserable compensation for his entire loss of liberty,—that they are petty things which weigh not at all against the mighty suffering of a life-long imprisonment."

How are you to know if a patient is capable of living beyond the walls of an asylum? The answer is simple; give him a trial: many unpromising cases I have known to benefit so much by the change that they would scarcely have been recognized. Few chronic lunatics are dangerous to others: these are easily known, and we should be slow to place in a private family any one who has ever committed a homicidal act, unless he is fully and perfectly recovered; suicidal patients require the protection of an asylum so long as any insanity remains, but there are scores of eccentric monomaniacs who are perfectly harmless, who only require surveillance and a limit to their supply of money, and can enjoy life thoroughly amidst the amusements of town or sports of the country, their eccentricities being greatly smoothed away by the constant society of educated ladies and gentlemen. As the last generation did away with the fetters and mechanical restraint used in asylums, so let the present release from the restraint of an asylum all those capable of enjoying a large amount of liberty and a freer atmosphere than that in which they now fret and chafe.

Yet, as Dr. Bucknill says, " There is another side to the question. If some insane person are kept in asylums who ought not to be there, certainly many others, perhaps as many others, are kept out of asylums who ought to be placed therein; and it is often at least as difficult to persuade the friends of a perverse, an intractable, or even a dangerous lunatic, or one who needs constant medical care, to place such a one in an asylum, as it is to prevail upon the friends of other lunatics who are harmless and docile to give them the indulgence, freedom, and happiness of domestic life. We are inclined to think that this difficulty is one of such magnitude and importance as to demand the interference of the Legislature; for if it exists¹ with regard to Chancery lunatics, who are under the immediate protection of the State, how much greater must be the evil with regard to those lunatics whose proper care and treatment are entirely dependent on the good intentions and right judgment of their friends. And the relatives of lunatics have, as a class, peculiarities which often render it a most difficult, and sometimes an impossible task to persuade and influence them to a right and rational discharge of their duties." [1]

If an asylum is inevitable, and thither the patient must go, the question will arise, How is a choice to be made? Various points must here be considered. Is the case likely to be of some duration? or is it acute and urgent, requiring immediate restraint? Is transit to an asylum likely to be difficult? Is the patient, when placed there, likely to be able to go beyond the premises? Is it desirable that he should be near, or at a distance from, his home and relatives? Are there circumstances about the case, such as sexual excitement, which make it essential that he or she should not come into contact with patients of the other sex? All these matters will guide us in the choice of an asylum. It may be important to have recourse to the nearest and most easy of access—to one within a cab or carriage drive. In many acute cases, cases especially of females, it may be most prejudicial to place the patient in contact with others of the opposite sex. In some cases a very small asylum, where the routine is domestic and home-like, is advantagous, but other patients may cause too much commotion in such a one, and may do better when merged in the community of a more populous institution. Much will, of course, turn upon the question of expense. As a rule, the cheaper the terms the larger is the asylum; but for some, a large and cheap asylum may act more beneficially, so far as cure is concerned, than a small one where the individual may be the object of even too much solicitude. Where there is no hope of cure and the case is chronic, the patient should be placed where he can have the greatest amount of occupation, amusement, and liberty compatible with his safekeeping on the one hand, and his peculiar tastes and idiosyncrasies on the other. It is too much the fashion to think that all asylums must be in the country. Green fields, though charming

---

¹ Psychological Medicine, p. 704.

at first to denizens of a town, are extremely monotonous, and many a patient would gladly exchange his country walks and muddy lanes for the shops of Regent Street or for Rotten Row.

I now must say a few words on certain aids or adjuncts, which are from time to time necessary either for the treatment of patients or the prognosis and diagnosis of the disorder. Certain patients refuse their food and require to be fed by force. Various authors advocate various plans of feeding, either through the nose, or through the mouth, with or without a tube passed down the œsophagus. Of these, I would say that no one is suited to all cases, that each has its merits and demerits; but that if I were compelled to choose one, and one only, I would select that of the ordinary stomach-tube passed into the œsophagus through the mouth. Many advocate feeding by a spoon through the mouth, the patient being held on his back by attendants, and his mouth forced open for the purpose. The objection to this plan, in my opinion, is the length of time it takes. We may be dealing with people who are extremely weak and prostrated, but who, nevertheless, resist violently; and if the struggle is prolonged for half an hour, the patient loses as much through fatigue as he gains by the food; yet when the struggle is not great, when food can be got down easily by this method, and when he swallows readily the food once placed in his mouth, there is no necessity for the stomach-pump, and he may be fed four or five times in the twenty-four hours. I feel bound to describe to you the various methods, that you may be prepared to adopt one or other as circumstances or opportunity indicate. Feeding may be accomplished by the mouth without the introduction of any tube. This plan is set out by Dr. S. W. D. Williams of the Sussex Asylum, in a paper in the "Journal of Mental Science," October 1864, and this is his plan of procedure:—

"With the aid of three attendants the patient is placed on his back on a mattress on the floor, and covered by bedclothes, being, as a *sine qua non*, in his night dress, as far as the armpits, the arm being free. The head rests on a well-filled bolster, an attendant kneels on each side on the bedclothes covering the patient, and thus easily but effectually secures the body. One hand is placed on the patient's wrist, and the other presses down the shoulder. By these means he is perfectly restrained in the least irksome way to both patient and attendant, and, which is of primary importance, but few if any bruises need be inflicted. Hold a person in any other part of his body, or by any other means, and he surely becomes covered, after a few operations, by a mass of bruises, which often leads to unpleasant recrimination and fancies on the part of friends and relatives, and tends to foster the prevailing ideas current among the many as to the management of institutions for the insane—ideas which it behoves every conscientious alienist physician to persistently endeavor to dissipate, if he would wish to hold any claims to philanthropy. The operator kneels at

18

the patient's head, and, if the patient is very restive, may steady his head with his knees, but this is seldom necessary. A third attendant takes his place at the operator's left elbow. It should be here ascertained that the patient's throat is quite free externally from any clothing. The next operation is to get the spoon into the patient's mouth: this, if the patient be a woman, is generally easily done by getting her to talk, and slipping it in when the mouth is opened to speak; this device failing, however, persistent but moderate pressure with the spoon against the teeth, aided, if necessary, by inserting a finger between the upper and lower gums behind the last molar, will soon effect our object. Of course, in putting a finger into the mouth, one must look out for being bitten; but if the spoon be firmly pressed against the teeth so as to slide between them immediately the masseters are relaxed, such an accident cannot readily occur. The best spoons to use are the small iron ones, to be found in most of our large asylums, with the handle straightened. This should be placed far enough into the mouth to command the tongue, care being taken not to excite the reflex action of the fauces. It should then be restrained by the thumb and index finger of the left hand, the palm and remaining fingers firmly grasping the chin and preventing any to-and-fro or lateral motion of the head. The third attendant now passes his right hand under the operator's engaged arm, and firmly closes the nostrils, leaving his other arm free for any emergency that may arise. The operator can now with his right hand pour the food into the patient's mouth, and, provided the spoon well commands the tongue, deglutition is easily and perforce obtained, even in the most obstinate cases; but the patients are really by this means so completely mastered, that the majority of them drink the food down easily, and often the spoon is not required at all, but the nostrils being closed, the lips may be separated and the food poured into the mouth without opening the teeth. Indeed, for the last three weeks of last January, I fed a young lady, in this way four times a day, although she was obstinately bent on refusing food. The most convenient instrument for containing the food, is a caoutchouc bottle or ball holding about half a pint, and having for a stopper a bone tube like the extremity of an enema tube. This bottle can easily be commanded in the hand, and the bone tube having been inserted into the hollow of the spoon as it is held between the teeth, after a little practice, by squeezing the bottle the quantity of fluid to be injected can be judged to a nicety, and the tube removed after each injection. Not more than half an ounce should be injected into the mouth at once, one good respiration being allowed between each mouthful. After an expiration there is a short pause before the next inspiration, and if this moment of rest be chosen for filling the mouth, there is but little likelihood of the larynx being irritated by particles getting into it and delaying the operation by causing a fit of coughing. By a careful compliance with these rules, and a little practice any one may administer in all ordinary cases at least a pint of liquid in from ten to fifteen minutes, without a possibility of any danger or harm accruing, which cannot be said of the various other modes in vogue."

This is, in other words, the ordinary attendants' mode of feeding a patient; place the patient on his back, support his head, hold it between the knees, force open the mouth and keep it open, usually with a "forcing stick," pour in a mouthful of liquid, and hold the nose till it is swallowed. It is a very simple plan, which succeeds well, as Dr. Williams says, "in all ordinary cases," especially where the patient is by it "completely mastered," and takes the food quietly when placed in his mouth, but it fails in extraordinary cases. If we meet with a very powerful patient, two attendants will never keep him quiet, and in the struggle his arms, to say nothing of body and legs, will present the mass of bruises Dr. Williams so rightly deprecates. Then as the mouth is held open all the time, the patient can eject by an expiration some, at any rate, of the fluid, and if he holds his breath as long as he can at each mouthful, the administration of a pint of liquid will occupy much more than fifteen minutes. The food being only placed in the fauces to be swallowed as respiration demands, some of it does always go the wrong way, and great is the coughing and choking produced thereby. Frequently more is wasted than taken in this way; nevertheless, it is well suited for many cases, and may always be tried before resorting to more instrumental means. Some use a funnel inserted behind the teeth; others a glass bottle with a valve controlled by a spring, on which the operator places his thumb, and by which he can let flow as much or as little as he likes. A bottle of this kind made by Coxeter, called Dr. Paley's feeding apparatus, has a flat metal mouthpiece which keeps down the tongue, and through which the fluid escapes. It is a most excellent contrivance, and by it many patients may be fed frequently and satisfactorily

Much the same plan has been recommended by Dr. Moxey, in the "Lancet," March, 1869, the only difference being that, instead of putting the food into the fauces through the mouth, he pours it through a funnel placed in one of the nares.

In the fourth volume of the "Journal of Mental Science," is an interesting paper by Dr. Harrington Tuke, who reviews the various methods of feeding, and gives the preference to an œsophageal tube introduced through the nose and reaching the stomach. Dr. Tuke speaks strongly against the plan of forcibly feeding by a spoon or funnel. "It is not only the violence that must accompany the administration of food in this manner that inclines me strongly to deprecate this mode of treatment, but I believe that it must sometimes be an exceedingly painful operation. The sensation of something going the wrong way is familiar to us all, and it appears to me that pouring soup into the pharynx of a screaming and violently resisting patient is very apt to induce spasms of the glottis, or even cause the passage of some of the fluid into the lungs.   I do not think that an exhausted patient could safely be submitted to such treatment." Another equally strong objection is argued by Dr. Tuke against this plan. "It involves the medical attendant in a sort of personal contest with his patient, which must engender feelings of hostility most detrimental to the exercise of moral influence.   The medical attendant, living on terms of intimacy with his patients, should never descend to the position of a rough nurse.   Feeding with a tube *secundum artem* is a painless surgical opera-· tion, which, if rapidly and skillfully done, will not give rise to the same feelings of degradation as I should imagine 'funnel' feeding must occasion."

Dr. Tuke has a great dread of the ordinary stomach pump tube, and prefers a small tube introduced into the stomach through the nose, if the patient will not open the mouth.

"The instruments I use for injecting food into the stomach are œsophageal tubes about seventeen inches in length, made of elastic gum like an ordinary catheter, and of various diameters from the size of No. 3 to a No. 6 urethral catheter.   One of these, if the patient will open his mouth, I pass down into the pharynx.   If there is resistance, and the mouth is obstinately closed, I send the tube best adapted to the size of the nostril, without any stylet, but well oiled, along the floor of the nasal passage, and so into the cavity of the stomach."

Various objections may be urged against this method, some of which Dr. Tuke admits and comments on.   First, the catheter strikes against the cervical vertebræ, and there remains fixed: to obviate this difficulty various ingenious but complicated contrivances have been invented by the French, with which I will not now trouble you.   "The remedy," says Dr. Tuke, "is simple.   Let the instrument be previously bent so as to give it a tendency to turn downward; and, at the moment it approaches the posterior nares, let the head of the patient be thrown back, so as to diminish the sharpness of the angle it must describe.   It is obvious that the operation should not be performed when the patient is in the supine position.

"The next problem, that of avoiding the entrance of the larynx or the opening of the fauces, is solved by bringing the patient's head forward and downward, which will send the point of the tube against the posterior wall of the pharynx, but to a practiced manipulator this will not be necessary, and this part of the operation will be as easily performed as the *tour de main* with which a good surgeon sends the sound below the arch of the pubes into the bladder.   The tube having thus far proceeded comes within the grasp of the constrictor muscles, and now glides down the œsophagus almost without aid from the operator.

"The next objection is the likelihood of the catheter entering the larynx, and the danger of the lungs thus receiving the fluid intended for the stomach.   This danger is common to tubes introduced either through the mouth or the nostril; perhaps in the latter case the smallness of the tube may render the accident more probable. A simple rule will prevent this mischance producing any serious result. The operator must never attach his injecting apparatus to the catheter before at least fourteen inches of the tube have been passed. If no violence has been used this will sufficiently indicate that its point has entered the cavity of the stomach."

I confess I think this argument fallacious. A tube might pass fourteen inches down the trachea, bronchus, and bronchial ramifications, if we reckon these inches from the exterior of the nose; and it is not easy to bend the head of a resisting patient backward and forward at our pleasure. It is also a disadvantage not to be able to feed a patient in the supine position.

Dr. Clouston feeds by the nose in ordinary cases, using a small funnel and six inches of india-rubber tubing from a baby's feeding bottle. If the patient blows the food out of the mouth, he uses a long red-rubber elastic tube. When this plan fails, he feeds by the mouth, opening the teeth by a suitable instrument, sometimes by two, one on each side, wrapped around with strong tape to protect the teeth. He passes an india-rubber tube of large size and feeds his patients lying on a bed or sofa.[1] I have known nose feeding have an excellent moral effect. One old lady who had no objection to be fed by the stomach-tube, gave in at once when fed by the nose, and has never refused her food since.

I have never seen any plan of feeding violent and refractory patients which equaled that of Dr. Henry Stevens, formerly Medical Superintendent of St. Luke's Hosital. Having had considerable experience of this method, I will here describe it, because by it many of the objections usually urged against the stomach-pump are removed.

Where a patient can be fed without extraordinary difficulty or exhaustion by Dr. Williams' method, I adopt it, and the sight of the stomach-pump apparatus laid out on a table at hand often produces a moral effect, and facilitates the operation, which is conducted by attendants in my presence. If they cannot easily succeed, I use the stomach-pump after the following fashion:—

In the first place, the patient is to be rendered incapable of sudden movements. Dr. Tuke says, "I have known one of the most expert surgeons in London pierce the thoracic aorta in consequence of the accidental movement of a patient while the tube of an ordinary stomach-pump was being passed down the œsophagus." Recollect that no grasping on the part of a number of attendants can hold a very powerful patient motionless, because they are not all the time acting together. The patient is to be placed in a wooden arm-chair, and his body, arms, and legs, are to be

---

[1] Clinical Lectures, p. 117.

swathed in sheets drawn through the arms and legs of the chair so as to render him immovable. By this means all sudden movements and consequent accidents are prevented; he cannot struggle, therefore there is no exhaustion, and bruising is prevented far more effectually than by Dr. Williams' method. When the patient is thus fastened, half of Dr. Tuke's objections disappear. If the teeth were firmly closed, they were slowly and gradually opened by Dr. Stevens by a silver-plated wedge, which is expanded by means of a screw; thus without the slightest violence or chance of breaking a tooth, the teeth were separated sufficiently to insert the hardwood gag, which was held by an attendant standing behind the patient. All chance of passing a tube into the glottis is obviated by using one of a size that will not enter it. Nothing is gained by using a very small tube; as we do not use the smallest-sized catheter to pass along an unstrictured urethra, so we need not use a very fine tube to pass down the œsophagus, through which a coiner bolts his bad half-crowns with perfect impunity. The tube, then, should be at least of a size that will not enter the larynx; it must be flexible to the end, and must not have the stiff wooden extremity which generally terminates the tubes in the ordinary stomach-pump cases. Passing it through the hole in the gag, we direct it, not straight at the vertebræ, but to the right, having previously oiled it. No force whatever is to be used. In all probability the patient will hold it with his tongue, preventing its descent; we are not to force it, but simply hold it steadily; in a few seconds he is obliged to take breath, he relaxes his hold, and the tube slides within the action of the muscles by which it is swallowed, and so passes into the stomach. No haste is to be used in pushing it down or drawing it back. We then affix our injecting apparatus, which may be an ordinary brass pump or an india-rubber bottle. If the tube is not too small the food need not be mere liquid drink, but may consist of a custard of milk and eggs, or a mess of finely-pounded meat and beef tea, thickened with potato or flour; it is important that there be an adequate quantity of farinaceous and vegetable material. Such things as brandy, wine, eggs, and medicines may be added at discretion, and the medicines may be mixed with the food in or out of the sight of the patient, according as we think fit. Frequently, when he finds that we can administer all we wish him to take, he gives in, and eats his food. The preparations for feeding often produce the same effect, but I have found that patients do not experience any great pain or inconvenience from this method, and sometimes will refuse their food merely to give trouble. Such persons often dislike being fed by the spoon method far more than by the stomach-pump, as it is a longer process. One patient would eat all the rest of his meals if I would give him his breakfast with the stomach-pump; otherwise he would take nothing all day. This he continued for a month. Another gentleman who had had considerable experience of feeding in various ways, fell to discussing the subject with the attendants one day after I had fed him, and stoutly maintained that the stomach-pump, used as described, was the least unpleasant of any. Only excessive violence can lead to such accidents as piercing the aorta. Even teeth can be got open without breaking, by tact and practice; and all danger of going the wrong way is at an end if the tube be of the proper size. Doubtless, some will talk about mechanical restraint, and so forth; to those I would say, compare a patient struggling for fifteen or thirty minutes in the hands of three or four attendants, with one fastened with sheets in a chair for five minutes. Let both be seen before judgment is passed. Of course, when a patient is weak and exhausted, or passive and unresisting, he may be fed on the bed without being raised to the upright position in a chair. In this case I assume that, without any great amount of force, he can be held on the bed by a few attendants. But when we have to deal with a very violent or very powerful man, who will struggle desperately with attendants whether he is in the recumbent posture or not, I think the method just described, of placing him in a chair and rendering him immovable by means of sheets, will be found by far the most efficacious.

It may occasionally happen that we are obliged to feed by force a patient who is not suffering from acute melancholia, but from some other variety of insanity—one who refuses food from sheer opposition, or because he thinks it poisoned. In such cases one or two operations generally work a cure; but in acute melancholia we feed because the patient's life is jeopardized by want of food; and, in spite of our feeding, such a one may sink, for his acutely melancholic state is often only the last stage of a melancholy which has been gradually reducing the strength of the individual for months, and which for want of vigorous treatment in the early stages has gone on to a point when cure is impossible. Nevertheless, we must not let a patient die of starvation, and as a long and exhausting struggle is not to be thought of, we must feed with the stomach-pump twice a day, or oftener.

LECTURE XVIII.

AFTER what has been said about forcible feeding, I may fitly draw your attention to what I consider an indispensable appliance in every asylum—viz., the weighing-chair. In the treatment of insanity this is frequently as useful as is the thermometer in acute diseases. And it is useful in both recent and chronic cases. There are but few patients who do not gain weight before they recover their reason. At St. Luke's Hospital in six years, 1852–57 inclusive, there were discharged 784 patients, these being, with very few exceptions, persons who had been admitted within a year. Of the whole number, there were only 104 who had lost weight, and of these 104, 43 were uncured. Some lose weight very rapidly if admitted at the commencement of an attack of acute delirium. A week or two of such an illness may reduce them so much that they do not recover their former weight while in the asylum, as the delirium often passes away, and sanity returns rapidly. But many, especially the melancholic class, have usually lost flesh before we see them, and as the restoration of the bodily strength is essential to recovery, the weighing-machine enables us to ascertain whether progress is being made or not. If the patient is fed by force, we learn whether the food we give is producing the effect we desire, and whether it need be increased. Many patients who take their food unwillingly, contrive in various ways to avoid or secrete it, if we are not present to see that it is swallowed. The loss of weight as revealed by the machine will often enlighten us on this point, and show us that the food ordered is not really taken. In cases of melancholia with a tendency to refuse food, we cannot afford to wait till the eye reveals the wasting of the body. We want a more accurate gauge of what is going on, just as we take the temperature by a thermometer and not by the touch alone. By weighing a patient weekly, we soon discover whether he is gaining or losing, and our feeding will be regulated accordingly. It may also enable us to detect the presence of other diseases in those who are unable to say what is the matter with them. A sudden loss of weight in a chronic patient should always arrest our attention and make us seek the cause. It not unfrequently happens that chronic demented patients lose weight in the winter, even if there is no falling off in the appetite. This is due to the cold weather, of which such persons are peculiarly susceptible, and indicates that they require an extra amount of warmth, warm clothing, bedding, and rooms.

It is an old remark that a large increase of weight without mental improvement is a bad sign, and betokens the approach of dementia. This is often true, and patients go on to stoutness, and even corpulence, the mind becoming more and more childish. We must not, however, despair, if at first increase of weight precedes mental restoration, nor must we on that account withhold food. Frequently there may be considerable delay before the mind begins to awaken, and the bodily health may have apparently advanced much further. Yet by degrees, as strength returns, the mental faculties and attention will take a fresh start, and once started will go on to recovery, and this cannot take place in an enfeebled and emaciated frame.

I must not omit to mention the ophthalmoscope, which has been used by certain physicians for the examination of the insane with a view to elucidate the pathological condition of the brain. The first, I believe, who in this country published the results of such examination was Dr. Clifford Allbutt, who observed the eyes of many patients in the asylums of Yorkshire, and has published the account of the changes noticed in the eyes of those afflicted with general paralysis, mania, epilepsy, idiocy, and dementia. Another observer is Dr. Aldridge, who, in the volumes of the West Riding Reports, has recorded the examinations made by him in epileptics, general paralytics, and in acute dementia, as well as in persons to whom various toxic agents had been administered. In epileptics he noticed during the stage of stupor succeeding the convulsions great paleness of the optic disk, showing emptiness of the capillaries, also smallness and attenuation of the retinal arteries, and lastly, a restoration of the circulation more or less complete. During the inter-paroxysmal period there was in the majority of instances a condition of hyperæmia of the retina and optic disk. In general paralysis the appearances indicated that the affection commences with inflammation and slight exudation, and ends in atrophy, the changes being at first a pink suffusion and hazy appearance, or a deep hazy red tint with slight swelling, while later the disks are white and atrophic. In acute dementia a state of retinal anæmia was observed. The optic disks are pale, but it is not the brilliant pallor of atrophy. Another distinguishing condition between this anæmic state of the disk and atrophy is to be found in the uniform grayish-white appearance of the disk, and the fact that one disk is in exactly the same condition as the other, and also that no partial anæmia of one optic disk can ever be seen.[1]

It will be noticed that these ophthalmic appearances tally with the pathology of the various disorders as usually received. In epilepsy there

[1] West Riding Asylum Reports, iv., 296.

is supposed to be a spasm and arrest of the circulation; in general paralysis the condition is one of slowly progressing inflammation; the pathological condition of the brain in acute dementia is one of anæmia.

Dr. Jehn compares the results of the observations of various physicians, Bouchut, Dubuc, Köstl, Allbutt, and Tebaldi, and shows that considerable differences exist in the results arrived at.[1] It is certain that ophthalmoscopic research is yet in its infancy, and it is to be hoped that by the multiplication of observers and observations something definite will ere long be attained.

Another instrument which has been employed by some is the sphygmograph. In Drs. Bucknill and Tuke's " Psychological Medicine "[2] are given sphygmographic tracings from papers by Dr. Wolff,[3] Dr. Hun,[4] and Dr. George Thompson.[5] The tracings were taken in cases of dementia, melancholia, mania (acute and chronic), general paralysis, and epilepsy. According to Dr. Hun, the pulse is nearly tricrotic in the early acute stage of insanity. When this passes away, and dementia and mental apathy succeed, the pulse becomes dicrotic, and at last monocrotic, as the result of paralysis of the sympathetic system. Dr. Thompson thinks that his experiments prove that there is a characteristic tracing to be found in general paralysis, the line of ascent being slanting and short, that of descent being gradual and prolonged, not displaying the usual aortic notch, but instead a number of wavelets, which, if counted carefully, will be found to have eight distinct rises and depressions. This he thinks is due to a loss of resiliency in the vessels, and a narrowing of the capillaries, and compares such a tracing with that taken from the wrist of a healthy individual who had been immersed in cold water for some length of time, and who was in consequence in a state of chill. Dr. Savage, however, has stated[6] that he had taken periodical pulse-tracings of twenty-five fresh cases in Bethlem, but had not been able to get any one which was characteristic of general paralysis. In such a progressive disorder, the tracings, if of any value, ought to vary according to the progress of it. But here, too, observations must be multiplied to a very large extent before we can come to any conclusions derived from this instrument.

Electricity has been employed to a considerable extent of late years, not for the diagnosis, but the treatment of insanity. In the year 1870 Dr. Arndt published the result of his practice with both the induced and continuous current.[7] In 1872 Dr. Clifford Allbutt gave, in the second

[1] Allgemeine Zeitschrift für Psychiatrie, xxx. Band, v. Heft.
[2] Page 309, etc.
[3] Allgemeine Zeitschrift für Psychiatrie, 1867–69.
[4] American Journal of Insanity, January 1870.
[5] West Riding Asylum Reports, vols. i. and ii.
[6] Journal of Mental Science, xxi., 149.
[7] Archiv für Psychiatrie, Band ii., Heft ii.

volume of the West Riding Asylum Reports, a series of cases treated with the continuous current, and the conclusion arrived at was that those in which marked improvement took place were patients suffering from acute primary dementia; those in which distinct improvement was noticed, but to a less degree, were cases of mania, atonic melancholia, and perhaps recent secondary dementia. Those in which no change was observed were cases of chronic dementia, and some of melancholia, while in hypochondriacal melancholia, and perhaps brain-wasting, the result was unfavorable. Electricity was also tried at the Sussex Asylum, and the results have been recorded by Dr. Newth.[1] The patients were fifteen in number, of which nine were cases of melancholia, one of acute mania, two of mania, one of dementia, one of progressive paralysis, and one of locomotor ataxy. His opinion is that in the cases where there seems to be a want of tone in the nervous system, the continuous current has in the majority of cases a most marked and beneficial effect. Dr. Newth believes that an almost unerring guide as to the probable result is the state of the pulse after a few applications; if this increases in force and slightly in frequency, there is a great chance of the treatment being successful. Dr. Beard, of New York, says[2] that electricity in any form acts as a stimulating tonic, with a powerful sedative influence, and an agent for improving nutrition in any condition, local or general, where improvement in nutrition is required. The result of his practice leads him to think that the first tentative applications should be very mild, the strength of the current and time of the sitting being increased as the patient proves himself able to bear it. And we must not look at the immediate result only, but must watch for the permanent effects that are observed after weeks or months of treatment. The latest researches on this subject are those of Drs. Löwenfeld and Tigges. The experiments of the former prove that the circulation within the cranium can be influenced by the application of the interrupted current. "It can," he says, "when transmitted through the moistened hand laid on the head, produce great relief in intense headaches;" and he quotes the opinions of others on the favorable results obtained in neuralgia and sleeplessness. Dr. Tigges observed that patients slept after the application of the constant current, and in cases of melancholy with stupor he found the pulse fuller and more frequent, the face pale, and the pupils more dilated and rarely contracted, after the constant as well as the interrupted current.[3] In my own experience, which is very limited, I have found it of most use in young persons suffering from primary dementia, *stupidité*, or stubborn mania. I am inclined to think, however, that in several of these cases the effect was moral rather

[1] Journal of Mental Science. xix., 79.

[2] Journal of Mental Science, xix., 355.

[3] Journal of Mental Science, xxix., 415.

than physical. In melancholia I have found it less beneficial. It frequently caused great terror, and had to be discontinued.

I must now say a few words concerning some of the drugs that are used in the treatment of insanity. Incidentally I have already mentioned most of them, but as great controversy exists about several, I wish to make one or two general observations. And first let me speak of chloral. There is such an extraordinary prejudice in the minds of some against this medicine at the present time, that one would think that dementia and even death resulted from the administration of a few doses. I can only give you my own experience, for to quote that of others for or against it would require a volume. My own experience leads me to think it one of the most valuable drugs we possess for the treatment of insanity generally. Of course the cases in which we give it must be selected. It is not to be supposed that it will affect every patient in an equal and like degree. Is there a drug in the pharmacopœia of which this can be said? But of all those which we give for the purpose of procuring sleep—and this I hold to be the great use of chloral—I am certain there is none which will bring it about with equal certainty or with less unpleasant consequences. And I will tell you very briefly the precautions which, in my opinion, should be observed in the administration thereof, by means of which you may possibly avert some of the symptoms which are said to accompany its use.

First, I never give it for any other purpose than to bring sleep to a patient whose sleep is insufficient. When it was first introduced, at any rate, it was often given two or three times a day for the purpose of quieting a noisy or violent patient. As I never adopted this plan, I cannot tell you the exact result, but I apprehend that given thus it failed to produce sleep at night, its effect wore off, and no good was done. I am as averse as any of the opponents of drugs to the administration of them as mere methods of keeping patients quiet and wards orderly. But it is very important to procure sleep at night in many cases. In acute delirium you may save a patient's life by causing him to sleep even two or three hours in a night. In acute mania the worst stage may be shortened by the same method. In chronic mania the patient may be saved from sinking into dementia by securing a due allowance of sleep.

Secondly, do not give more than is necessary. I have often seen a large dose of chloral ordered when a much smaller one would have been sufficient. This must of course be ascertained by trial, but it should be your aim to attain the amount of sleep desired by the minimum dose that suffices. Sometimes it is advisable to give it when a patient goes to bed. But some will go to sleep at first and wake in two, three, or four hours' time; here you may allow them to have what natural sleep they can get without a narcotic, and give the latter when they wake, and so a smaller dose may suffice. I often find that the effect of a dose of chloral is prolonged over more than one night, so that the patient will sleep the next night without any. Sometimes a dose every three or four nights is enough. Trial should be made at intervals in every case, and it should be discontinued as soon as it can be dispensed with, and sleep can be procured by dint of exercise or other means.

Thirdly, I find that the effect of chloral is greatly augmented by combination with other substances. I have given it without effect in acute delirium, but when combined with bromide of potassium it produced sleep with certainty. In melancholia it often seems to produce but little sleep, and to have little or no beneficial effect, yet in combination with opium it renders the hypnotic action of the latter more sure and speedy, and from the two the patient gets more good than from either given singly. Frequently the action is quickened and increased by the administration of it in or with a stimulant, and sometimes the addition of an alkali assists.

That a drug which can bring about profound sleep in a few minutes is a powerful medicine is self-evident, and, like all powerful medicines, its effect may be good or evil; but I am convinced that the evil effects complained of by some are often due to the use of impure samples. For this reason I never order any of the syrups of chloral which are largely advertised, and may be made of cheap samples of the drug. Dr. Oscar Liebreich, whose name will ever be remembered in connection with chloral, has put this very strongly in the *Practitioner*,[1] and I cannot do better than quote his words:

" I have good reason to believe that a large proportion of the chloral, both solid and liquid, which is at present employed in medicine is not trustworthy in respect of its purity. Its impurities are calculated to have both a directly and indirectly mischievous effect. Directly, because some of these impurities contravene the simply hypnotic and anæsthetic power of chloral, and confer upon the so-called chloral with which they are mixed irritating and exciting properties; indirectly, because by rendering the chloral in certain cases feebly hypnotic, they render the practitioner or the patient uncertain of the proper dose, lead to multiplication of doses, and so tend to fatal results. The accidents which have occurred frequently in England are, I am satisfied, largely due to the impurities and uncertainty of the many preparations, and especially of the many solutions in vogue." Dr. Liebreich then speaks of the nature of these impurities, and reminds us " that crystals of chloral hydrate which are not dry and transparent, are not at all trustworthy." " It has happened to me more than once that my professional colleagues have brought to my notice cases in which chloral hydrate has produced considerable nervous excitement; the state of excitation combating, and in some cases overcoming the hypnotic effect. In investigating them I found the chloral was

---

[1] June. 1877.

of the impure kind. On opening the bottle irritating fumes arose. It was acid, and contained certainly a good deal of chloral, but with it other compounds, which give rise, by their decomposition, to acid and irritating substances of complex character, which have a very injurious physiological action.

"The hypnotic effect of chloral is best developed in a normally alkaline state of the blood: its effect is heightened when the blood is excessively alkaline, as in typhoid or typhus, and counteracted by the opposite chemical conditions as in uric arthritis. Here chloral in large doses may produce great excitement without giving sleep."

If, as has been said, a river of chloral has flowed through the land, it is not too much to say that a sea of bromide of potassium has overwhelmed all who come under the designation of "neurotic." I see not a few who have never taken chloral, but rarely does any one come under my care who has not gone through a course of bromide. Bromide is now given to all indiscriminately, just as opium was formerly, and as no immediate ill result follows, the indirect evil results are ascribed to something else. I have seen a patient, an old lady of seventy-five, who had been taking 60 grains of bromide daily for weeks, whose speech was unintelligible and who was more or less hemiplegic. All these symptoms vanished so soon as the bromide was discontinued. You will find bromide a most valuable medicine, as valuable as chloral or opium. But it is no more applicable to every case than are the others, and you will have to select the patients to whom you give it. Bromide of potassium, if given for any time, is most lowering in its effect. A certain amount of quiet may be produced by it, and for this reason it is often given and continued. But this is a kind of quiet such as is rightly denounced by those who object to all sedative drug treatment; and in many cases you will find that so far from tending toward cure, the patient under such treatment will become worse and worse. Especially is this true of the large class of persons who suffer from depression; although at first a certain amount of apparent relief may be obtained from this medicine, I believe that if it is given for any time, it will certainly add to the physical prostration, which is almost the constant concomitant of this form of mental disorder, and thus increase the emaciation and melancholic symptoms. On the other hand, if you have to deal not with depression but with riotous or hilarious mania occurring in young and "sthenic" individuals, you may find bromide most efficacious when given alone, and the effect of chloral when given as a hypnotic will often be much increased by the combination of the former drug. Dr. Clouston, who has studied most carefully the action of this and other medicines, says that he should place the bromides of potassium, sodium, ammonium, and iron, at the very head of the list of neurotic drugs, and of these the bromide of potassium is by far the most efficacious and the least hurtful. He also remarks that in the preliminary stages of insanity, before the symptoms have actually developed into decided psychical aberration, when they are chiefly sleeplessness, irritability, and commencing want of control, there is no drug equal to the bromide of potassium. Dr. Clouston also gives the results of combinations of bromide and other drugs, especially cannabis indica, and tells us that by giving bromide and cannabis indica together, not only is the effect of either given separately immensely increased, but the combination has an essentially different action from either of them given alone.[1]

I have incidentally spoken of opium already, and what I am about to say concerning it will be very brief. Of its value in insanity, as in disease generally, there can be no doubt; but it is not suited to every case of insanity any more than it is suited to every sane person. I believe if you were to take the first hundred persons you meet, and give each of them a dose of opium at bedtime, the effect in a large number of cases would be not the production, but the prevention of sleep. This must be borne in mind in treatment of the insane; not only is it unsuitable in certain forms of insanity, but you may find patients suffering from every kind of mental disorder, who may be intolerant of opium, and to such it is useless to give it. For this reason it is necessary to experimentalize, so to speak, on each individual, and determine by the result of a few doses whether the medicine is likely to be beneficial or not. The form of insanity in which, in my experience, it does least good and most harm, is acute delirious mania in "sthenic" patients, where there are great excitement and heat of head. The cases in which its benefit is greatest are those of quiet melancholia, the very opposite of the former. In the cases intermediate between these two varieties you will find it sometimes beneficial, sometimes not. As I have said, it must be tried. In chronic insanity, chronic and recurrent mania, and paroxysmal dementia, it is often of great service, not only producing sleep at night, but calming the general condition of the patient and improving the bodily health and appetite. Wherever it is given, the effect must be carefully watched, and if but little sleep is produced by a full dose, it is not to be inferred that the dose was insufficient, but we should suspect that the patient is one to whom opium in any quantity will do harm rather than good. Then arises the question, what preparation is to be given? Many speak most highly of the hypodermic injection of morphia, and I would refer you to a paper on this subject by Dr. M'Diarmid.[2] He affirms that, given hypodermically, it does not cause constipation or loss of appetite, two common effects when it is administered by the mouth. But he admits, as I have found, that it often causes vomiting. My favorite solution of the bimeconate of morphia, which I always give by the mouth, does not cause vomiting or constipation, and

[1] Fothergillian Prize Essay, 1870. Brit. and For. Med. Chir. Review, Oct., 1870, and Jan., 1871.

[2] Journal of Mental Science, xxii., 18.

where medicine can be administered in this way without difficulty, I use nothing else. The dose is so small that it can be easily mixed with food or drink, and patients will take it for months without in the least suspecting that they are taking a sedative. In some cases it may be advantageous to administer the subcutaneous injection, but of course this entails the attendance of the medical man to perform the operation.

Cannabis indica has been highly advocated by Dr. Clouston when given in combination with bromide of potassium, in order to subdue maniacal excitement. Given together, he says, in doses of a drachm of each, they are more powerful to allay such excitement than any other of the drugs or stimulants which he tried, more uniform and certain in their effects, more lasting, interfering less with the appetite, and to produce the same effect the dose does not require to be increased after long-continued use. Dr. M'Diarmid [1] says that the tincture of cannabis indica in doses of ℞ xx. to xxx. is a useful supplement to an insufficient or delayed dose of morphia, and that, when a patient is liable to vomit after the injection of morphia, or suffers from a weak heart, a safe plan is to give a small dose subcutaneously, and then to administer a small dose of cannabis. The objection to cannabis in my experience is that the specimens of the medicine vary immensely, and also, it is so nauseous that there is great difficulty in getting patients to take it.

Calabar bean, *Physostigma venenosum*, I have already spoken of as advocated for the treatment of general paralysis, the theory being that this drug oppresses and slows the heart, so counteracting the great hypervascularity found in this disease.[2] So far as I am aware, Calabar bean has not been used in other forms of insanity, though, if this is its effects, it ought to be equally useful in many kinds of acute mania. My own experience of it is very limited, but it does not lead me to think that the course of general paralysis is materially affected by its use.

Ergot of rye is decidedly beneficial in some cases of epilepsy, and arguing from the analogy of bromide of potassium, we might infer that, in insanity also, it would be a useful medicine. Dr. Crichton Brown has advocated its use, on the theory that it produces contraction of the vessels. It is, he says, eminently useful in certain varieties of recurrent mania, chronic mania with lucid intervals, and epileptic mania. "In these forms of cerebral derangement I have found it almost uniformly efficacious in reducing excitement, in shortening attacks, in widening the intervals between them, occasionally in altogether preventing their recurrence, and in averting that perilous exhaustion by which excitement is so often succeeded."[3]

Hyoscyamus is a very old remedy in insanity, but given in the ordinary

---

[1] *Loc. cit.*

[2] Dr. Milner Fothergill, West Riding Asylum Reports, v.. 186.

[3] Practitioner, June, 1871.

pharmacopœail doses, its effect is almost *nil*. Given in larger doses, however, it is decidedly useful, and may, like the cannabis indica, be combined with bromide of potassium, and the effect of either be increased by the combination. I wish now, however, to draw your attention to the powerful effects of that which has received the name of hyoscyamine. Two substances have been so called,—a white granular body, which has also been called hyoscyamia,[1] and the extractive hyoscyamine of Merck. It is important to distinguish these, as of the latter a dose of 1 gr. may be given, while of the former the dose is from $\frac{1}{30}$ to $\frac{1}{20}$ of a grain. In noisy and destructive mania, particularly that of chronic patients, this drug often produces a great effect, rendering such persons quiet and manageable. Its operation, however, appears to be extremely unpleasant. The patients dislike it intensely, and accuse the doctor of poisoning them; and nearly all those who use it have cases to relate where most alarming symptoms have followed its administration. Dr. Browne[2] advises it to be given hypodermically, as, owing to the presence or absence of food in the stomach, its effect, if given by the mouth, varies greatly. If given hypodermically it is more constant, and the dose he generally used was from $\frac{1}{30}$ to $\frac{1}{18}$ of a grain of Merck's crystalline hyoscyamine dissolved without heat in glycerine and water. Beyond all question this very powerful remedy must be used with great caution, and is better fitted for asylum than private practice, as it requires watching after administration.

It is much to be desired that extended trials of these last named drugs should be made at our large asylums, and that the result should be communicated to the profession. If we had the testimony concerning Calabar bean, ergot, and hyoscyamine, that we possess with regard to chloral and bromide of potassium, we should be better able to appreciate their usefulness, and decide upon the cases in which they should be administered or withheld.

Next, I must say a few words upon feigned insanity. A disordered mind has been simulated from the earliest ages—witness the dementia which David successfully feigned, and the pretended imbecility which saved the life of Lucius Junius Brutus—and it will be assumed perpetually by those who have a motive for shifting from their shoulders the responsibility of their acts. Fortunately, few know how to feign insanity. It is only a Shakespeare who can depict the assuming of a Hamlet or an Edgar. The majority of simulators are clumsy performers, whom you will detect without difficulty; but here and there you may chance to see a case which is not so easy to decide, and which, though eventually you may be satisfied as to its character, cannot be recognized at a moment's notice. Doubtless they who have the insane ever before their eyes will

---

[1] Dr. Savage, Journal of Mental Science, xxv., 177.
[2] British Medical Journal, Nov. 25, 1882.

19

most readily detect the sham disorder, yet there are certain points which
will enable you to come to a conclusion respecting the greater number of
cases. If we except the instances of hysteria, catalepsy, pretended fasting, and the like, which can hardly be called feigned insanity, we shall
find that most persons who simulate the malady have an obvious motive
for so doing. Therefore we do not meet with feigning in ordinary private
practice; but if any of you become surgeon to a jail or to the army, you
will not seldom be called on to see malingerers who adopt this as a means
of getting to comfortable asylum-quarters, or obtaining a discharge from
duty. As, however, men do really become insane under the same circumstances, you will have carefully to discriminate between the real and the
affected symptoms.

Uneducated as the mass of such persons is, the attempt will generally
be clumsy and easy to detect; but here and there an educated man, who
brings himself within reach of the law, may with greater success carry on
the cheat.

The first remark to be made is, that the insanity simulated may be
transitory or persistent. The individual may pretend that he was in a
delirious or unconscious state at the time the criminal act was committed,
or he may be apparently insane at the time we see him. Secondly, he
may be in an acute and active state of feigned excited mania or melancholia, or may pretend to be in a quiet and apparently chronic condition
of monomania or dementia. Thirdly, he may put on this appearance soon
after the commission of his wrong act, to make it appear that it was committed by him while insane; or he may feign insanity while in prison to
get away to the better fare and idle life of an asylum.

If you are told by a prisoner that at the time he committed the act he
did not know what he was about, or that he has no recollection of it, he
virtually simulates the form of transitory mania which is seen occasionally
in conjunction with epilepsy, or taking the place of the latter. You will
recollect, however, that such attacks are extremely rare, that they are not
usually so transient as to be unnoticed by others, or so severe as to take
away all recollection of what was done in them. Here you will inquire
into the previous history of the individual as regards former attacks of
insanity, epilepsy, blows, or cerebral affections.

He may tell us that he suffered from an irresistible impulse to commit
the act, from some sudden and overwhelming idea; he may simulate the
so-called *impulsive* insanity. Applying what I have said already concerning this, you will look for other symptoms, for a history of previous
attacks or previous head affections; you will not consult the individual
about these, for you may easily put such into his mouth, but you must to
the best of your ability arrive at an account of his past life, and consider
the character of the deed not only as regards its enormity, but also its
senselessness, want of motive, or eccentricity. Such acts, if really com-

mitted under the influence of either of these forms of temporary insanity, are usually violence against self or others. If this plea is put forward as an excuse for small and petty thefts, or forgery, acts of indecency or exposure of person, we may reasonably suspect it. The latter are committed by madmen, but not by those whose insanity passes off so soon as the act is over.

More commonly, however, we are called to see a person who is apparently insane at the time of our inspection. Comparing a true with a feigned case, I may say, generally, that a real lunatic, when approached by a stranger, appears at first rather better than worse, and more on his guard; he tries to bring his wits together and understand what is going on. But a sham lunatic, when we go to him, redoubles his efforts to seem insane; he is more energetically noisy, idiotic, and maniacal. A sham lunatic, recollect, always wishes to be thought a lunatic. If we ask him whether he is out of his mind, he tells us at once that he is. In fact, he dare not say the opposite; whereas a real patient rarely confesses it, unless he be in the depth of melancholia.

Feigned insanity is almost always overdone. As there is no subject on which such erroneous notions prevail among people in general, so the imitation is, with rare exceptions, a bungle. If noisy and violent mania is the form assumed, detection is easy. The malingerer, unlike the true maniac, will tire himself out and go to sleep. No sane person can maintain the incessant action, singing, and shouting of a genuine maniac for any but the shortest time. No genuine maniac would, in the middle of all this, at an early stage of the attack, go to sleep, and sleep many hours. Watch such people without their knowledge, and you will have little doubt as to the case.

If a less excited maniac is feigned, and the feigner will talk and answer questions, he generally overdoes his part by pretending to have lost all reason and memory. He will not give one correct answer to the simplest question; he will not know his own name; but will display an ingenuity in evading answers and in talking nonsense entirely at variance with the loss of mind he pretends to have suffered. Or he will answer questions correctly about everything that does not concern himself, but so soon as we question him as to his crime or history, he becomes suddenly demented and entirely deprived of memory and intelligence. Loss of memory is not common among insane, except in cases of dementia, primary or secondary. It may be feigned, but will rarely be a clever simulation. Here we must look for an absence of mind; and if we see a presence of mind, and a sharpness and quickness displayed in many ways, the notion of dementia is incompatible. Dirty habits may be adopted to further the deceit, and malingerers will daub themselves with, or even eat, their fæces; but in conversation we may generally discover that they are not so lost as they seem. Frequently something casually mentioned in their presence is done in consequence of the hint, showing that their attention has been fixed on all that has been said.

A man may feign melancholy, or sit silent and desponding, and say nothing. Here some knowledge of the insane may be requisite to guide us to an opinion. He may refuse his food, and say that poison is put in it. We must watch him, and look for physical symptoms. Is his tongue clean, his skin cool, and pulse normal? Does he sleep well at night? Does he alter his conduct according to that which is said in his presence? Does he dress and undress himself? Melancholia, or *melancholia cum stupore*, is a form distinct from others, from mania especially, and his distinction is not likely to be carefully preserved by a malingerer. Neither mania, melancholia, nor primary dementia comes on in patients, full blown, in an hour. The history of the previous days is almost conclusive in the majority of cases, especially if there is a knowledge of the sleep the pretended lunatic has enjoyed.

The detection of feigned insanity is, and ever will be, difficult, when we have to examine men and women in whom madness and badness are so intermingled that observers cannot determine which it is that regulates their conduct. Amidst our criminal population are hundreds who can hardly be said to be sane and responsible, but who, in the lower ranks of life, commit a succession of crimes, perhaps of no great magnitude, which render them the almost perpetual inhabitants of jails. Some of them are so violent, outrageous, and destructive, so silly in their motiveless fury, and childish in mind, that we may call them imbeciles or insane, and have good grounds for our opinion. Such there will ever be on the border-land of insanity. But each of these must be judged by himself. My purpose here is not to speak of doubtful, but of feigned, insanity—insanity feigned by those of whose sanity at other times we have no doubt.

We cannot depend on any physical signs for the certain detection of simulation. We find among the insane the pulse neither slow nor quick, a cool head, and normal urine; and he will be bold who shall affirm that he can detect insanity or its absence by the sense of smell, though such men are to be found. Nevertheless, there is almost invariably such a disturbance of the health in a person whose insanity is just commencing, that our suspicions should be aroused if this be wanting. Want of sleep, a coated tongue, constipation—all, or some, are nearly always to be found; so that if a man suddenly feigns insanity, we look for them and for symptoms of recent and acute mental disorder. If the insanity simulated is that of quiet and apparently chronic monomania or dementia, we know that these forms do not come on suddenly, and that there must have been a previous stage.

Various plans have been advocated for the purpose of making the simulator confess the imposture, and give up his acting. Speaking in his presence of remedies which will probably have to be used, such as the actual cautery, and the sight of its preparation, may frighten some pretenders. The sight of the stomach-pump may make a man take his food; but then a lunatic will take it for the same reason. In accordance with the truth, *in vino veritas*, feigned insanity has, it is said, been detected by the opening influence of an intoxicating amount of wine; this, however, is hardly applicable to the inmates of jails. Little is to be gained from drugs, unless it be from a good dose of tartar emetic, which may make a man confess rather than have another. This, of course, is to be given only when our mind is made up concerning the case, and we want to put an end to the play. A cold shower-bath may cure another, but probably nothing is so efficacious as the application of a galvanic battery. In a very interesting paper, Dr. David Nicolson, one of the medical officers at Portland, has related the valuable aid derived from this instrument. His remarks on feigned insanity among prisoners are well worth perusal.[1] When you are convinced that a person is shamming, you will probably effect a rapid cure by a few turns of the machine, or a repetition of it twice a day for a few days.

There are cases on record where skillful cheats have deceived for a long period even alienist physicians, but such are rare. Consider if there is a strong motive for feigning insanity; if there has never been anything of the kind prior to the motive arising, and if the insanity is violent and acute in character, we may reasonably suspect it, and close observation will generally leave no doubt of the deception.

I alluded a minute ago to a belief not altogether uncommon, that the insane possess a peculiar odor, and that insanity may be detected by the nose. I will not relate to you the various opinions and modifications of opinion on the subject. It is one of those matters which can hardly be brought to a definite test, for the sense of smell is strongly subjective, and a preconceived notion may go far to help a person to discover an odor. Doubtless many lunatics smell offensively. I have already told you that in acute mania there is often an intolerable effluvium, especially from women. Many patients can with difficulty even in a chronic state, be kept sweet, and, if very stout, their odor may be perceptible enough. Many of the poorer classes wear their clothes a long time, and thus acquire a stale and disagreeable smell. But that there is a smell peculiar to the insane, which emanates from every insane person, I myself have failed to discover. It may be that my sense of smell is not so acute as that of others, though of this I am unaware; but certainly I believe that I have seen insane ladies and gentlemen who, washing and dressing like other people, were as free from smell as the sane who sat with them at table unconscious of their presence. Unfortunately, we are not likely to advance beyond mere theories and opinions on the subject, and my opinion is all that I will advance at the present time.

---

[1] Journal of Mental Science, xv., 586.

LECTURE XIX.

The Law of Lunacy—Private Patients—Order and Certificates—Single Patients—
Notice of Discharge or Death—Leave of Absence—Order of Transfer—Pauper
Patients—The Property of Patients—Commission of Lunacy.

It now becomes my duty to tell you something about the legal
methods of dealing with persons of unsound mind. Legislation has again
and again, during five hundred years, regulated the manner in which the
persons and property of such people are to be cared for; and although the
statutes relating to the subject are not less than forty in number, I hope
to be able to put before you in brief that which it is essential for you to
remember while practicing your profession. And I may as well say at the
outset that I am not lecturing for those who have, or are to have, the care
and charge of an asylum. Any of you who undertake this duty will learn
the details, legal and medical, by special study. My present object is to
teach to those who are not specially concerned with this branch of practice
that which they require for the purpose of sending a patient to an asylum,
attending one who does not require the restraint of an asylum, and giving
evidence before a commission in lunacy, or on other occasions when a
man's insanity is called in question.

The subject naturally divides itself into two parts, one which relates to
the person, the other to the property of a lunatic; and the former may be
subdivided into one portion relating to the person of private lunatics, and
that which is concerned only with the custody of paupers. I therefore
shall speak of it under these heads.

I. { 1. The care and custody of private lunatics.
{ 2. The care and custody of pauper lunatics.

II. The care of the property of lunatics.

Here I would remark that a man does not necessarily come under the
cognizance of the lunacy laws because he happens to be a lunatic. He
may be a lunatic for years, and may be tended and restrained in his own
house, or in that of a relative or friend, provided that his own friends or
relations take care of him, and take care of him properly. It is the common law of the land that a man's friends may restrain him from harm,
or protect him, if he is unable to protect himself. But if the lunatic is
not taken care of by his own friends, or if they neglect him, and he is
found to be wandering at large or improperly confined or maintained,
then the Lunacy Acts reach him, the Lord Chancellor or Home Secretary
may order him to be visited in the friends' or his own house, and necessary steps to be taken for his amelioration.

The Lunacy Acts define with tolerable accuracy the persons who may take care of lunatics without legal supervision. They must be persons "who derive no profit from the charge." Any one deriving profit, whether as proprietor of the house or lodging, or as companion, nurse, or attendant, must comply with the statutes I am about to describe.

There is, however, one exception to this. The committee of a person found lunatic by inquisition may take charge of such person, or may commit him to the charge of another, without medical certificates, upon his own order, having annexed to it an office-copy of his appointment.

With these exceptions—viz., the care of a patient by his own relatives or friends, or his own committee or committee's agent—all private lunatics are to be restrained and kept only after the due execution of three legal documents, which are called the "Order and Medical Certificates." Although you, as medical men, are chiefly concerned with the latter, it is right that you should also be familiar with the "order," that you may be able to instruct the friends of a patient.

Here is the order in the statutory form. Generally speaking, we fill up printed forms, but the whole may be in writing, if no printed form is at hand.

ORDER FOR THE RECEPTION OF A PRIVATE PATIENT.

I, the undersigned, hereby request you to receive *John Jones*, whom I last saw at 20 *Smith Street, Paddington*, on the *twenty-first* day of *March*, 1884, (*a*) a (*b*) *person of unsound mind*, as a patient into your house.

(*a*.) Within one month previous to the date of the order.

(*b*.) Lunatic, or an idiot, or a person of unsound mind.

Subjoined is a statement respecting the said *John Jones*.

Signed,   Name, *Mary Jones*.
Occupation (if any),
      Place of Abode, 20 *Smith Street, Paddington*.
Degree of Relationship (if any),)
      or other circumstances of } *Wife.*
connection with the patient,)
Dated this *twenty-first* day of *March* one thousand eight hundred and *eighty-four*.

To *Robert Brown, Esq.*,
      (*c*) *Proprietor*, (*d*) *Bath House Asylum.*

(*c*.) Proprietor or superintendent of.
(*d*.) Describing the house or hospital by situation and name.

I have here filled up the order with the name of an imaginary patient, *John Jones*, the other names being, of course, equally fictitious. Now, observe that the person signing the order must have seen the patient within a calendar month. This is a most proper regulation. Formerly a person might sign an order for the reception of one whom he had never seen in his life; but now he must have seen him within a month, must state where he saw him last, and affirm that he is of unsound mind. In the marginal notes you will see that the patient may be described as a *lunatic, idiot,* or *person of unsound mind.* One of these he must be called, and it is usual to adopt the last as the least painful to friends, and, at the same time, most comprehensive. "Subjoined is a statement." This must accompany the order, and to it I shall come immediately. Who may sign the order and who may not? I suppose, in my imaginary case, that the wife signs it. It should, in my opinion, be signed by the nearest relative; but frequently there is a great objection to so doing on the part of relatives, and the statute allows any one to sign who can show any sort of reason for interfering, as a friend, a magistrate, or the minister of the parish. He or she must, however, have seen the patient within the month, and this the date at the bottom will indicate. But certain people may not sign the order. First, no person may sign who receives any percentage on, or is otherwise interested in, the payments to be made by, or on account of, any patient received into an asylum or other house. Secondly, no one can sign the order who is the medical attendant, or the proprietor, of the asylum into which the patient has to go. Thirdly, no one can sign who is the father, brother, son, partner, or assistant, of either of the medical men who sign the certificates, or who himself has signed one of the certificates. The order must be directed to the person under whose care the patient is to be placed, whether it be the owner of a private house or lodging, or the owner or superintendent of a private lunatic asylum or hospital.

This order, you are to recollect, will authorize the reception of a patient during one calendar month from its date, and no longer. If a month has expired a fresh order will be necessary.[1]

Underneath the order on the printed form is placed the "Statement," which I will fill up with supposed particulars.

STATEMENT.

If any particulars in this statement be not known, the fact to be so stated.

Name of patient, with Christian name at length, ...	*John Jones.*
Sex and Age, ...	*Male,* 35.
Married, single, or widowed, . . .	*Married.*
Condition of life, and previous occupation (if any), ...	*Clerk.*
Religious persuasion, so far as known, .	*Church of England.*
Previous place of abode, ...	20 *Smith Street, Paddington.*
Whether first attack, ...	*Second.*
Age (if known) on first attack, . .	*Thirty.*
When and where previously under care and treatment, ...	*Bath House Asylum* in 1872.

[1] 25 and 26 Vict. cap. 3, sect. 23.

Duration of existing attack, . . .	*Three weeks.*
Supposed cause, ...	*Unknown.*
Whether subject to epilepsy, . . .	*No.*
Whether suicidal, ...	*Yes.*
Whether dangerous to others, . . .	*No.*
Whether found lunatic by inquisition, and date of commission or order for inquisition, ...	*No.*
Special circumstances (if any) preventing the patient being examined, before admission, separately by two medical practitioners, ...	*None.*
Name and address of relative to whom notice of death to be sent, . . .	*Mary Jones,* 20 *Smith Street, Paddington, W.*

(e) Where the person who signs the statement is not the person who signs the order, the following particulars concerning the person signing the statement are to be added.

Signed, Name, (e) *Mary Jones.*
Occupation (if any),
Place of Abode,
Degree of Relationship (if any),
or other circumstances of connection with the Patient.

This statement, which is the appendix, as it were, to the order, needs little explanation. It is a statement of the facts of the case for the guidance of the proprietor of the asylum, and for the information of the Commissioners in Lunacy. As in the order, the name of the patient must be stated in full, Christian and surname, and every other detail must be filled up in some way. No space must be left blank. There are certain points on which friends are very reluctant to give accurate information, and yet it is important that we should have it. They are very apt to give the duration of the existing attack as being very short, when it may turn out on inquiry that the patient has been insane for a long period, though possibly only dangerous or excited during a few weeks or days. Then we rarely get the true cause assigned. Frequently this is hereditary transmission—a fact which friends are most loath to mention. And they do not like to describe a patient as suicidal or dangerous, and yet it is of great importance to those who are to have the charge that this should be stated, and if there be any doubt, it is better to state the suspicion than to give a direct negative to the question. With regard to the last question but one, "special circumstances," etc., I must say a word. Inasmuch as it is often very difficult for a medical man to gain access to a patient, and it may be of the utmost consequence that such a patient should be at once deprived of the power of doing harm to himself or others, there is a clause in the Act—16 and 17 Vict. c. 96, sec. 5—which provides that "any person (not a pauper) may, under special circumstances preventing the examination of such person by two medical practitioners, be received as a lunatic into any house or hospital, upon such 'order' as aforesaid, and with the certificate of one physician, surgeon, or apothecary alone, provided that the statement accompanying such order set forth the special circumstances which prevent the examination of such person by two medical practitioners; but in every case two other such certificates shall, within three clear days after his reception into such house or hospital, be signed by two other persons, each of whom shall be a physician, surgeon, or apothecary, not in partnership with or an assistant to the other or the physician, surgeon, or apothecary who signed the certificate on which the patient was received, and not connected with such house or hospital, and shall within such time and separately from the other of them have personally examined the person so received as a lunatic."

Such is the meaning of the question commencing with the words "special circumstances."

The statement is commonly, but not necessarily, signed by the person who signs the order. It may be signed by any one having the knowledge requisite, and he must state his relationship or connection after the signature.

We now pass to that which more immediately concerns ourselves, viz., the medical certificates.

And first, who may and who may not sign these? Any physician, surgeon, or apothecary may sign a certificate, if he be a person registered under the Medical Act passed in the session 21 and 22 Victoria, cap. 90. Not only must he be legally qualified, he must also be registered. This is not generally known; but the reception of a certificate from a non-registered practitioner would, in my opinion, lay the proprietor of an asylum open to the charge of illegally receiving.

Certain medical men, however, are precluded from signing the certificates.

1. The two medical men must not be professionally connected, must not be in partnership, nor may one be the assistant of the other.

2. Neither of them must be the proprietor of the house or asylum into which the patient is to be received, nor must he receive any percentage on the payments to be made for the patient, nor must he be the medical attendant after reception of such patient, whether in a private house or an asylum.

3. No medical man who, or whose father, brother, son, partner, or assistant, is wholly or partly the proprietor of, or the regular professional attendant in, a licensed asylum or hospital, shall sign a certificate for the reception of a patient into such house or hospital.

4. No medical man who, or whose father, brother, son, partner, or assistant, shall sign the "order" already spoken of, shall sign any certificate for the reception of the same patient.

Thus, you observe the various persons—the person signing the order, and those signing the certificates—are to be entirely independent one of another, and all three are to be independent and unconnected with the proprietor of the asylum, or the medical attendant of the patient, if he is not in an asylum. So the co-operation of four independent persons, of whom three must be medical men, is requisite for the restraining of any one under the Lunacy Acts, and each of the two medical men who are to sign the certificates must examine the patient separately. This you must recollect, because in all probability it will happen that you will be called to meet another practitioner to consult with him as to the propriety of placing some one under legal restraint. Although you together make an examination for the purpose of consultation, you must again visit and question the individual separately, and, repeating the examination, you must elicit that which you are about to write down in your certificate. Otherwise, if at any future time the alleged lunatic were to bring an action against the proprietor of the asylum for false imprisonment, the certificates would be invalidated by neglect of this rule.

I now pass to the consideration of the form of the medical certificates, one of which I will fill up, as I filled up the order, with imaginary details.

MEDICAL CERTIFICATE.

(a) Set forth the qualification entitling the person certifying to practice as a physician, surgeon, or apothecary, ex gra.:—Fellow of the Royal College of Physicians in London, Licentiate of the Apothecaries' Company, or as the case may be.
(b) Physician, surgeon, or apothecary, as the case may be.
(c) Here insert the street and number of the house (if any), or other like particulars.
(d) Insert residence and profession or occupation (if any), of the patient.
(e) Lunatic, or an idiot, or a person of unsound mind.
(f) Here state the facts.
(g) Here state the information, and from whom.

I, the undersigned, being a (a) *Fellow of the Royal College of Surgeons of England,* and being in actual practice as a (b) *Surgeon,* hereby certify that I, on the *twentieth* day of *March* 1884, at (c) 20 *Smith Street, Paddington,* in the county of *Middlesex,* separately from any other medical practitioner, personally examined *John Jones,* of (d) 20 *Smith Street, Paddington, clerk,* and that the said *John Jones* is a (e) *person of unsound mind,* and a proper person to be taken charge of and detained under care and treatment, and that I have formed this opinion upon the following grounds, viz.

1. Facts indicating insanity observed by myself. (f) *He is under a delusion that he has committed some unpardonable sin, that he is Antichrist, and that his name is mentioned in all the newspapers. His appearance denotes great agitation and depression.*

2. Other facts (if any) indicating insanity communicated to me by others. (g) *I am informed by his brother, Robert Jones, that he has attempted to jump into the river, and out of window.*

Signed, Name, *William Green.*
Place of Abode, 10 *Richmond Street, Paddington.*

Dated this *twenty-first* day of *March* one thousand eight hundred and *eighty-four.*

Now, if you consider this form of medical certificate, you will notice that, according to the directions appended in the margin, you are first of all to state your legal qualification. Not merely are you to say that you are a physician or surgeon; you are to give the name of the diploma you hold. In addition to this, you must assert that you are in actual practice; a retired practitioner, or a medical man who has given up the profession and is otherwise occupied, cannot sign either of the certificates. Then comes the date, and this is important. The dates of the certificates are quite different from those of the order. The order may be signed and dated by any one who has seen the patient within a month, and is valid for a month from the signing thereof. But the medical certificate is only valid for seven days, not from the signing, but from the examination of the patient. The date of the examination, the first date in the certificate, is the important part: within seven days from this the reception of the patient must take place, or the certificate expires. It may be signed and dated at any time between the examination and the reception. The date of the day of the month and the year must be given, and also the place of examination. And you are to specify the street and number of the house, if it has one, as well as the county. Also your examination of the patient must take place without any medical man being present, as I have already explained to you. Other people may be present, but they must not be medical men practicing. The names, Christian and surname, of the patient must be written at length, together with his residence, profession, or occupation. You then affirm that the said patient is one of three things—a lunatic, idiot, or person of unsound mind; and, as I said in the case of the order, it is better to use the last expression, which comprises every variety. You also affirm—and this, too, is important—not only that the individual is of unsound mind, but that he is "a proper person to be taken charge of and detained under care and treatment;" in other words, to be taken care of as a lunatic under certificates of lunacy. There may be many patients afflicted with unsoundness of mind, temporary or other, for whom we might hesitate or refuse to sign certificates of lunacy for the purpose of restraint. Formerly the medical man merely stated his opinion that the patient was of unsound mind, without giving reasons, and upon such a certificate the patient was received. The same practice still continues in Ireland; but in England and Scotland you are obliged to state your reasons for coming to such a conclusion; and the Commissioners will reject the certificate and release the patient if they do not consider the reasons strong enough. Now, the reasons are divided in the form into two parts—the facts observed by yourselves, and those communicated by others; and I need not tell you that those observed by yourselves are the most important, the others being necessarily hearsay reports, which frequently you may have reason to disbelieve. Now, these facts are supposed to be observed by you on the day of examination, the day mentioned as the date, and when they consist of the result of conversation carried on upon that day, there can be no doubt about the matter. When, however, your opinion is based not upon a particular delusion, but upon the general conduct of the individual, there is often great difficulty in getting enough on one particular day to warrant your signing a certificate. And the Commissioners in Lunacy insist on this being done. In their Fifteenth Report (1861) they say: "It would, of course, be impossible that any examining medical man should exclude from his consideration facts known to him of the antecedents of the patient, immediate or remote; these are entitled to their full influence; but the Legislature has been careful to guard against such facts exercising undue influence in the certificate he is called on to give, by requiring that this certificate shall be directly deducible from examination on a particular day and at a specified place; and that the opinion expressed therein as having been formed on such particular day shall be set forth as the result of his having observed at that time in the person under examination some specific fact indicating insanity."

You will, therefore, have to connect that which you may have observed previously with what you observe on the particular day. If a patient justifies his past conduct, and defends it in an insane manner, you may elicit sufficient for your purpose; or, without asserting delusions, he may admit that he has entertained them previously, or otherwise indicate that he has not given them up. Frequently, when you anticipate that you will have to examine a patient for a certificate, and have reason to think that he will deny his beliefs, it is as well not to subject him to any cross-examination upon them till the actual day arrives. But I shall have something to say concerning the examination of patients subsequently: here I am only speaking of the requisite formalities.

It is not necessary that any facts communicated by others should be inserted. Where those observed by yourselves are plain and unmistakable, it rather weakens than stengthens a certificate to supplement them with others received on hearsay. But frequently that which you observe is explained and illustrated by what the patient has said to others; and acts committed by him, acts of attempted homicide, suicide, or other violence, may not have been witnessed by you, yet may be valuable indications of insanity.

Two certificates complete the formalities requisite for placing a patient under restraint. Each must be the independent opinion of a registered practitioner, who, in signing this legal document, does so under grave responsibilities. If he does it negligently or fraudulently, he is liable to an action at law, and to be mulcted in heavy damages. Upon such order and medical certificates, a proprietor or superintendent of an asylum may receive a person, as a lunatic, pleading them in justification; but he must send a copy of them to the Commissioners in Lunacy within twenty-four hours; and then, after the expiration of two clear days, and before the expiration of seven days, he also transmits to the Commissioners a "statement," containing his own observations upon the mental and bodily state of the patient. The same thing is done by the proprietor, if the patient is removed, not to an asylum, but to the house of a private individual, becoming what is called a "single patient."

You will have noticed in the newspapers reports of prosecutions, instituted by the Commissioners in Lunacy, against various persons, for wrongfully receiving and taking care of people of unsound mind; and from the phraseology adopted, you may think, as many do, that they were prosecuted for receiving these patients *without a license.* But this is not so. No license is required for the reception of one patient. When two are received, then a license becomes necessary. What is requisite is that these single patients should be received upon an "order" and two certificates, just as if they went to an asylum, and that copies should be sent to the Commissioners, thus registering the patients on their records.

As you may have occasion to send patients to reside in this way with a family, and may wish to attend them while there, I will briefly describe the regulations to be observed. You wish to send one to the house of some private individual, male or female. The order and certificates are procured in the usual way, and copies of them are to be sent by the proprietor of the house to the Commissioners in Lunacy within twenty-four hours of the admission of the patient, together with a notice of the admission signed by the said proprietor. *If you are to be the medical attendant, you must not sign either of the certificates.* After two clear days, and before the expiration of seven days, you will send to the Commissioners a "statement" of the mental and bodily condition of the patient. Then once a fortnight at least you will enter in a book, to be kept at the house for the inspection of the Commissioners, an account of the patient under the following heads:—

Date.	Mental State and Progress.	Bodily Health and Condition.	Restraint or Seclusion since last Entry. When and how long. By what means and for what reason.	Visits of Friends.	State of House, Bed and Bedding, etc.

This is called the "Medical Visitation Book," which will be inspected and signed by the Commissioners when they visit the patient. When the patient leaves, a "Notice of Discharge" must be sent to the Commissioners by the proprietor in the following form:—

FORM OF NOTICE OF DISCHARGE.

I hereby give you notice, that   a single
patient received into this (^{a})   on the   day of
18 , was discharged therefrom (^{b})   by the authority of
       on the   day of   18 .

         Signed,_____

           (^{c})_____

Dated this   day of   one thousand eight hundred and
eighty
     To the_____

(a) House.
(b) Recovered *or* relieved *or* not improved.
(c) Superintendent *or* proprietor of ——— house *or* hospital at———.

If he dies, a " Notice of Death," signed by the medical attendant, must be sent to the Commissioners and also to the Coroner of the district, who may hold an inquest if he thinks fit. There is a special form for the notice of death.

NOTICE OF DEATH.

I hereby give you notice, that   a single
patient, received into this (^{a})   on the   ·   day of
18 , died therein on the   day of   18   ; and I further
certify that   was present at the death of the said
       and that the apparent cause of death of the said
       (^{b})   was

         Signed,_____

           (^{c})_____

Dated this   day of   one thousand eight hundred and
eighty
     To the Commissioners in Lunacy.

(a) House *or* hospital.
(b) As ascertained by post-mortem examination, *if so.*
(c) Medical attendant of ———.

I give you the forms of this and of the notice of discharge.

Blank forms like the above may be purchased, but it is not absolutely necessary that the order and other documents should be on a printed form. The whole may be in manuscript if a printed form is not procurable, provided that the wording is the same.

If it is thought advisable to send the patient for change of air to the seaside or elsewhere, or to allow him to go home upon trial, " Leave of Absence " may be obtained from the Commissioners.

OFFICE OF COMMISSIONERS IN LUNACY.

       19 WHITEHALL PLACE, S. W.,
           18 .

By virtue of the power vested in us, by the 86th section of the Act 8 and 9 Vict. c. 100, we hereby signify our consent to the removal,

under proper control, of   a certified patient in   House,
to   for the period of   calendar month   from
                                         } Commissioners
                                          in Lunacy.

To

NOTE.—In forwarding "the approval in writing" required by the above section. It should be stated, whether it is "of the person who signed the order," or "of the person who made the last payment."

If it is necessary to remove him from one place of residence to another, or from one asylum to another, this may be done by obtaining an "Order of Transfer" from the Commissioners, in which case fresh certificates will not be required.

TRANSFER OF PRIVATE PATIENT.

CONSENT.

We, the undersigned, Commissioners in Lunacy, hereby consent to the removal, on or before the   day of   18 , of   a private patient in   House
      Given under our hands this   day of
      in the year of our Lord one thousand eight hundred and
                                   } Commissioners
                                   in Lunacy.

ORDER.

I*   the undersigned, having authority to discharge a private patient in   House,   hereby order and direct that the said   be removed therefrom to   House
      Given under my hand this*   day of
      in the year of our Lord one thousand eight hundred and
               Signed,
               Place of abode,

NOTE.—This order must be signed and dated subsequently to the consent of the Commissioners; and must be signed by—

Generally {

1. The person who signed the order for the patient's admission:
2. If such person be incapable (by reason of insanity, or absence from England, or otherwise), or if he be dead, then by the husband or wife of the patient:
3. If there be no husband or wife, then the patient's father:
4. If there be no father, then by the patient's mother:
5. If there be no father or mother, then by any of the patient's nearest of kin: or by the person who made the last payment on the patient's account.

If a patient escapes, he may be recaptured within fourteen days upon the original order and certificates: if fourteen days have elapsed, a fresh order and certificates must be obtained. Notice of the escape and recapture must be sent to the Commissioners. If not recaptured, notice of the escape must be sent within two clear days.

All these enactments apply equally to private patients in asylums, and to single patients. In England persons can be received in any registered hospital for lunatics as voluntary boarders, and may contract by bond or agreement to conform to the rules and regulations of such hospital. Persons can only be received as voluntary boarders in licensed houses if they have been within five years preceding certified patients in any asylum, or under care as single patients, and leave in writing must be obtained from the Commissioners or visitors. In Scotland any one can be received as a voluntary boarder in any asylum.

I will now say a few words as to the method of proceeding when we desire to place a pauper in an asylum. Of the management of public asylums I say nothing, but it may fall to your lot to send thither poor people who have been under your care.

The law enacts that the medical officer of a poor-law district, on becoming aware of a lunatic, shall give notice thereof to the relieving officer, or, if there be not one, to the overseer. In the same way any person may give notice of the same to the relieving officer or overseer. The latter is in turn to give notice to a justice of the peace of the county or borough, who within three days shall cause the lunatic to be brought before him, or shall visit him at his house, and shall examine him, with the aid of a medical man. If the latter gives a medical certificate, and the justice is satisfied that the pauper is a lunatic, and a proper person to be taken charge of and detained under care and treatment, he shall make an order for his reception into an asylum. If two medical certificates are given, one by the medical officer, and a second by any other medical man, the justice *must* make the order without any option.

If the pauper cannot be taken before a justice or be visited by him, he may be visited by an officiating clergyman, together with the relieving officer (or overseer), and their joint order may be given for his removal, after the medical certficate or certificates are signed. The medical certificate is in precisely the same form as that I have already given. It must not be signed, however, by any medical man who is the medical officer of the asylum nor by any one whose father, brother, son, partner, or assistant shall sign the order.

If the relieving officer cannot at once take the lunatic to the asylum, he may take him to the workhouse, and in point of fact a greater number of patients are taken there first; but it is enacted that " No person shall be detained in any workhouse, being a lunatic, or alleged lunatic, beyond the period of fourteen days, unless in the opinion, given in writing, of the medical officer of the union or parish to which the workhouse belongs, such person is a proper person to be kept in a workhouse, nor unless the accommodation in the workhouse is sufficient for his reception; and any person detained in a workhouse in contravention of this section shall be deemed to be a proper person to be sent into an asylum within the meaning of section sixty-seven of the Lunacy Act, chapter 97; and in the event of any person being detained in a workhouse in contravention of this section, the medical officer shall, for all the purposes of the Lunacy Act, cap. 97, be deemed to have knowledge that a pauper resident within his district is a lunatic and a proper person to be sent to an asylum; and it shall be his duty to act· accordingly, and further to sign such certificate with a view to more certainly securing the reception into an asylum of such pauper lunatic as aforesaid."—(25 and 26 Vict. c. 3, sec. 20.)

This section enacts that any medical officer having knowledge of a lunatic being in his district, being a proper person to be sent to an asylum, shall give notice of it in writing to the relieving officer or overseer.— (*Vide ante*, p. 305.) .

The foregoing remarks apply to pauper patients resident in a parish or district.   But patients are often found at large—wandering lunatics, as they are called—and the law deals with them in this capacity.   It is enacted (16 and 17 Vict. c. 97, sec. 68) that every constable, relieving officer, or overseer, who shall have knowledge that any person wandering at large within the parish is deemed to be a lunatic, shall immediately apprehend and take such person before a justice, calling to his aid a medical man, and obtaining from him a medical certificate, may make an order for the lunatic's reception into an asylum or hospital.   Or the justice may act on his own knowledge, and may examine the lunatic at his own abode or elsewhere.

This is to be done whether the patient is a pauper or not.   Patients may be found wandering at large and be taken care of in this way till their friends can be communicated with, or they may be taken to an asylum, if paupers, and thence transferred to the asylum of their own parish.   But in this manner they are to be dealt with according to the law.

There are other patients for whose amelioration the law makes provision.   These are people not wandering at large, but ill-treated or neglected by their relations or friends.   Not unfrequently do we read in the newspapers of lunatics found caged in cellars, attics, or outhouses, and more or less neglected or cruelly treated.   Or, short of this, a lunatic may be allowed by his relatives to remain in his own house in a state in which he is dangerous to himself or others.   Here the enactment is in some respects similar to the last mentioned.   The constable, relieving officer, or overseer of any parish, having knowledge of there being such a lunatic not under proper care or control, or being cruelly treated or neglected by any relative or other person having the care or charge of him, shall give information on oath within three days to a justice of the peace, who shall visit and examine such person, or direct some medical man to visit and examine him; and shall then require any constable or relieving officer to bring the lunatic before any two justices of the county or borough, and they shall call upon a medical man to examine him, and, with his certificate, send him to an asylum.   They may, however, suspend the removal for a period not exceeding fourteen days; and they may hand the patient over to his friends, if satisfied by them that he will be properly taken care of.

You are not to forget, however, that it is lawful for any one to restrain a lunatic who is dangerous to himself or others, by virtue of the common law, apart from the lunacy statutes. This has been decided more than once. In Scott *v.* Wakem, an action of trespass was brought against a medical man for placing the plaintiff under restraint while in a state of delirium tremens; and Baron Bramwell ruled that a medical man may justify measures necessary to restrain a dangerous lunatic. The same opinion was held by Chief-Justice Cockburn, in Symm *v.* Fraser and another, in 1863. Here Mrs. Symm, a widow, had been restrained while in a state of delirium tremens. It is done, in fact, constantly; certificates of lunacy are not signed for patients whose malady only lasts for a few days. We use the measures necessary for their safe custody, as we should for those delirious from fever or other diseases. And in the case of dangerous lunatics, you will recollect that you are justified in restraining them by force from doing mischief, till the order and certificates necessary for placing them in an asylum are signed. Do not be timid in taking such steps. Do not, as is so often the case, let the patient go on till something dreadful occurs. The bench of judges will take care that you are held blameless in such a case, whatever prejudiced juries may think. The Chief-Justice, in the latter of the actions I have named, desired the jury "to consider the case not only with reference to the interests of the individuals committed to the care of medical men, but also with a view to their interests in another sense—taking care not to impair or neutralize the energy and usefulness of medical assistance, by exposing medical men unjustly to vexatious and harassing actions."

In Scotland the procedure is different. Instead of a patient being placed in an asylum on the "order" of a relative and two medical certificates, the order is given by the sheriff. A petition is presented to this official by a relative or other person, accompanied by a statement and two medical certificates, asking for an order that the alleged lunatic may be received into a specified asylum. If the asylum is a public asylum, the patient a pauper, one of the certificates may be signed by the medical superintendent, or physician, or assistant physician thereof. If a private asylum, the certificates must be signed by medical men having no pecuniary interest in it. If the sheriff is satisfied, he grants an order which is valid for fourteen days. A copy of the order, petition, statement, and certificates, has to be sent to the Commissioners after two clear days, and before the expiration of fourteen days, with a notice of admission and report of the bodily and mental health.

The person who places a patient in an asylum can authorize his discharge. If another relative or friend wishes to procure his discharge, he must obtain an order from the sheriff, together with two certificates signed by medical men approved by the sheriff, and must give eight days' notice in writing to the person on whose petition the patient was sent to the asylum. The Commissioners in Lunacy may release a patient on the certificates of two medical men appointed by them.

The same provisions may apply to the reception of a patient in a private house, but the ordinary method in such a case is to get authority for such reception from the Board of Lunacy, which renders the sheriff's order unnecessary. The legal penalty is twenty pounds, if a patient is received without such order or authority. But any person suffering from incipient or transitory mental disorder may be received and kept for profit for six months without any such order or authority, if the doctor in attendance grants a certificate that the case is not confirmed, and that it is expedient that the patient should be so placed in a temporary residence, with a view to his recovery. This is an excellent measure, and it is much to be regretted that it is not in force in England.

There is one other legal procedure on which I must say something. Hitherto I have been speaking of the legal methods of restraining the person of a lunatic, private or pauper. But the law, by another process, makes provision for the guardianship of the property of a patient.

In old times the King was held to be the natural guardian of idiots and lunatics, and committed the care of them to whom he chose; but now the Lord Chancellor is directed by the Crown to perform this office, and such people become wards of the Court of Chancery. There is a numerous array of statutes relating to "Chancery lunatics," as they are called—statutes which have grown up alongside of those I have already mentioned, and which in some respects clash with them. There is a separate Board of Commissioners to look after such patients, and the consequence is that in many details confusion exists.

For, although a patient may have been for twenty years a certified patient in an asylum, visited regularly by the Board of Commissioners in Lunacy, at the head of which Board is, nominally, the Lord Chancellor, yet so far as his property is concerned, the said patient is considered of sound mind; and to deal with it on his behalf a commission must be issued by the Lord Chancellor to try whether he be of unsound mind—a fact which may have been known to one Board of Commissioners for a long period.

Not to go into details which do not concern you, I may say, that the present practice is for some one or more persons interested in the patient to petition the Lords Justices to direct that an inquisition shall be held as to the state of mind of the said patient. This petition must be accompanied by affidavits of the mental condition, and you may be called upon to give such an affidavit. The patient must have notice given him of the presentation of the petition, and, if he chooses, he may, within seven days of such notice, demand a jury. If the Lords Justices direct an inquiry, it is held, generally speaking, by one of the Masters of Lunacy. But if the Lord Chancellor think fit, he may direct the issue to be tried by one of the judges of the Supreme Court. When the property of the alleged lunatic does not exceed in value the sum of one thousand pounds, a commission of lunacy may be avoided. By the Act 25 and 26 Vict. c. 86, sec. 12, in order that the property of insane persons, when of small amount, may be applied for their benefit in a summary and inexpensive manner, it is enacted as follows:—" Where, by the report of one of the Masters in Lunacy or of the Commissioners in Lunacy, or by affidavit or otherwise, it is established to the satisfaction of the Lord Chancellor that any person is of unsound mind and incapable of managing his affairs, and that his property does not exceed one thousand pounds in value, or that the income thereof does not exceed fifty pounds per annum, the Lord Chancellor may, without directing any inquiry under a commission of lunacy, make such order as he may consider expedient for the purpose of rendering the property of such person, or the income thereof, available for his maintenance or benefit, or for carrying on his trade or business: provided, nevertheless, that the alleged insane person shall have such personal notice of the application for such order as aforesaid as the Lord Chancellor shall by general order direct."

The alleged lunatic may demand a jury, and the demand must be complied with, unless the Lord Chancellor is satisfied by personal examination that the individual is incompetent to express or form a wish on the subject. Practically, we find that many patients are in this condition; no jury is demanded, and then the issue is tried by one of the Masters without a jury.

Whether there is a jury or not, you may have to be examined on oath as a witness, and, it may be, cross-examined, and it behoves you to form a very clear and accurate conception of the opinion you are going to give, and the grounds on which you will uphold it. Counsel will try to entrap you in every way, and ask you to define insanity or unsoundness of mind. Do not, however, be tempted into discussing any abstract questions; confine yourself to the case before you, the state of mind of the alleged lunatic, and that which he has said or done. You will be assailed with questions as to whether you think this or that act indicative of insanity. Such an act may possibly be done by a sane person, but a number of such acts may be conclusive as to the insanity of any one, or one act may at once stamp the particular individual as insane.

Your opinion may be asked as to the advisability of holding a commission of lunacy, for it is not expedient to take this costly step if the patient is likely to recover soon, or to die. Solicitors often fancy that a commission of lunacy is to be taken out as soon as a patient is put under legal restraint, but this is not so. Unless his affairs urgently demand it, such a step should be deferred until it can be seen whether he is likely to recover in a reasonable time or not.   I have known a patient nearly well before the commission was held; and if he is likely to recover within a few months, it is most unfair to subject him to the expense and stigma of a commission, and throw upon him the trouble and cost of superseding it. Your prognosis will be based upon the principle enunciated throughout these lectures, which I need not repeat here.   Time is in this your great auxiliary; though patients do recover after years of insanity, they do so but seldom.   If a patient has been under care and treatment for a twelve-month, and does not show manifest signs of real improvement, his case is sufficiently unfavorable to warrant at all events an inquisition.   For, as I have said, this may be superseded on recovery.   The patient will petition the Lord Chancellor or Lords Justices to supersede the petition and set free himself and his property, and he must support his petition by medical affidavits.   Here the questions of recovery or partial recovery, or apparent recovery, will arise, and you will recollect what I said on these heads in a former lecture.

The proceedings in Scotland analogous to a commission in lunacy differ from the latter in some important respects.   A *curator bonis* may be ap-pointed by the Court of Session upon the petition of a relative supported by two medical certificates. If this is done, the patient's property is taken care of, but his person is not interfered with by such proceeding.   To deprive him of his personal liberty the sheriff's order is necessary, as before mentioned.

LECTURE XX.

THERE remains one subject on which I must say something. I have spoken of the legal formalities necessary to be observed when a man or woman is placed in confinement, and have mentioned that you will be called upon to sign medical certificates and affidavits of the unsoundness of mind of a patient. I propose to say a few words concerning the way in which you are to examine such people with a view of testing their mental condition. Very general must my observations be, for it is not possible to lay down rules for the performance of such a task with anything like strictness. Yet some hints may be useful to those who are quite without experience in the matter. You have two things to decide before you sign a certificate; first, whether the individual is or is not of unsound mind; secondly, whether he is a fit and proper person to be detained under care and treatment. These are distinct questions, and it is clear that the Legislature, by thus distiguishing them, allows to medical men a certain judgment in deciding whether or not a person, who may be of unsound mind, is a proper person to be detained under care and treatment as a lunatic protected or restrained by certificates of lunacy. Many patients during acute illness may be for a time of unsound mind, yet can in no sense be called proper persons to be detained as lunatics under care and treatment; and there may be some of feeble mind, yet gentle, harmless, and docile, who do not require the protection of the lunacy laws, and are not proper persons to be detained. As I have said elsewhere, it is not always easy to sign a certificate for a patient concerning whom we may make a general declaration in an affidavit, for the Commissioners in Lunacy insist that all that is alleged of the patient shall have been observed on a given day. It must not be the outcome of an acquaintance extended over some years, and although you have a general opinion that the individual is weak-minded or insane, it may be based rather on what you have heard than on what you see.

Concerning the cases of acute disease in which the mind is temporarily disordered, little is to be said. You will not think of signing certificates here. And in acute insanity, where medical assistance is urgently needed, there will be little difficulty in appreciating the state of mind, and signing

a certificate.   In these cases the real difficulty experienced is more frequently in gaining access to the patient, and engaging him in conversation.
This done, his malady is revealed, and our end is accomplished.   In gaining admittance to a patient, our difficulties may come from the patient
himself, or from ill-judging or ill-meaning friends, who, because they
think that all doctors are leagued together to shut every one up in a mad-
house, or because they have an interest in keeping the patient where he is,
frustrate the endeavors which perhaps his nearest of kin are making for
his safety or cure.   Such persons resist the inspection of the patient, on
the plea that he is not insane, but only a little excited, and requires rest
and quiet.   They will insist that he is not dangerous, and to the best of
their ability they will keep him from doing anything very outrageous.   I
suppose, that scarcely one lunatic has ever been placed in an asylum without some of his friends or acquaintances denouncing the sinfulness of the
proceeding.   There is, however, little danger, though there may be some
difficulty, in visiting such a patient.   There is more to be apprehended
from one who himself dreads and avoids you, and who, from a fear that
you are coming to do him some harm, may resist to the uttermost, using
murderous weapons.   In such a case it is not possible to lay down rules
which are universally applicable.   You have to converse with the patient,
to assure yourself of his insanity, to sign a certificate.   Here, if at all, it
may be justifiable and necessary for you to resort to stratagem, to invent
an excuse for an interview, to feign to be other than a doctor.   Such
measures are to be avoided when it is possible, and they often can be
avoided, by tact or by open and straightforward plain speaking.   They
often lead to great difficulties, cause the patient to distrust all about him,
and give him occasion to make great complaint.   But I am not prepared
to say they can always be dispensed with.   If a madman has armed himself with a revolver, and vows that he will not be shut up, and if he has,
by previous experience, found out that doctors are a necessary item in
the process, he will be a bold man who will go in a strictly professional
capacity to sign a certificate.   One thing is certain, that stratagems are
better left alone in many cases where friends urge their adoption, especially the devices invented by friends, which frequently are so clumsy that
you may by them be absolutely debarred from having the requisite conversation with the alleged lunatic.   I have, on arriving at a house, been
shown suddenly into a patient's room, and introduced to him as some
person of whose name, occupation, or relationship I was utterly ignorant.
If you are introduced, not as a doctor, but as a lawyer, man of business,
or the like, you cannot discuss the patient's health, mental or bodily; and
questions which you may wish to put will sound impertinent or absurd,
or will make him suspect you to be a doctor in disguise, and he may then
refuse to hold any conversation with you.   In most cases go as a doctor,
and as nothing else.   You have then a reason, whether he admits it or

not, for cross-examining him closely as to his bodily and mental health. If stratagem is absolutely necessary, consider it well before hand, its probable direction and consequences, and be sure that those in league with you play their parts faithfully. I am assuming now that the insanity of the patient is not doubted, but that conversation with him is difficult. The peculiar features of the insanity will furnish suggestions for your plan of proceeding. One man has invented a marvelous scheme for enriching himself and all belonging to him. You are come to treat with him for the purchase of his patent, or a partnership in his business. Another is going to buy houses and lands. You have houses and lands to sell. There is little difficulty in dealing with such, or in gaining access to them. But if a man is suspicious, fears a conspiracy, and shuts himself up against police, bailiffs, or the like, he may resist strenuously all efforts to observe him. Such a patient is, however, by the nature of his case, fearful; and if, accompanied by sufficient assistants, you boldly confront him, he will probably not be able to escape entering into conversation with you. If access is denied to you, not by the patient, but by others, you must consider how the law stands. A man's own house in this country is his castle, and, sane or insane, he cannot be removed thence except for some good reason, and after lawful proceedings. The law allows a man's relatives or friends to remove him from home for treatment and cure upon a legal order and certificate; but if a husband chooses to keep his insane wife in his own house, or a wife her insane husband, no one can order his or her removal unless it can be shown that he or she is improperly treated or neglected. Cases of this kind often arise, and the lunacy authorities are appealed to and requested to give an order for the patient's removal; but they have not the power, and the only method of effecting it is to lay information before a justice or justices, as I have mentioned in my last lecture.[1] If a patient is properly treated in a relative's house or his own, and has medical advice and care, no magistrate would feel called on to order his removal, even if other relatives desire it. But if a person who is no relation takes charge of and detains a patient against the wishes of all the family in his own or the patient's house, it is probable that a magistrate's order might be obtained, and access demanded. The Commissioners in Lunacy have little power over patients until they are brought under their jurisdiction. When they prosecute persons for illegally receiving and detaining patients, they do so only when they can prove that the patient is taken into the house, or taken charge of, "for profit," and taken charge of "as a lunatic," that is, by one who must have known him to be a lunatic. A friend taking charge of a lunatic without profit, for friendship's sake alone, would not be reached by his portion of the Act, and the Commissioners could not order the removal, which is only to be effected through the intervention of a magistrate.

[1] *Vide*, p. 306.

Passing from cases where our difficulty lies in gaining access to the patient, I come to those where opportunities of observation and conversation are afforded, but where the insanity of the individual is doubtful or difficult to detect, or is denied by himself or certain of his friends or relations. The difficulty may lie in the slightness of the insanity, or in the ingenuity with which the patient baffles our endeavors to detect it.

If the alleged lunatic has been previously under our care, and is known to us, we shall need little information from others; but we are often consulted as medical men, by the friends of patients with whom we are previously unacquainted, and of whose sane condition we are entirely ignorant. We are consulted by friends who wish us either to say that the patient is insane or sane, according as they themselves think, and they wish, of course, to enlist our assistance and evidence to support their own view of the question. Now, do not be led away by the *ex parte* statement which you will receive about a doubtful or disputed case of insanity from those who first consult you. Do not be induced to be an *ex parte* witness, retained like a barrister on one side or the other. Receive all you hear as matter requiring proof, and recollect there are two sides to all such questions. Before you go into the presence of a patient, find out as much as you can concerning him from people who differ in the opinion they hold, and, if it be possible, from as many persons of all ranks as you can, —relations, acquaintances, servants,—and consider whether their accounts agree or not. And if they disagree, and one side represents him to be sane, and the other insane, consider which is likely to be the better informed, the least prejudiced, the least interested, and the more reliable. One party, who does not wish the patient removed or interfered with or taken out of its hands, will say that he is not violent or dangerous, but only a little "excited;" that all his so-called delusions are not delusions, for they are all based on facts. Others will justify what he has done, or say that he was provoked to do it. "Excited," "excitable," "excitement," are words which are bandied about in an extraordinary manner. One person is said not to be insane, but only "excited," while "excitement" in another case is alleged as the chief evidence of insanity. It is a vague word meaning nothing, and I advise you not to employ it in writing certificates, or giving evidence concerning insane patients. If friends use it, request them to explain what they mean, whether excitement of speech, excited acts and gestures, or what? As for delusions not being delusions because they are based on something true, we know that the greater part of delusions, like dreams, arise out of some fact or combination of facts, and have these as their groundwork; but they are none the less delusions. And if, on the other hand, friends allege that the patient is under delusions when he holds certain opinions, or asserts certain facts, which, though improbable, are possible, you will have to consider whether they may be true, even if the friends would wish to persuade you that they are all phantoms, or whether a man may not hold extraordinary, or even extravagant opinions, without their proceeding from insanity. All the information that can be got you will receive and weigh, and will then test it by personal conversation and examination of the patient.

I now suppose that you are brought face to face with the alleged lunatic. If you are shown into a room where he is with other people, you may not be able to say at a glance which is the individual, and may not be at liberty to ask. It is something more than awkward to commence a conversation with the wrong person, so that I strongly advise you to make sure before you enter the room that you will have no difficulty in fixing on the right one. You can ask such questions concerning the number of people there, the appearance of the individual, or the distinguishing marks of his dress, so as to render any mistake impossible. This may seem a piece of trifling advice, but I have known the difficulty to occur. If you can, get a friend to introduce you and open the conversation, and there can be no better way of doing this than by inquiring after the patient's health. Frequently you can be introduced by his ordinary medical adviser, or you may say that you have come as his substitute. There will be little difficulty in discovering the insanity of the melancholic. Though he thinks he is past all human aid. he will freely tell you his woes and fancied misfortunes. The gay, exalted, and hilarious paralytic and maniac will disclose his malady readily enough. But when you have to deal with the suspicious monomaniac, the man of concealed hallucinations and delusions, with patients who have been shut up already, or with people who are merely weak-minded, or whose insanity is displayed in acts rather than in words and delusions, you may converse. for a very long time without being able to detect the hidden disorder, or to satisfy yourself that what you have heard is true, and that what has been done has been done from insanity, and not from depravity or wanton mischief. There is no better way of commencing the conversation than by inquiring after the patient's health, because it is a conversation on a point in which he is concerned. You may talk to a man forever on points which do not concern him, on the weather or the crops, on politics and the topics of the day, and he may converse freely, rationally, and like an ordinary being, if he has no delusions concerning such matters. You must bring round your conversation to himself, for this is the point on which he will display his insanity. He is the subject of whom all his delusion are predicated. You may talk over an enormous range of ground and an infinity of topics, you may even talk of matters concerning which he has delusions, but if you do not connect him with them, your labor may be in vain. And now you see the advantage of appearing in your own character of doctor. You assume the right of questioning the patient about everything which directly or indirectly affects his health, such as occupation, residence, mental work or worry, habits—in fact, his daily existence.

He may assert that you are not his medical adviser, that you have no
business to question him, and that he wants none of your advice; but you
will assure him that you have been requested by his family or his own
medical man to see him, and will tell him that they have been alarmed
at his symptoms, at what he has said, or done, or threatened to do, and
this he must explain away or deny.   And in his justification, explanation,
or denial of insane sayings and doings, he will generally open up the real
state of his mind.   If you gain his confidence, and he enters into conversation, it is as a rule not difficult to detect the insanity.   But he may
refuse to talk, and, without keeping absolute silence as a melancholic or
demented patient, may yet tell you nothing whatever.   Such answers as
he does give are pertinent and correct, but he will not converse concerning himself or any one else.   When this is the case, you cannot sign a
certificate, and there is nothing to be gained by pressing a patient beyond
a certain point; it is better to take your leave and see him again on a subsequent day.

I warned you that in many cases it is time lost to talk on indifferent
topics which do not concern the patient; for the same reason there is
nothing to be gained by going into his presence, ostensibly to talk to
some one who is in the room with him.   You cannot turn from such a
person, and suddenly commence to question the patient; and if he, upon
your entry and assumed business with another, gets up and leaves the
room, you have no excuse for detaining him.   Your visits must be to him
and to no one else.   I mention this because it is a plan often proposed by
friends, which as often fails, and the failure of your first attempt often
involves the second in greater difficulties.

You may possibly observe something in the patient's appearance, dress,
or occupation, which may form a topic of conversation and afford a clue
to his mental peculiarities.   It may be so extraordinary as at once to indicate insanity, or it may suggest delusions which you may by its aid extract.   Nor will you forget during the interview to survey the apartment,
supposing it to be the patient's own, and notice anything that is *bizarre*
or startling.

Much has been written concerning the physiognomy of the insane.   It
is supposed that insanity stamps itself in the countenance of a man, and
is recognizable there.   In many cases it is, but it is recognizable in some
only by those persons who know the patient in his sane state.   Nevertheless, when we approach a patient for the first time, we shall generally find
that his emotional state shows some peculiarity.   He is not perfectly easy,
unconstrained, and void of all undue emotion.   Either he is a little too
gay, or a little too dull.   On entering into conversation we may find him
in high spirits, jocund, hilarious, and boisterous toward a stranger; or dull,
suspicious, and snappish; or decidedly depressed and melancholy.   And
these various moods may of themselves, without further information, aid

us in discovering the corresponding delusions. The gay and hilarious man will have exalted ideas concerning himself, his personal strength, beauty, and prowess, his wealth, rank, and prospects; the depressed man will have all the delusions of melancholia, will think his soul is lost, his fortune and business ruined, or his body a prey to all manner of disease; while the irritable and suspicious man will think that there is a conspiracy to ruin him, or that people are accusing him of unnatural crimes, that every one looks at him in the streets, and that the newspapers refer to him in all they report. From noting the general manner and demeanor I have often discovered delusions which have been unknown to the friends, and hitherto carefully concealed.

It is always satisfactory to discover delusions. Although insanity may and does exist without them, yet there can be no doubt of its presence when we discover them. But you must make perfectly sure that what you think or are told is a delusion, and one beyond all reasonable question. Certain statements admit of no doubt. A patient told me that he had the devil in his inside, who had converted everything there into cinders, and he produced, in support of this, some black powder, which he said was his fæces thus converted. But many assertions we only know to be delusions from information derived from others—information not always forthcoming, and not always credible. A very common delusion is that which a man entertains concerning his wife's infidelity. But if this is all you can discover, and you have no informant but the man on the one side and the wife on the other, are you necessarily to believe her statement that it is a delusion? You will consider the man's general state, his mode of asserting the fact, and the grounds he gives for his belief, and you may very likely, without further testimony, convince yourself from his whole story that he is insane. You will in a certificate state the delusion, your belief that it is a delusion, and your grounds for the belief. Always say that such and such a fancy is a delusion, if it is a thing possible to happen or to have happened; or that you believe it to be a delusion, and your grounds for such belief. Thus, " The patient tells me that he is ruined, which I am assured by his wife, or son, or lawyer, is an entire delusion;" or, " The patient asserts that his wife is unfaithful, but he cannot tell me the name of any man, or give any grounds whatever for his belief, which I look upon as altogether a delusion."

Friends may have an interest not only in shutting up an individual, but in making it appear that statements made by one who is undoubtedly insane are all of them delusions. Madmen have an unpleasant way of revealing family secrets, and it is convenient to call all such revelations delusions. Here you must, if possible, derive information from others who are not primarily concerned—old servants, medical men, acquaintances, and the like. But for them I should certainly have been disposed to accept as delusions some of the facts that have been told me by patients, and even now I am in doubt about some, never having been able to arrive at the truth. You must well consider who the person is who makes a statement to you concerning a patient and his delusions. Is he or she the person who wishes to place the patient under restraint? Has he or she any interest in so doing beyond the welfare of the alleged lunatic? Fathers and mothers do not often wish to confine their sons and daughters as lunatics, unless they really are such, but it sometimes happens that fathers would like to shut up as mad an unruly son who is vicious and bad. I receive with the greatest caution that which I hear from husbands concerning wives, and *vice versa*. Then beside the interest which one person may have in confining another, we must make all due allowance for the possibility of our informant being frightened, prejudiced, or ignorant. We must also admit of the patient being an ignorant person, or one holding very extraordinary opinions on certain points, as religion or politics. Others who differ with him may look on such ideas as quite sufficient to warrant their assertion that the man is mad. We must also make due allowance for the class of life of the alleged patient, especially if we have to base our opinion upon violence of conduct and language. If a gentleman or lady of exalted station and refined manners uses blasphemous and obscene language and vile epithets, we may reasonably question his or her sanity; but if one of the *plebs* call his wife filthy names, and threatens to beat her, it does not follow that he is under a delusion as to her fidelity, or that he is insanely dangerous or homicidal. All due allowance must be made for the rank, station, and previous habits of the individual. Counsel may ask us whether we consider swearing, indecent or profane language, proof of insanity? Of course it is not, looked at *per se*, but when uttered by this or that person it might be as direct evidence of the state of mind as anything could be. A lady once told me that she knew her husband must be mad when he met her at the station in a white hat. Wearing a white hat is not usually considered evidence of insanity, but such an act on the part of this gentleman—a grave clergyman—convinced his wife that something was very much amiss. So, when you hear people talk of a patient having delusions, you must request them to state exactly what these are, and it may be that you will have to cross-examine them pretty closely before you examine the patient. You will then examine him upon them, either bringing round the conversation to such points, or telling him that you have heard such things alleged of him. You need not be supposed to believe them, but you can inform him that such things are said of him, and beg for, or at any rate hear, his explanation. And if he denies the whole that you say you have heard, confront him, if possible, with those who have told you. I say, if possible, for friends are very reluctant to assist in such matters, and wish that the lunatic should not know whence we get our information. We are constantly requested not to mention the name of this or that informant. Do not be hampered by any promises whatever. If a matter is to be mentioned, it must have come to us from some one, and patients of the partially insane class can see through the shallow subterfuges so often invented by friends, and their distrust and dislike are only increased and strengthened thereby.

Still greater will be the difficulties which you will have to encounter in signing certificates for patients who have no delusions. You may be called on to do this in the case of a man who is said to be "morally insane," who is altogether altered in character and habits, and who in various ways behaves absurdly or outrageously. Everything here will depend on the opportunities you may or may not have of recognizing the change in the individual. If I can say of my own knowledge that a patient is totally altered, it is perhaps the strongest assertion that I can make in support of the opinion that he is insane; but if I am called in to see for the first time one who is said to be thus changed, I cannot compare his present with his former state. I can only compare the present with what I hear of the past. Now, these patients have generally sense enough to behave themselves decently while in the presence of a stranger, especially if they know him to be a doctor. We do not see their eccentric or insane acts, and they may absolutely deny them. They will justify or explain away a thousand little sayings and doings which, though they may sufficiently illustrate the change which has come over them, may, nevertheless, when taken singly, sound trivial, and are not enough to constitute insane acts. It is not likely that you will be able to sign a certificate in any such case after one single interview: probably you will have to visit and examine such a patient several times, and a comparison of what has passed on the several occasions will greatly aid you in coming to a conclusion. If the disorder is acute and rapidly advancing, you may discover delusions in a patient in whom a week previously there were none; in others delusions may have existed and have passed away, yet the man has not yet recovered, but remains changed, eccentric, restless, excitable, insane, obviously unfit to take charge of himself or his affairs. In many of these, though we can discover no delusion, there is obvious intellectual defect. The patient rambles from subject to subject, and do what we will we cannot keep him to the point. Ask him half-a-dozen simple questions, and you will not get a plain and direct answer to one of them. He will display not an incoherence of words, such as we find in the babble of delirium or acute mania, but an incoherence and inconsequence of thought, which is quite unnatural to him and incompatible with the proper conduct and occupation of men in general. Then if we question him closely as to the extravagance or absurdity of his acts, his attempts to justify them are ofttimes ridiculous and childish. Before we come into the presence of one of these patients we may obtain all the information we can upon the following points from as many friends as possible. Is there anything which they consider to be the unmistakable origin of the disorder? Has he had any epileptic, or epileptiform, or apoplectiform attack, or anything at all of the nature of a fit? Has he ever had a blow or fall on the head, or any bodily disorder which at the time affected the head? Has he undergone any serious loss, worry, anxiety, overwork, or loss of rest? If they tell you that he is altered, make them state distinctly in what respect he is altered. Are his altered acts those of commission or omission? Does he neglect his business, forget his appointments, forget the dinner hour? Or does he buy useless things, or articles at a price beyond his means? Does he keep loose company, ill-treat and abuse those nearest to him, profess dislike or indifference to those he has hitherto most dearly loved? Has he become bold, noisy, and talkative, instead of being shy and reserved? Has he made any extraordinary changes in his personal appearance, his dress, hair, or beard? Does he now take more drink than he ought, having previously been sober? Is he excessively restless, never settling to anything, but constantly coming and going? From the answers we receive to these questions we may, if our informants are credible, satisfy ourselves of the patient's insanity before we enter his presence. It is probable that such questions will in the main be answered truly. We may then ask one more, which is as likely to be answered untruly. Has any relative of the patient's ever been insane? In our examination of the individual we shall bear in mind all that we have heard, and shall consider the mode in which he gives his answers, their coherence, his justification or denial of his acts, and the reasons for his treatment of or estrangement from his family and friends, if such exist. And herein, as I have before reminded you, our questions must be applicable to the condition, education, and station of the individual, be he gentle or simple. We must avoid the discussion of abstract right or wrong, and the question of whether such acts as his are right or wrong. We have only to consider whether they were done by him because he was insane, and would not have been done by him in his sane mind. We are concerned not with things wrong or right, moral or immoral, but with things irrational. As men have held the most extravagant opinions, so have they committed the most diabolical crimes, while sane. But the manner and method of their deeds and opinions we have to criticise. The arguments used to uphold them may bring to light an insanity not visible in the acts, and may indicate the disorder of the mind from whence they sprang. There is a strong love of argument and controversy in many such patients, and they often defend themselves in an exceedingly ingenious, yet no less insane fashion.

You may have to examine, for the purpose of signing a certificate, one who is not and never has been insane, in the ordinary sense of the word, but whose unsoundness of mind is of the nature of weak-mindedness or imbecility, defective rather than insane mind. You have here no former condition of healthy mind with which to compare the present. You must compare it with an assumed standard of average humanity, and the question will be, at what point will you decide that soundness ends, and unsoundness of this description begins? In conversing with such a person you will find no delusions, no loss of memory; there will be no definite commencement of the defective state, nor any assignable cause. All that you have to guide you is of a negative character. It is the absence or negation of mind, rather than the positive presence of symptoms of insanity. There may be positive vice, vicious habits and propensities, fondness for low company, thieving, lying, and the like; but your difficulty will be, first, to recognize this for yourselves, for such offenses will not be committed in your presence; and, secondly, to decide whether they proceed from depravity and badness, or from imbecility, which renders the individual irresponsible. There are many of our criminals who, were they higher in society, would be considered irresponsible and placed in asylums; being what they are, they fill our jails. But I have already spoken of these patients; I here merely wish to make a few suggestions for your guidance in dealing with this most difficult class. Unless the individual has been under your observation for a long time, you will scarcely be able at once to give a decided opinion in a doubtful case. The friends will come to you strongly biassed. Either they wish to shut up the alleged lunatic because he disgraces his family by his vice and iniquities, or they wish to prevent him from squandering his property; or, on the other hand, they wish to make out that there is nothing the matter with him, if, perchance, the Commissioners in Lunacy consider that he ought to be protected by certificates. You will not therefore be able to receive that which you hear from friends without considerable caution and discrimination. You are brought face to face with a youth of this class, and proceed to question him. If you tax him with his vices, he either denies them, or, admitting them, confesses and acknowledges that they are wrong; but probably he can mention many others who do the same things. You proceed to test his knowledge; and this, in my opinion, is the only test, especially when vice is absent. Is he teachable, capable of receiving instruction and profiting by it? The chances are that, being hard to teach, dull and disinclined to learn, his education has been neglected, and he has been placed with tutors to be kept rather than taught; or put to learn farming, or sent to sea, or to some other calling for which he is equally unfitted. So that his ignorance on special subjects may be due to his preceptors rather than to himself. But a gentleman's son ought to have learned by the time he is twenty-one to read, write, and spell; and if at this age his spelling and letter-writing are those of a child of eight, we may reasonably think that deficiency may be the cause, if he has had fair opportunities. The ignorance and inability to learn may be a good ground for making an affidavit for a commission in lunacy that a patient is incapable of manag-

21

ing his affairs; but in the matter of signing a certificate, I think it also ought to be shown, and you ought to be able to say, that in your opinion the individual is a proper person to be detained under care and treatment as a lunatic, because he is not fit to be at large unless closely watched. Many of these people will run off in a vague and purposeless way from any place in which they are living. They cannot be trusted alone, for they would get into mischief, and they cannot be allowed to have the control even of a small sum of money. Indoors they may destroy furniture or their clothes, wantonly set things on fire, or practice horrible cruelty to children, dogs, or cats. Where we find this to be the case, we may reasonably think and say that such a one is a proper person to be detained under care and treatment, if it appear from the mental deficiency that his whole condition is that of an imbecile or idiot.

Far more easy is it to sign certificates for those who, from old age or brain disease, are fatuous and demented. You will here have signs and symptoms of a more certain weight and significance than are to be found in the last-mentioned class. You can compare the present condition with the past. Whereas a man was once vigorous in intellect, clear in conception, and of good memory, we now have but the wreck of mind remaining. Chief of all we find a failing memory. There is no recollection of what happened yesterday or the day before—no recollection of your last visit, possibly not even of your name. If a patient drifts into dementia from a state of mania or melancholia, we may find still some of his old delusions or the traces of them. But in many who are demented from paralysis or epilepsy, there may be no delusions, but an absence of mind and intellect—a repetition of the same question, or the same story, or the same sentence, with entire forgetfulness of having asked such a thing before, and an equal forgetfulness of the answer received. You will be asked to sign certificates concerning such patients to legalize their residence in an asylum or family, or to deprive them of their rights by a commission of lunacy, and you will, as I have said, have little difficulty in coming to a decision. Loss of memory is a morbid state far more perceptible and appreciable than the defective intelligence of the weak-minded youths I have last spoken of, and such patients demand care and personal attention in a way that many of the others do not. Of course, it will be for you to consider the extent and the constancy of this impairment of memory, and how far it would necessarily render a man incapable of managing his affairs. As, on the one hand, no person's memory is perfect, as we all at times forget a name or date, so, on the other hand, every man whose mind is not completely obliterated, or who is not unconscious, can recollect by the eye, or in some way, something of past events, or those persons nearest and dearest to him, therefore, when you meet with imperfect memory the result of disease, and no other symptoms, you will have to decide whether the imperfection is sufficient to render the patient unsound of mind in the eye of the law. "Sound mind, memory, and understanding," is the phrase lawyers use to indicate capacity. Does the patient before you possess all of these? Clearly a man cannot manage his property who cannot recollect its nature or amount. If he cannot recollect whether he has a wife, or the number of his children, he cannot be considered competent to make a will. If he has so forgotten places that he does not know whether the house he is in is his own or another's, nor the name or situation of it, he cannot be said to be dwelling there of his own free choice, or to be able to arrange such matters for himself. The facts in these cases speak for themselves. An old man may dwell in his own house, may sign his name to a piece of paper as he is directed, may hand to any one a sum of money previously prepared and put into his hand, and so may be said to manage his own affairs; but you may find that to-day he has totally forgotten every occurrence of yesterday, that he knows nothing about his property, thinks people are alive who are long since dead, and that all his management consists in signing everything that is placed before him. Patients of this feeble mind are, by reason of their feebleness, contented and happy. They put up with any treatment, however bad. They do not run away, and are said to remain voluntarily, whesever they may be. But their volition is as feeble as the rest of their minds, and their submission and surrender of all independence speak for themselves. And they are hopelessly incurable. They have to be taken care of for their natural lives, whereas the weak-minded youths may, in many instances, be improved. Hopes may be entertained of them, but of the demented there is no hope. They have possessed a mind, but have lost it past recall. Therefore certificates may be signed with much less hesitation than in the last mentioned cases.

Most unquestionable your aid and advice will be asked for the purpose of placing under legal restraint the inveterate drunkards so often called *dipsomaniacs.* The question of restraining them by special legislation, and in special inebriate asylums, is attracting much attention in the present day; but so far nothing has been done in this country, and I have only to speak to you of the existing law, which is the law of lunacy. It is not easy to advise legislation on the subject. We have at present a machinery consisting of Lunacy Acts, asylums, and Commissioners in Lunacy, by which any one can be put under restraint who is of unsound mind. What further addition can be made? Inebriate asylums have been instituted, in which people may voluntarily place themselves for the cure and eradication of the habit; but it is difficult for legislation to sanction the forcible incarceration of every person who may be called by his friends an habitual drinker. For those who are really of unsound mind, the present means are sufficient; and in the examination of one alleged to be a dipsomaniac, you will have to ascertain not only that he is a fit person to be taken care of and detained, but also that he is of unsound mind.

Although I cannot hope to give you any very definite or precise rule for your guidance in coming to a decision on this most difficult point, yet some few hints may be of service.

I abolish dipsomania from the varieties of moral insanity, together with such monomanias as erotomania, pyromania, and kleptomania. The unsoundness of mind which exists in connection with habitual drinking must be estimated like unsoundness in any other individual. Not every drunkard is insane, nor can he be confined because he ruins his health and property, any more than a confirmed gambler or opium-eater. There is an insanity, the marked feature of which is a craving for drink, but it is not the condition of a man who, after his work is over, goes to the public-house and gets drunk, whether he does so nightly or occasionally. He may squander his money and wreck his constitution, but during his working hours he is a sane and intelligent man. It is not the condition of a man who periodically drinks himself into delirium tremens. During the delirium we may, of course, sign certificates of insanity, if it is absolutely necessary for his protection to do so; but when he has recovered, he cannot be detained because at some future time he will most probably drink himself into another attack. In the interval he may be a perfectly sane man.

We shall have no difficulty in signing a certificate when unmistakable insanity has been produced by drink, whether it presents the symptoms of mania, melancholia, or dementia. Such will be recognized and dealt with as any other case of insanity.

There are, however, certain persons who seem impelled to drink, as others are impelled to murder or suicide. And this impulse is so strong, that they are rendered entirely unfit to take care of themselves or their affairs. If left to themselves, they would drink continuously till they reached the stage of delirium tremens or alcoholic paralysis. Closely examining them, we find them to be people who, from congenital or acquired weakness of mind, are unable to exercise any self-control, and are practically of unsound mind. They may have suffered from blows on the head, fits, previous attacks of insanity, or they may have by inheritance an insane neurosis. They probably desire to place themselves under control, and will voluntarily enter an asylum if it be possible. Here the drinking is most frequently the result of the insanity, which is, however, aggravated by the perpetual alcoholization.

Except in view of a particular case, one is obliged to speak in very general terms of such patients; but, as I have said concerning other doubtful forms of insanity, there is generally to be discovered some mental defect or peculiarity other than the act or habit of drinking, if we look for it carefully and have sufficient opportunity for its discovery. It is for the most part quite impossible to sign a certificate upon a single examination of one of these, but a longer acquaintance may remove our doubts, and enable us to say that he is not of sound mind, memory, and understanding. It is, of course, essential that we see him free from the influence of recent drink: always inquire into that which an alleged lunatic has had to drink since his last sleep; serious consequences might ensue were we to sign a certificate for a man who was only drunk.

This drinking insanity may be periodical or permanent. In the former case the patient must be liberated, as any other who has recovered from his insanity. He may be liberated upon trial, and if his recovery is only apparent, he may be readmitted. Very frequently it is permanent, and, in fact, is only the commencement of a more marked degeneration of mind.

Lastly, in your examination of a patient be careful, above everything, that he has fair play. You are about to do that which will deprive him of that we all hold most dear, and he is in some respects in the position of an accused person. You are examining him to satisfy yourself of the real state of his mind, not to trip him up and extract that which will sound well in a certificate. If he is a person inferior to you in intellect, you may puzzle him by cross-examination, so that he may seem really wrong in his head; but you know that a man in the witness-box may in the same way "lose his head," and, without the slightest intention of doing wrong, swear black is white. If the case is not urgent, and admits of doubt, always see a patient twice before you sign. Some vary considerably, especially women, at different times, and if possible you should see them at their best and at their worst.

You are not to omit the inspection of a patient's letters and writings. Many, who are very shy and reticent when brought face to face with a medical man, will, in that which they write, reveal the delusions and fancies under which they are acting. The whole style of a patient's letter, the signature and direction, may show the idea predominating in the mind; and defect of intelligence, the imbecile and childlike weakness, the failure of mental power, and forgetfulness of dementia, may be all displayed in written characters. Letters will assist you in finding out concealed ideas, or may illustrate points on which you have only been able to gain imperfect information from the patient or his friends.

Once more I must warn you, that you undertake a serious responsibility when you sign a certificate of lunacy. You are performing an act which deprives another of his liberty and rights, and are signing a legal document which you may have to defend in a court of law. Never put anything in a certificate which you cannot justify in the witness-box. You will not be absolved from responsibility because your certificate has been received by the Commissioners in Lunacy. A man may even be insane, and yet if your certificate be not true, and if all the requirements of the law be not carried out, you will be held responsible. Let that which you say in it be as short as possible, if only it be strong. And do not add to the facts which really indicate insanity others which do not indicate it at all. Be accurate in the filling up of the whole certificate. Scarcely a single certificate is ever sent in from a medical man that has not to go back to him for the correction of some error, or the insertion of something omitted. In your statement of facts, see that you state fully what you observe and what you mean. Do not write down a series of single words, such as " great excitement, delusions, refusal of food," but say, " I found the patient looking so and so, doing this or that, and he said that," etc. And if what he said was not unquestionably a delusion, you must add that it is a delusion, as you believe or are informed. And when you speak of refusal of food, or excessive drinking, or the like, take care that you speak of what you have yourself observed, or else that you state that such a fact has been communicated to you by others. I often find that the two sets of facts are intermingled by medical men. Do not use vague terms as incoherence, excitement, fatuity; but reduce your statements, as far as possible, to the enunciation of concrete facts.

Here, Gentlemen, I bring these lectures to a close. They are but brief sketches of some among the many topics which I might discuss. Of their imperfection no one is so aware as myself; nevertheless, they may serve as suggestions for further reading and observation; and I venture to hope that you will find, that from them you have derived some hints, when at a future time you have to treat insanity, or take the necessary steps for placing under legal restraint an insane patient.

TYPES OF INSANITY.

BY

ALLAN McLANE HAMILTON, M.D.

INTRODUCTION.

As we progress in our study of insanity, we are constantly reminded of the physical changes that take place in the patients committed to our charge. Disease of the brain makes itself known by well-marked bodily symptoms, that are in themselves almost as important as the many variations of disordered mental action. In the present work it has been my aim to put into simple form a few suggestions that may prove useful to medical men who, from time to time, meet with cases of insanity in their practice. The plates are drawn by Mr. T. J. Manley from instantaneous photographs; the subjects were selected from many hundreds of patients, and I believe them to be typical.

I wish to express my obligation to Drs. A. E. Macdonald and Franklin, as well as the gentlemen of the medical staffs of the male and female insane asylums of the city of New York, and to the various superintendents of asylums throughout the country who kindly sent me abstracts of State laws. I am especially indebted to Dr. A. Trautmann, of the Ward's Island Asylum, for the accurate sphygmographic plates, and to Dr. G. D. Smith, of New York, for valuable assistance.

CONTENTS.

CHAPTER I.

THERE are various changes in the appearance of the insane that are
almost as important in their way as the evidences of mental trouble displayed in conversation. Not only is the unbalanced mind evinced by
alterations is facial expression, and by departures from former habits in
the matter of gesticulation, postures, and dress, but physical alterations
as well are presented, which are the outcome of disease of the brain, and
are sometimes so trivial as to escape ordinary observation, but nevertheless
should be always looked for. Especially is such the case in those examples of insanity which are masked or concealed.

When one walks through the wards of any asylum for the insane, he
will be immediately impressed with the repulsiveness of the faces about
him, for the general appearance of the insane patient is in no sense pre-
possessing, and this is especially the case in the female. Women of beauty,
as writers upon insanity have observed, rapidly lose their good looks with
the establishment of mental disease, and plainness or downright homeli-
ness is the rule among asylum patients, whether of high or low social
station. What with slovenliness in dress, filthiness in habits, changes in
the color of the skin, and the condition of the hair, much of the romance
that is supposed to belong to insanity disappears. There are few Lears,
and fewer Ophelias.

The physiognomy of the insane consists not only in the portrayal of
inharmonious types of expression, but in transitory and intensified manifestations of dominant feelings. The latter is often the case in commencing insanity, and in forms of mental disorder that have stopped short
of dementia. It is well in all cases to systematically study the condition
of the organs of expression themselves, and ascertain if there be functional
derangements as well as general structural changes which may be the
result of defective innervation. Such study should be careful and continued, and not only the manifestation or absence of expression should be
taken into account, but the possible existence of paresis of certain facial
muscles, the condition of the eyes and hair, the coloring and appearance
of the skin, and the general muscular tonus should be noted as well. Relaxation and rigidity of the muscles are conspicuous factors in the expression of insanity, and in states manifested by lowered emotional activity we
find the former to be nearly always present. Such is the case in melan-

cholia and dementia, and in the atonic stages of other forms of asthenic disease. Rigidity, on the other hand, is the rule in mania and in conditions attended by excitement, as well as in certain sthenic forms of melancholia. The *melancholic* patient dramatically expresses mental distress by the position assumed, which is the embodiment of utter resignation to the worst; the facial muscles are relaxed, the mouth sags at the corners, and the eyelids droop, leaving exposed a small portion of dirty white sclerotic. The color of the skin is muddy, and in appearance greasy, as the sebaceous secretion is abundant; the nose and ears may be red or else livid, and the lips swollen and ill-defined. It is not rare to find spots of acne upon the forehead or back, or herpetic patches about the mouth. When the patient raises the head, which is usually bowed, it is to look wearily into vacancy (Plate III.), and a position of this kind may be assumed and kept for hours at a time. The hands hang listlessly in the lap, are dusky and swollen, and the fingers are intertwined or engaged in picking imaginary particles from the clothing. When the back of the hand is pressed, a white mark remains, slowly disappearing, however, as the sluggish capillaries refill. The nails are pale, or have a bluish tinge, and often there are hang-nails which are idly picked. These latter, in association with acne upon the forehead, are very common in sexual insanity, especially among masturbators. Such melancholics are not disposed to pay much attention to what goes on about them, and beyond an occasional deep-drawn sigh they give little indication of their feelings, but seek to avoid interference or notice of any kind. In lighter grades of melancholia the expression is of a much more sthenic and active character, and this is especially the case in forms of depression alternating with excitement. The patient is loquacious and communicative, as well as restless. His anxiety and anguish are evinced by certain forcible actions, such as pressing his hands over the face or head, by appealing gestures, a supplicating expression, or one of fear or remorse, by rolling up of the eyeballs; by bending the body usually forward, the patient assuming a crouching attitude, and by other evidences of an intense play of the more active of the depressing emotions.

In *melancholia*, when there is a complicating hysterical element, it is not rare to find libidinous gestures and postures, which are, however, more marked in mania of an hysterical form.

In *mania* everything indicates the play of ambitious feeling. Under the sway of pride, self-satisfaction, inordinate vanity, rage, hate, and certain dominant and all-absorbing passions, the bearing and demeanor of the patient suggests only excitement, restlessness, and irregularly expended energy. Muscular rigidity succeeds relaxation, but there is an exhibition of power which is entirely disproportionate to that needed for the performance of any special act. The movements made by the patient are rapid, cumulative, and startling. He paces to and fro, and his emotions are of a kind that must find vent in muscular action. The elated sense of importance is shown by his pompous deportment; his smile is supercilious and constantly plays about the mouth, the upper lip being raised to expose his teeth. As he rapidly strides through the ward or room in which he may be placed, his body erect, and his face turned upward and usually to one side, he presents a striking picture. The maniac gesticulates in a way that is not to be forgotten: he pats his breast, smoothes down his clothing with both hands, strokes one hand with the other, points to himself, raises both hands with their palms toward the visitor, and he does all this in a rigid and puppet-like manner. In the midst of his rapid walk he commonly turns and strikes an attitude (Plate VI.). The same patient at another time manifests an extravagant expression of rage, which is no less actively displayed, the brows being corrugated, the teeth covered by compressed lips, the eyes widely opened and the balls fixed, but it is rare for such a patient to look a person squarely in the face, and the intense expression may be rapidly succeeded by one entirely different.

In *chronic mania* we find that the dominant features of the patient's insanity have left indelibly marked traces. The suspicious, violent maniac (Plate V.) glares at the passer-by, with averted head and sinister expression, while the brow is contracted and the lines about the mouth are deep and sharply drawn. Such a patient suddenly starts up to swear and curse, and shake her fist violently in the face of the spectator, while her scowl may be succeeded in a moment by a contemptuous sneer or a malicious grin. For days together there may in such cases be little variation in the play of expression.

Certain sub-varieties of insanity are manifested by peculiarities in the behavior and appearance of patients which have more than passing interest. In hysterical mania, during the attacks the patient often presents the appearance of transfiguration alluded to by Charcot in his writings upon hystero-epilepsy, there being a condition of ecstacy, the excitement displaying itself in fixed attitudes, in which she remains for several hours. Varieties of moral insanity of sexual outgrowth in young people of both sexes, but especially in males, are expressed by great shyness, timidity of manner, or an effeminate appearance which is highly suggestive. In certain young women with sexual insanity a restless manner and the existence of a morbid self-consciousness and vanity are constantly present.

In *dementia* the facial change consists in an absence of expression of any kind, and the muscular atony often gives to the countenance a masklike vacancy and immobility. Under stimulation a meaningless smile may be brought to the lips, but it is not the reflex of any intelligent mental action. The lower lip is often relaxed and dependent, and from the corners of the mouth drools a stream of saliva which the patient makes no attempt to arrest (Plate VII.). The eyes are cold, fishy, suffused, and expressionless, and in advanced cases betray no indication of intelligence.

Particles of food collect in the interstices of the teeth and the breath is offensive and peculiar. The demented patient ordinarily is slow in his movements, remains in fixed attitudes, and his circulation is defective, the extremities being cold and livid; the lids are red, and sometimes there is a tendency to lachrymation, the person crying without apparent provocation.

In many of these cases there are exacerbations of feeble excitement, usually short-lived, and accompanied, in old people, by restless movement and incoherent loquaciousness and irritability. Loss of memory being one of the most important mental defects in dementia, we frequently find that the dement does not recognize any one with whom he may have come in contact since the development of his condition, although in cases not far advanced he may be able to remember old friends, but cannot call them by name.

A form of dementia which is rare is known as *primary dementia*, and affects young people, as a rule under the age of thirty, is manifested by a self-absorbed stupid manner, and by what Browne describes as "a perplexed vacant expression." The movements of the patient are slow, and, like older dements, he assumes a position of utter dejection and rarely changes it. He may make movements of an automatic character when such are suggested to him. When his hands or feet are placed in a certain position they maintain that position with a sort of cataleptic fixation, though there is little or no rigidity. His hands are cold, and the heart's action is weak. When he does talk it is in a garrulous manner, and like the echolalic idiot he repeats the last phrase he may have heard, or one word over and over.

The *imbecile* usually presents changes in appearance which are very marked. As possessors of inherited taint and the victims of early cerebral disease, we find defective development of various parts of the body, such as misshapen, though not necessarily atypical heads, evidences of early hydrocephalus, distorted and contracted limbs, the result of infantile paralysis, and secondary degeneration and atrophy. In many of these cases there are ocular defects and various errors of nutrition. The expression of the imbecile is repulsive in the extreme, and we find varying indications of intellectual change (Plate II.). As a rule, the countenance indicates a low order of brutality, the eyes are small, furtive, and cunning, and the movements are quick and cat-like. Imbeciles are often deaf and dumb, and pantomime may be a striking feature. The facial symmetry which often exists in the imbecile, and is due to early unilateral disease of the brain, is detected by drooping of one corner of the mouth, absence of one nasal fold, flatness of the nostril on the same side, and unevenness of the palpebral openings. It will be found also in many cases that the tongue is not protruded in a straight line, or is the seat of hemiatrophy, and there may be in connection with this a drooping of one arch of the palate, and a deviation of the uvula. Many imbeciles are epileptic, and during examination may have attacks of petit mal, or localized spasms of various kinds.

The appearance of the *idiot* is so familiar, even to the lay observer, that not much need be said on this subject. Nevertheless a word of caution may be given to those who are liable to confuse imbecility with a congenital condition of non-development, which is idiocy (Plate I.). The physical defects of the idiot are symmetrical, and the defective development is always of a type which can be duplicated. The body is generally undersized, the arms are sometimes long and there is a general tendency to flexion. In low grades the head is, as a rule, much smaller than normal, and out of proportion to the size of the body; the facial angle is often very great and the upper jaw has an advancing alveolar process, and may contain irregular and carious teeth, which usually protrude, presenting a rodent-like appearance. Cleft palate and other osseous defects are often suggestive accompaniments of deeper errors of development. The mouth of the idiot is usually large, the lips are thick, the eyes prominent and surmounted by bushy brows; the hair is coarse and bristling and inclined in the centre to grow well over the forehead. The physiognomy of the idiot betrays a slight degree of intellectual activity, but usually emotional excitement of an inconstant kind is all that we find. The grimaces and facial contortions are exaggerated and, as a rule, are suggestive of pleasurable feelings; or, on the contrary, we find passing expressions of rage or sorrow, which follow the most trivial provocations. So monkey-like is he in his behavior and motions that the diagnosis should never be difficult. In other cases of idiocy, not so pronounced, there is little to indicate the mental condition except certain vacuity, which shows how inconsiderable is the interest taken by the patient in things about him. Such patients are amiable and tractable, and the cranial atypy may be very slight.

Among certain idiots there are certain physical peculiarities which are the result of defective development of the lateral and posterior column of the spinal cord. Among them is spastic paralysis. The feet may present various deformities, there being talipes valgus, varus, or equinus. In cases of idiocy it is not rare to find supplementary fingers or toes.

The *general paretic* manifests his disease more in disorders of motility than by any alteration in facial expression, if we may except the appearance of elation which accompanies the delusions of grandeur, or the flatness and immobility of the facial muscles, or the local pareses, which are features of the stage of dementia. In the early stages we are furnished with tremulousness of the lips and tongue, and fibrillary tremor of the facial muscles, difference in the size of the pupils, drooping of the eyebrows, a staggering walk which does not exactly resemble that of any other form of spinal or cerebral disease, and which indicates rather an uneven expenditure of power than a loss of muscular strength. There

22

may possibly be incoördination of the upper extremities in advanced cases.
The patient, when he attempts to speak, uses his lips and tongue in a
way that is peculiar, and his speech is explosive or shuffling, and this is
especially noticeable when he uses words which contain many consonants.
The manner of the paretic is especially pronounced, and he is fond of
attracting the attention of any one who will listen to his extravagant
delusions, and rarely misses any opportunity of seeking notoriety. At a
later stage of the disease he loses all his energy, and may present the
appearance of an ordinary dement, there being, however, in addition, the
special motor troubles and the pupillary alteration.

In the physical diagnosis of insanity it is well, especially in cases with
well-marked history of heredity, previous mental trouble, or cerebral
disease, to carefully examine the configuration of the head, to determine
as nearly as possible the capacity of the cranium, the existence of evidences
of premature closure of the fontanelles, and to look for marks of injuries,
or syphilitic bone or aural disease. Measurements of the head may be
taken by means of a flexible lead pipe or tape, which should be of sufficient thickness not to lose its shape when removed. When such moulds
are made they should be fastened to a smooth board and carefully
measured.

For the purpose of measuring the facial angle we may avail ourselves of
either of the instruments described by Broca in his work upon craniom-
etry, or more simply by the use of three ordinary rulers which may be
joined by adjustable screws.

Records should be kept of the bi-aural or transverse, circumferential,
and antero-posterior measurements. If, however, there is reason to
suspect irregularities, moulds may be taken in different regions and
measurements compared.

The head of the insane is more often abnormally long (*dolio-cephalic*),
but occasionally a short (or *brachy-cephalic*) configuration is found. With
one-sided atrophy it is not rare to find imbecility, and lead moulds should
be taken at several points in such cases to determine the inequality. In
cases of hemiatrophy of the brain dependent upon disease of early origin,
unilateral bone atrophy frequently results.

In our observations and craniometric investigations, we are to avoid
the mistake of attaching too much importance to simple irregularity and
distortion, for every hatter's collection will show that men of brightest
intellect are possessors of heads of decidedly irregular shape. After all,
we are to look for atypical crania, which are either disproportionate in
size, or present facial angles so great as to suggest at once a small or
undeveloped brain.

CHAPTER II.

THE eyes of the insane undergo changes which are often of the greatest
importance, and should never be disregarded in making an examination.
We should take into account, first, the condition of the pupils; second,
the mobility of the eyeball; and third, the condition of the fundus by
means of the ophthalmoscope. In melancholia, as a rule, the pupils are
dilated and sluggish, while in mania, except in the active stages, they
are moderately contracted, or present no apparent change, and respond
readily to the light. In epileptic insanity they are mobile and usually
dilated, but this is by no means invariably the case. If both pupils are
found to be much reduced in size, and local reflex action is abolished or
impaired, the condition is highly suggestive of the first stage of general
paresis, or of complicating disease of the pons; but care should be taken
not to mistake the contraction that is the result of opium, and which may
be a feature of the insanity. Unequal dilatation of the pupils is of great
significance, as it is so common a feature of general paresis. Such
unequal dilatation is by no means always confined to one side. Mickle
found that pupil variation bears a decided relation to the changes in
mental symptoms. In patients presenting alternating depression and
elation, the condition of the pupils is alike in the two mental conditions.
In several cases he saw they were dilated and sluggish, but differed slightly
in size. He noted in the confirmed disease, in the stage following excitement and expansive delirium, that there was always a difference; at first
the left pupil was usually the larger, and afterward the right. They were
always irregular and sluggish, while in the quiet stage preceding extreme
dementia the pupils were commonly small. He, as well as others, has
noticed that after unilateral convulsions there is temporary dilation of
one pupil. It would seem that the left pupil is more frequently dilated
than the right, the pupillary changes, however, are not constant in their
method of appearance. In cases of insanity of syphilitic origin we commonly find changes in the color of the iris and irregularity of the pupil,
that suggest old iritis.

In cases of insanity directly traceable to coarse diseases of the brain,
there are to be discovered, as a result of paralysis of the various muscles
moving the eyeball, a variety of visual defects, the most important of
which is diplopia or double vision.

Paralysis of the third nerve results in ptosis, dilatation of the pupil, immobility of the eye, except in the outward direction, followed by divergent squint, and crossed diplopia, which is produced when the patient is directed to look at an object held in front, above, or on the side opposite the affected eye; he will then see two images, one above, below, or at the side of the other. If the patient be directed to hold his hand over the sound eye, and he is told to touch a specified object in front of him, he is utterly unable to do so and is apt to become dizzy. Of course, it is rare to find complete paralysis of all the fibres of the third nerve and the appearance of all these symptoms conjointly. *Paresis of the internal rectus* causes the patient to look toward the other side, in order to overcome the diplopia; a divergent squint results with crossed diplopia, the lateral distance between the true and false images widening as the object is moved in the direction of the sound side, away from the affected eye. With this form of paresis, when the object is held obliquely upward and inward, the images will be divergent above, that of the affected eye inclining to the opposite side. With the reverse position the images will converge above, that perceived by the affected eye inclining toward the impaired side. *Paresis of the external rectus* is symptomatized by homonymous diplopia. When the object is placed directly in front of the affected eye, at a distance perhaps of five feet, no diplopia exists; but when moved laterally so that the paralyzed muscle cannot be exerted to bring the eyeball to follow it, homonymous diplopia results—the patient turning his head toward the affected side.

Paresis of the superior rectus is manifested by a diplopia shown in the upper half of the visual field. The patient holds his head backward so that the objects may be brought into the lower half of the field. If the sound eye is covered and the patient is told to place his finger upon a certain object, he will invariably shoot above the mark. *Paresis of the inferior rectus* gives rise to diplopia opposite to that of the last named variety. *Paresis of the superior oblique,* is difficult to diagnose, because of its slight character. Objects below the horizontal median line appear to be double and irregular, while above no diplopia whatever is produced. In the double vision that occurs with this form of paresis the images appear at different distances from the patient, that seen by the affected eye being nearer to him.

Limited space will not permit me to go into this subject as fully as I could wish, and I will refer the reader to works on ophthalmology, where he will find much that relates to the mechanical defects of the motor apparatus of the eyeball. Tests should be applied in all cases of organic insanity, especially when there are visual hallucinations. In syphilitic insanity ocular paralyses are common and early manifestations, and in idiocy and imbecility various motor defects of this kind are to be found.

The ophthalmoscope has been used extensively as a diagnostic agent in determining the existence of organic insanity, but, so far, the appearances found differ widely.

Optic neuritis with atrophy of both kinds have been discovered in the eyes of general paretics, dements, and the subjects of epileptic insanity. In general paresis there is a progressive neuritis, which passes into a peculiar atrophic condition, observed by Loring and others.

Enlarged veins, shrivelled arteries, and choked disk may be detected in one or both eyes of the insane, and it is not rare to find atrophy of the disk associated with incoördination, pains of the lower extremities, and other symptoms of associated spinal trouble.

In many cases of mental disease no impairment of the visual power is associated with neuritis. Ophthalmoscopic appearances have been found in acute and chronic dementia, idiocy, imbecility, and syphilitic insanity, but rarely, if ever, in simple melancholia or mania. One of the first symptoms of atrophy of the optic nerve is impairment of the color sense. The failure is met with most frequently in dementia, hysterical insanity, and general paresis, and, among men, more often in the latter disease. The power of seeing red and green is lost first, as a rule, and afterward the other colors. To apply the color test, the examiner should supply himself with a number of skeins of different-colored worsted, which the patients are asked to match. Hemiopia is an occasional feature of insanity, depending on gross cerebral disease; consequently it is more common in secondary dementia, general paresis, and syphilitic insanity than in other psychoses. In addition to the defects mentioned, we may find clonic spasms of the orbicular muscles (nystagmus), diseases of the lids, and a tendency to lachrymation; and this latter is a very common accompaniment of dementia. In examining the eyes of the insane we are furnished with diagnostic suggestions of the greatest value. Especially true is this with regard to *expression.* Bucknill and Tuke lay stress upon the absence of expression of the eyes in the delirium of fevers, in contradistinction to the intensity which exists in mania; and they call attention, on the other hand, to the prominence of the eyeball and the bloodshot appearance which characterizes the excitement dependent upon cerebral meningitis. In mania there is simply an intensification of emotional expression.

The "*insane ear,*" or *otheotoma,* which has been described by a number of observers, is probably the result primarily of trophic disorder, and may arise from a slight injury or some such trivial exciting cause. The auricles become the seat of violent inflammatory process, which goes on to suppuration and may entail a considerable destruction of tissue.

The appearance of the ear in the acute stages of such inflammation is quite striking. It becomes hot, engorged with blood, and swollen to an extraordinary degree, so that the normal folds and indentations are lost in the general tumefaction, and there is closure of the external meatus.

The affected ear is exquisitely painful, the patient shrinking from the slightest touch. It is not long before there is an increase in the violence of the inflammatory process and the formation of one or more abscesses, which, if not opened, burst and discharge a large quantity of bloody pus. An abscess may sometimes form behind the ear. Extensive separation of the cartilage from the other tissues often occurs from burrowing of the pus, and when reparative process takes place a conspicuous deformity remains, due to contraction, so that the affected ear is shrivelled, crenated, and often flattened. The "insane ear" may be of slow origin, and the result of a low inflammatory process, or it may arise in a single night. It is common in cold weather, and may follow exposure to cold or pressure. This condition is sometimes met with in people who are not insane, and is considered by some authors to be very rare and a different affection, but I can see no difference, considering the pathological condition in both to be a perichondritis.

An appearance of the ear is occasionally met with which is misleading, and should not be confounded with that of the disease under consideration. I allude to the deformity produced by insane patients who constantly pull their ears. Not only do we find elongation of the lobule, but ulceration and diffused redness as well. In certain cases of congenital insanity the auricles may be either abnormally large, pointed at their extremities, and stand out prominently, or else they are unusually small and flat.

The *mouth* undergoes changes in configuration which have much to do with the insane physiognomy, and is perhaps the most expressive organ, with the exception of the eyes, in the portrayal of mental variations. Its appearance in repose and in excitement should be noted, and the coloring and formation of the lips should be likewise. In certain forms of insanity the latter are tumefied and often dry and cracked. The buccal mucous membrane is pale and sometimes insensitive. The tonicity of the oral muscles undergoes considerable diminution, especially in such forms of chronic insanity as dementia and general paresis. In the former it is common to find a drooping of the lower lip, and in advanced cases there is inability to prevent the escape of saliva, entirely independent of the patient's mental disregard of its accumulation. There may be a paretic condition, which manifests itself in unevenness of the mouth. In general paresis there is tremor of the lips, which is especially noticeable when they are slightly parted, or when the attempt is made to speak, and seems to be increased by the effort of the patient to control it.

The *tongue* also trembles in general paresis, and when protruded is not only agitated by vermicular movements but is suddenly retracted. We also find this tremor in chronic alcoholism, but it is coarser and is not associated with the peculiar speech defects of the former disease. In varieties of insanity due to organic disease it is not unusual to find that the tongue, when protruded, points to one or the other side, or that it is impossible for the patient to bring the tip in contact with the roof of the mouth. The tongues of certain idiots and cretins are unusually large and swollen, of pale color and decidedly flabby. The appearance of the tongue in insanity as an index of various bodily states is also of great importance, as melancholia and diseases of like character are connected with digestive disorders, especially of an hepatic nature, and it will be found that this organ is usually coated with a heavy white or brown fur, and the breath is foul. In acute mania we may expect to find a red and glazed tongue. Various peculiarities in the appearance of the teeth are found among idiots, as well sometimes as among those who are of the "insane neurosis" or temperament.

In certain rare cases it is possible to find two rows of teeth in each jaw, one set being the permanent and the other the milk teeth, which emerge at different points. A tusk-like development is frequently found, and it is not rare to find a large canine or incisor jutting out from the anterior surface of the alveolar process.

The *nose*, according to Hofling, in shape and appearance undergoes noteworthy changes, which he regards as important. We should therefore note the condition of the nostrils, whether they are distended or compressed, together or singly, and their mobility.

Evidence of general mal-nutrition is the rule in the early steps of all forms of mental disease, and may arise from insufficient food, many patients refusing to eat, or from the constant wear and tear incident to excessive excitement. When dementia follows chronic insanity it is quite usual for the patient to become much improved in appearance. In fact, the sudden increase in size and improvement in the color of the skin often leads the friends of patients to believe in a great improvement, while, on the contrary, this change is one that makes the prognosis bad.

CHAPTER III.

THERE are temperature and pulse variations in insanity which are valuable evidences of structural changes. These should be studied in every case if possible by means of surface and deep thermometers, and by the sphygmograph. The asthenic mental disorders which are grouped under the head of melancholia are usually attended by lowered surface and deep temperature, and in dementia the same condition of affairs is found to exist. In all forms of insanity attended by slowness of muscular movement this diminution of temperature is notable. The surface circulation is extremely sluggish, and it is with difficulty that the extremities are flushed or made warmer by energetic rubbing. When bulbar symptoms are present, a unilateral lowering is by no means uncommon. In alternating insanity (folie circulaire), or melancholia attended by transitory attacks of excitement, there is often a sudden rise of temperature with the beginning of the irritability. In general paresis the elevation is constant and important, even in the melancholic stage toward the latter part of the day, and is most decided during the first and last stage of the disease. In the first stage, however, the increase is connected with the maniacal attacks, in the second stage it is lowered, but rises again in the third stage. The increase in temperature continues with excitement, and a very great and sudden increase of the body heat is a forerunner of death. In mania the elevation is very conspicuous, and bears a direct relation to the muscular irritability and restlessness. In the mania of debility, the temperature may continue two or three degrees higher than normal for some days. In phthisical insanity there is an evening rise, which is attended by flushing and distention of the temporal vessels.

In puerperal insanity there is a primary elevation with quiet small pulse. In patients suffering from insomnia, and who are violent, there is often a rise of two degrees at night. Macleod believes this to be the rule in all cases where there are destructive tendencies. As in various other diseases, an important point to have in mind is the difference between the morning and evening temperature.

In all forms of insanity the intercurrent complications are marked by sudden and conspicuous variations of temperature; bed-sores and typhoid states are evinced by increasing body heat, and in convulsive seizures, which may occur from time to time, the temperature is higher. After an attack of hemiplegia, whether following an epileptiform discharge or not, we find an increase of heat upon the paralyzed side. The surface-temperature is increased in mania, and the head, especially, is hot, while in some cases it is possible to detect local elevation of temperature. The sphygmograph in atonic conditions shows indications of lowered arterial tension. The tracings in melancholia vary, but are usually of an asthenic character. In complicated cases with cardiac hypertrophy the pulse is rapid, hard, and gives a tracing in which the first event is exaggerated and the diastolic line is marked by the absence of valvular breaks. In other forms of melancholia the heart impulse is weak, the tracing is almost straight, broken only by a feeble systolic elevation and tremulous valvular indentation.

On the page following are tracings taken by Pond's sphygmograph at the New York City Asylum for the Male Insane.

Absolute indications cannot be relied upon as the result of sphygmographic examination. There are general characteristics that are of great significance, and for this reason a number of tracings should always be taken.

In chronic and advanced insanity, the pulse is soft and compressible, and especially is this the case in melancholia. In diseases of this kind, circulation is exceedingly defective, and we find venous stasis in distal parts. As a result, it is found that inconsiderable injuries or exposure to cold, which in ordinary persons would have little effect, are apt to give rise to slowly healing wounds and sloughing. Chilblains are common, among demented patients especially, and dry gangrene is by no means uncommon, not only in old but in young people as well.

In some cases of asthenic insanity the pulse is found to be abnormally slow.

Heart disease has been found to exist in connection with nearly every form of insanity, and Berman has found that thirty-six per cent. of five hundred patients who died at the West Riding Asylum presented evidences of cardiac disease. We should, therefore, be on the outlook not only for cardiac murmurs, but for the signs of hypertrophy of the left ventricle. The cases in which we find heart complications most frequently are those of melancholia, impulsive insanity, and among patients who are sullen, morose, and suspicious. In general paresis the second aortic sound is accentuated, which is also the case in mania. In both of these diseases we find increased arterial tension, and it is advisable always to use the sphygmograph. In many cases of secondary dementia obstruction murmurs are to be detected.

The condition of the *skin* and its appendages is worthy of study. The cutaneous surface is usually dry, harsh, and presents evidences of malnutrition. In rare instances there is profuse sweating, notably in acute mania, but the action of the sweat-glands is feeble. In some forms of disease,

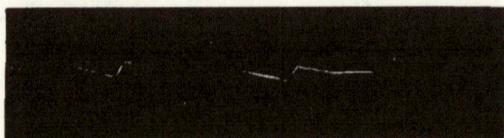

General Paresis. Right Radial Pulse; 78. (Most common.)

General Paresis. Right Radial Pulse; 70.

General Paresis. Right Radial Pulse; 76.

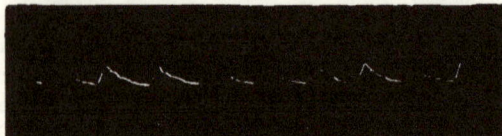

General Paresis. Right Radial Pulse; 96.

Acute Melancholia. Right Radial Pulse; 84.

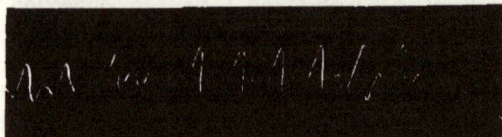

General Paresis, Maniacal Stage. Right Radial Pulse; 128.

Acute Mania. Right Radial Pulse; 82.

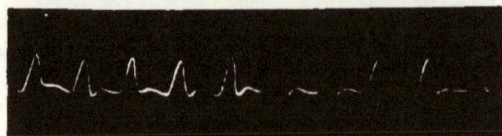

General Paresis; Maniacal Stage; 78.

acne, herpes, and certain bullous eruptions play a crisogenic part and disappear after each exacerbation. Moles and staining are frequently a feature of chronic insanity, especially among women, and changes in the hair are also found, and have been commented on by various authors: In mania the hair is peculiarly coarse and bristling and with every attack of excitement it becomes erect or crinkly. In some patients I have looked upon this as a prodromic sign of a developing attack of violence, and I have found such to be the case. In melancholia it is in appearance sodden and limp, and rarely curls. In many insane people premature or uneven blanching of hair occurs. In cases of hysterical insanity and in chronic insanity in women, there is a tendency to the appearance of hair upon the face—upon the upper lip and chin especially, the growth amounting to a beard in some cases.

There is occasionally found among children of weak mind a puffing or pseudo-œdema of the skin, which is associated with atrophy of the thyroid gland, and spots of staining. The face, in particular, is swollen and the lips and tongue are thick and enlarged. The voice is muffled and harsh and speech is slow. This condition is known as *cretinism*, and is quite rare in England and in this country. Of several hundred idiots I have examined, I have found but one case.

The *electric excitability* of muscles is not often affected. In a series of carefully made experiments, Lowe was unable to find any diminution in mania, but in general paresis and organic forms of disease, there was much loss of excitability. In even the first and second stages of general paresis no lowered reaction was found, but in the last stage he found that both in the arms and legs the muscular response to faradic excitement was considerably lowered.

The activity of the *tendinous reflex* depends very much upon the form and stage of insanity. In uncomplicated mania and melancholia it is rarely affected, but in all affections where there are symptoms indicative of affection of the posterior columns of the spinal cord it will be found to be diminished. In secondary degeneration of the lateral column, the particular reflex is of course exaggerated. In general paresis it may be normal or absent, depending of course upon the lesion.

Disorders of motility are occasionally present among the insane. The existence of fine fibrillary tremor is a common indication, especially in chronic insanity. It may be noticed in the face particularly, and a vermicular contraction may be detected upon close examination or by lightly striking the face. Allusion has already been made to the disorders of motility so conspicuous in general paresis. The tremor of alcoholism is of a different character, affecting the hands and lower extremities as well, and is usually connected with anæsthesia; besides, it is more active in the early part of the day. In organic disease with mental symptoms we find various grades of tremor and paralysis, which depend upon the region of cerebral substance involved. The gait of the insane is sometimes a valuable indication of the form of insanity. In general paresis the walk of the patient is uncertain and unsteady, and a true defect of coördination causes him to advance with widely spread feet and a tottering method of propulsion. In various forms of dementia there is a shuffling gait due to loss of power, and in secondary dementia we find quite often an old hemiplegia and its embarrassments.

Sensory symptoms are usually of an anæsthetic character, and vary greatly. In melancholia there is sometimes a general cutaneous anæsthesia of a profound nature, and in general paresis the same state of affairs is found, but most marked in the last stage of the disease, when, besides diffused loss of sensibility, there may be anæsthesia of the fauces and larynx. It is often difficult to determine the state of sensibility, owing to the mental obtuseness of the patient and his perverted perception. In rare forms of hysterical insanity there is pronounced hemianæsthesia, with color-blindness upon one side. In dementia the loss of cutaneous sensibility is markedly lowered and severe injuries or burns give rise to little complaint. The electric sensibility is occasionally increased in general paresis.

Loss of smell and taste are met with in general paresis, during the last stage.

In chronic alcoholism, hemianæsthesia, with anæsthesia of mucous membranes has been pointed out. With this there is color-blindness in the anæsthetic eye. In this organ the cornea is insensitive and may be touched without annoyance to the patient.

Involuntary discharges of urine and fæces may occur at various times in the course of mental disease. In mania the patient is so occupied with his delusions that he is apt to void the contents of his bowels and bladder, while in melancholia, according to Luys, there is a certain anæsthesia of the lining membrane of the intestines which prevents the patient from perceiving the distention of the lower gut by substances accumulating therein, and finally there is a mechanical escape. In dementia and advanced insanity, actual paresis of the sphincter prevents retention of the contents of the rectum and bladder.

In many cases of commencing insanity I have observed that it is common for female patients to void their urine, even though they are perfectly conscious of their weakness. In early melancholia and in hypochondriasis constipation is the rule.

Examination of the *urine* of the insane shows that there are great variations both in the amount and in the component parts. In melancholia and in conditions attended by slow organic changes, the quantity of urine is greatly diminished and the proportion of urea and chlorides excreted is diminished.

In conditions of excitement in mania, in the expansive stages of gen-oral paresis, the reverse is true, and in melancholia with excitement it is not rare to find abundant urine. In mania the quantity of urine may be very small.

Merson, Beale, Sutherland, and Lindsay have found that there is a plus amount of the phosphates in the urine, in acute mania, while, in the stage of exhaustion in mania, the third stage of general paresis, and in the feeble stage of acute dementia they are reduced The presence of albumen in the urine of the insane is occasionally found, especially in puerperal insanity, when the mental excitement often appears and disappears with the presence and absence of this substance. In the urine of general paretics it is often found, and in epileptic insanity it may be detected after the paroxysm.

The appearance of sugar in the urine may be determined sometimes in cases in which excessive thirst is a feature, and in which slight maniacal outbreaks follow inconsiderable excitation. I have frequently found it in the urine of paretics.[1]

In certain general paretics the urine presents an excess of alkaline carbonates.

In all cases of insanity it is well to inquire into the condition of the *menstrual functions*. Not only is insanity, as Falret and Esquirol have pointed out, very often caused by uterine and ovarian disorders, but there is a very important variation in the function of menstruation. Idiots and cretins menstruate very scantily or not at all, and puberty is delayed. In disorders of the asthenic type there may be amenorrhœa, though, so far as my experience goes, the development of the insanity has been preceded by excessive, protracted, and debilitating flowing. In mania and other conditions of excitement there is greater mental disturbance at the catamenial period; and attacks of epilepsy, when they have been a feature of the insanity, are apt to be more numerous and violent at this time. It is a well-known fact that forms of sexual mental disorder are much aggravated by menstrual disorder, and with any abnormality the patient is inclined to indulge in disgusting practices and foul conversation.

Sutherland holds that general paralytics undergo change of life much earlier than other women.

In certain varieties of insanity, especially in the early stage of general paresis, there is an excitement of the genital function, which manifests itself, besides lewd behavior, in frequent erections, masturbation, and ungovernable lust. Luys reports the case of a young man who indulged during the day in masturbation whenever he recounted his hallucination of the women who followed him, soliciting him to have intercourse with

---

[1] In the light of Magnan's theory of the origin of the disease in the fourth ventricle, this circumstance is an additional confirmation of the pathological and experimental production of diabetes.

them during the night. In dementia the tendency to masturbation is often constant, and it is found necessary to provide tight-fitting clothing in one piece, with sleeves sewed down to the sides, but even then the patient often manages to gratify his desire.

In nymphomania the behavior of the patient is perhaps more conspicuous than in the satyriasis of the male patient.

The salivary secretion is commonly increased, and in dementia very decidedly so. With accumulation the patient is apt to make what Luys calls "automatic attempts" to eject it, expectorating forcibly, with some degree of regularity (Plate VII.).

The breath and bodily odor of the insane are often very unpleasant, and by some authors the former is supposed to be as characteristic in its way as that emanating from the small-pox patient.

CHAPTER IV.

It is never wise to gain access to your patient by means of any ruse, and it is preferable that the medical man should appear before him in his own true character. The object of an examining physician is not to extort communications from the patient by misrepresentation or deceit, for in such cases the avowals of the alleged insane person are falsely based, and his motives are declared under a false impression. If the examiner has not sufficient tact to draw out the person whose sanity is doubted, he had better deputize some one else to do the work. It is only in the rarest cases, when the patient is violent and threatens actual harm to every one, that subterfuge is to be resorted to. Go to your patient, then, as you would to any other, and engage him, if possible, in conversation. If the occasion offers itself, ask him in relation to his feelings regarding his immediate family, business associates, and friends. His religious beliefs, if any, should be inquired into, and the possibility of any change in sentiments discussed. If there are morbid ideas, which show themselves in a disregard of the present or a dread of the future, it will be well to follow up the line of examination and ascertain the possibility of suicide or contemplated violence. If he has imaginary enemies it may be well to inquire who they are.

His business capacity and plans for the future are important considerations. He should be asked as to the extent of his holdings, both of personal property and real estate, and of his ability to perform certain duties. He should be put through certain tasks, regarding his competency to execute business instruments and deeds if occasion arises. His memory should be tested both as to recent and remote events, and any speech defects, aphasic or ataxic, noted. His handwriting should be examined and compared with specimens of older dates. Moral changes in the demeanor of the patient are important. With slight promptings the insane person will often indulge in salacious outbursts, and especially is this the case in hysterical insanity and general paresis of the insane. Peculiarities in dress, habit, and mode of life, as well as the changes already described, should be investigated.

After getting as much as possible from the patient himself, his friends —as many as possible—should be interrogated regarding his behavior and the truth or falsity of certain communications he may have made.

Hereditary disease, bad habits, and other factors of disease should be also noted. After due care, and repeated interviews if necessary, the physician may safely state an opinion, but he should never do so hurriedly or without deliberation.

A most important duty is the taking of memoranda, which should be kept for possible litigation that may arise.

In entering the house or room of the suspected lunatic, the physician should observe any peculiar or eccentric arrangement of furniture or decoration, for the patient is apt to surround himself with unnecessary objects, or to cover his walls with gaudy trash. In itself this tendency may amount to nothing more than harmless eccentricity, which has always existed, but when it is a new thing, and in contrast to the person's previous tastes, it will be found to be the result commonly of some delusion. The person with a delusion of grandeur will provide himself with worthless imitations of royalty, the insane woman who believes herself to be the wife of the President will cover the walls of her room with woodcuts of that dignitary from illustrated papers, or the young woman who becomes a subject of melancholia is apt to surround herself by a multitude of pictures of saints and martyrs and by relics and religious emblems.

An important indication of the deep-seated character of certain delusions in chronic insanity is manifested in laborious yet useless industry, shown in the manufacture of certain peculiar objects with which the patient is surrounded. One man with whom I am familiar has spent several years in the preparation of a curious astrological apparatus, constructed of refuse material and rags found about the asylum, while others have been diligently occupied in the manufacture of flying machines, and other objects requiring great time and labor in their construction. We may often find in the books read by the subjects of impending insanity pencilled comments and additions which show the drift of their minds. The value of these methods of expression cannot be too highly estimated.

An early and conspicuous indication of mental disorder is the *change in personal attire* made by the patient. Gaudy finery, "loud" colors, and peculiarly made clothing take the place of quiet dress. Bright and glittering gew-gaws are affixed to various parts of the hat, coat, or dress, and buttons, pieces of looking-glass, feathers, and gay pieces of colored rags are pressed into service (Plate VI.). The fondness of self-adornment is found among maniacs, dements, general paretics, melancholics, imbeciles, and idiots, and in such patients is connected with delusions of grandeur and excessive self-satisfaction.

Disregard of appearance and untidiness in dress are early and suggestive symptoms of insanity, both of melancholia and mania. The ordinarily neat and well-dressed person may neglect his razor and comb, and become slovenly, dirty, and careless; on the other hand, we sometimes observe an extraordinary neatness and personal cleanliness quite at variance with the patient's former habits. Many insane believe that they are contaminated by some foul substance, and therefore will frequently wash their hands or cleanse themselves in different ways.

It is not unusual for the insane person to remove all clothing either as a result of the seeming discomfort which their contact produces, or as a result of some delusion. In violent cases, and in many subjects where chronicity is being established, we find great destructiveness. They will

23

not only tear their ordinary clothing into strips, but will destroy the coarsest and strongest fabrics.

Much stress is laid upon the peculiarities which are found in the handwriting of the insane, and there can be no doubt that it possesses much that is interesting from a diagnostic point of view. In nearly every case a departure from the normal mental state is displayed by change, not only in the method of written expression, but in the chirography itself. This is especially noticeable in general paresis, and if a series of

letters be compared, it will be found that the more recent present various irregularities. Certain words are imperfectly ended, their terminal letters being absent, or they are extended in a scrawl, while at a later stage of the disease we find the omission not only of syllables, but of whole words; and in the letters of persons of precise habit before the development of mental trouble, it is rare to find an "i" dotted or a "t" crossed after the disease has made its appearance. At a later stage it is impossible to decipher anything that the patient may write.

One of the peculiarities of the letters of the insane consists in the use of illustrative diagrams, keys of explanation, and strangely coined words, and in forms of mental disease symptomatized by religious exaltation there are constant suggestions of the delusions of the individual, which are shown in maps and plans in which figure astronomical and theological symbols. In some cases there is a veritable *cacoëthes scribendi*, which is a feature in many maniacal patients. We find exacerbations of this form of mental trouble are preceded by vigorous letter-writing. The first example is a specimen of the handwriting of a young woman suffering from acute mania, who spent entire days in scribbling like rubbish, and

inditing numerous letters to persons she did not know. In her case, and in many others, it will be found that there is a disposition to use capital letters to an extraordinary degree and to cross-write, so much sometimes as to destroy legibility. In passing judgment upon the letters of doubtful cases of insanity, we must carefully read them through, bearing in mind that with the insane any sustained effort is impossible, and it is probable that in a long letter we shall find some manifestation of disordered mental action before the end is reached. The specimens upon this and the preceding page are examples of penmanship written by three general paretics, in different stages of the disease.

In some cases we have but little difficulty in making a diagnosis by the letter alone, because of the striking incoherency, which the individual may restrain in conversation but which he indulges in when left to himself; and in suspected cases, where patients are on their guard, it is well to ask them to write a letter.

In medico-legal questions one should not be too ready to express an opinion upon any document or letter that may be put in evidence, for ordinary bodily weakness may give rise to a tremulousness in the handwriting, and it will not do to make a hasty diagnosis upon this feature of a document. In other cases legal instruments may be presented to the expert witness for his opinion whether there are intermissions and interlineations. Care must therefore be exercised in taking into account the pertinency of the interlineations and the presence of marginal notes and corrections.

CHAPTER V.

THE laws of the different States regarding the commitment of lunatics vary greatly. In all cases, however, judicial endorsement is imperative, and in the State of New York it is necessary that the certificate of the examining physicians shall be approved by the judge of a court of record. In other parts of the country the formalities are more or less rigid, and in Canada a lunatic who is not dangerous may be received into an asylum by the approval of the superintendent. Besides the legal steps to be taken by the friends of the insane person, there are various local regulations pertaining to the asylums themselves. It is always necessary to give bonds, or to show that the alleged lunatic is without means, and proper blanks are prepared for the purpose. When such is the case, the pauper lunatic may be committed and cared for, after representation has been made to a local police magistrate or to the officer of the poor. In *New York City* the lunatic may be placed under arrest, and he is then transferred to a jail for examination by the medical officers of the Department of Public Charities and Correction.

It is necessary, when the patient is sent to a pay asylum, for two physicians to be appointed examining physicians. The *New York* laws are as follows:

SECTION 1. No person shall be committed to or confined as a patient in any asylum, public or private, or in any institution, home, or retreat, for the care and treatment of the insane, except upon the certificate of two physicians, under oath, setting forth the insanity of such person. But no person shall be held in confinement in any such asylum, for more than five days, unless within that time such certificate be approved by a judge or justice of a court of record of the county or district in which the alleged lunatic resides, and said judge or justice may institute inquiry and take proofs as to any alleged lunacy before approving or disapproving of such certificates, and said judge or justice may, in his discretion, call a jury in each case to determine the question of lunacy.

SEC. 2. It shall not be lawful for any physician to certify to the insanity of any person for the purpose of securing his commitment to an asylum, unless said physician be of reputable character, a graduate of some incorporated medical college, a permanent resident of the State, and shall have been in the actual practice of his profession for at least three years, and such qualifications shall be certified to by a judge of any court of record.   No certificate of insanity shall be made except after a personal examination of the party alleged to be insane, and according to forms prescribed by the State Commissioners in Lunacy, and every such certificate shall bear date of not more than ten days prior to such commitment.

SEC. 3. It shall not be lawful for any physician to certify to the insanity of any person for the purpose of committing him to a asylum of which the said physician is either the superintendent, proprietor, an officer, or a regular professional attendant therein.

In the State of *Maine* the certificates of at least two respectable physicians are necessary.   Sections 16 and 17 of the revised statutes, 143, §§ 11, 12.

SECTION 16. Parents and guardians of insane minors, if of sufficient ability to support them there, within thirty days of an attack of insanity, without any legal examination, shall send them to the hospital, and give the treasurer thereof the bond required, or to some other hospital for the insane.

SEC. 17. All insane persons, not thus sent to any hospital, shall be subject to examination as hereinafter provided.   The municipal officers of towns shall constitute a board of examiners, and, on complaint in writing of any relative, or justice of the peace of their town, they shall immediately inquire into the condition of any insane person therein, call before them all testimony necessary for a full understanding of the case, and if they think such person is insane, and that his comfort and safety, and that of others interested, will be thereby promoted, they shall forthwith send him to the hospital with a certificate stating the fact of his insanity, and the town in which he resided, or was found at the time of examination, and directing the superintendent to receive and detain him till he is restored or discharged by law, or by the superintendent and trustees.   And they shall keep a record of their doings, and furnish a copy to any interested person requesting and paying for it.

The *Vermont* laws, approved November 28, 1882, are as follows:

SECTION 1. Section 2906 of the revised laws is hereby amended so as to read as follows:

No person, except as hereinafter provided, shall be admitted to, or detained in an insane asylum, as a patient or inmate, except upon the certificate of such person's insanity, stating their reasons for adjudging such person insane, made by two physicians of unquestioned integrity and skill residing in the probate district in which such insane person resides, or, if such insane person is not a resident of the State, in the probate district in which the asylum is situated; or, if such insane person is a convict in the State prison or House of Correction, such physicians may be residents of the probate district in which such place of confinement is situated. And the two physicians making such certificates shall not be members of the same firm and neither shall be an officer of an insane asylum in this State.

SEC. 2. The next friend or relative of a person whose insanity is certified to, as above provided, may appeal from the decision of the physicians so certifying him to be insane to the supervisors of the insane, which appeal shall be noted on the certificate. The supervisors shall, when such appeal is taken, forthwith examine the case, and if, in their opinion, there was not sufficient ground for making such certificate, they shall avoid the certificate, otherwise they shall endorse their approval upon it. Such examination by the supervisors shall be had in the town where the appellant resides.

SEC. 3. When the next friend or relative of such a person takes an appeal, as above provided, he shall not be received in an insane asylum while the appeal is pending before the supervisors. And a trustee, or other officer, or employee of an insane asylum who receives or detains a person in such asylum whose insanity is not attested by a legal certificate which has not been appealed from or by a certificate duly approved by the supervisors on appeal, shall be imprisoned in the State prison not more than three years.

SEC. 4. Idiots and persons *non compos*, who are not dangerous, shall not be confined in any asylum for the insane. And if any such persons are so confined, the supervisors of the insane shall cause them to be discharged.

The *New Hampshire* laws provide (Secs. 12, 13, 18):

SECTION 12. If any insane person is in such condition as to render it dangerous that he should be at large, the Judge of Probate—upon petition by any person, and such notice to the selectmen of the town in which such insane person is, or to his guardian, or to any other person, as he may order, which petition may be filed, notice issued, and a hearing had in vacation or otherwise—may commit such insane person to the asylum.

SEC. 13. If any insane person is confined in any jail, the Supreme Court may order him to be committed to the asylum, if they think it expedient.

SEC. 18. No person shall be committed to the asylum for the insane, except by order of the court, or the Judge of Probate, without the certificate of two reputable physicians that such person is insane, given after a personal examination made within one week of committal; and such certificate shall be accompanied by a certificate from a judge of the Supreme Court, or Court of Probate, or mayor, or chairman of the selectmen, testifying to the genuineness of the signatures. and the respectability of the signers.

The more important laws regarding the commitment of the insane in *Massachusetts*, are appended:

SECTION 11. A judge of the Supreme Judicial Court or Superior Court, in any county, where he may be, and a judge of the Probate Court, or of a Police, District, or Municipal Court, within his county, may commit to either of the State lunatic hospitals any insane person, then residing or being in said county, who in his opinion is a proper subject for its treatment or custody.

SEC. 12. Except when otherwise specially provided, no person shall be committed to a lunatic hospital, asylum, or other receptacle for the insane, public or private, without an order or certificate therefor, signed by one of the judges named in the preceding section, said person residing or being within the county as therein provided. Such order or certificate shall state that the judge finds the person committed is insane, and is a fit person for treatment in an insane asylum. And the said judge shall see and examine the person alleged to be insane, or state in his final order the reason why it was not deemed necessary or advisable to do so. The hearing, except when a jury is summoned, shall be at such place as the judge shall appoint. In all cases, the judge shall certify in what place the lunatic resided at the time of his commitment; or if confinement is ordered by a court, the judge shall certify in what place the lunatic resided at the time of the arrest, in pursuance of which he was held to answer before such court; and such certificate shall, for the preceding section, be conclusive evidence of his residence.

SEC. 13. No person shall be so committed, unless in addition to the oral testimony there has been filed with the judge a certificate signed by two physicians, each of whom is a graduate of some legally organized medical college, and has practiced three years in the State, and neither of whom is connected with any hospital or other establishment for treatment of the insane. Each must have personally examined the person alleged to be insane within five days of signing the certificate; and each shall certify that in his opinion said person is insane and a proper subject for treatment in an insane hospital; and shall specify the facts on which his opinion is founded. A copy of the certificate, attested by the judge, shall be delivered by the officer or other person making the commitment, to the superintendent of the hospital or other place of commitment, and shall be filed and kept with the order.

SEC. 14. A person applying for the commitment or for the admission of a lunatic to a State lunatic hospital, under the provisions of this chapter, shall first give notice in writing to the mayor, or one or more of the selectmen of the place where the lunatic resides, of his intention to make such application; and satisfactory evidence that such notice has been given shall be produced to the judge in cases of commitment.

Sec. 15. Upon every application for the commitment or admission of an insane person to a hospital or asylum for the insane, there shall be filed with the application, or within ten days after the commitment or admission, a statement in respect to such person, showing, as nearly as can be ascertained, his age, birthplace, civil condition, and occupation; the supposed cause and the duration and character of his disease, whether mild, violent, dangerous, homicidal, suicidal, paralytic, or epileptic; the previous or present existence of insanity in the person or his family; his habits in regard to temperance; whether he has been in any lunatic hospital, and if so, what one, when, and how long; and, if the patient is a woman, whether she has borne children, and, if so, what time has elapsed since the birth of the youngest; the names and address of his father, mother, children, brothers, sisters, or other next of kin, not exceeding ten in number, and over eighteen years of age, when the names and address of such relatives are known by the person or persons making such application, together with any facts showing whether he has or has not a settlement, and if he has a settlement, in what place; and if the applicant is unable to state any of the above particulars, he shall state his inability to do so. The statement, or a copy thereof, shall be transmitted to the superintendent of the hospital or asylum, to be filed with the order of commitment, or the application for admission. The superintendent shall, within two days from the time of the admission or commitment of an insane person, send, or cause to be sent, notice of said commitment in writing, by mail, postage prepaid, to each of said relatives, and to any other two persons whom the person committed shall designate.

Sec. 16. After hearing such other evidence as he may deem proper, the judge may issue a warrant for the apprehension and bringing before him of the alleged lunatic, if in his judgment the condition or conduct of such person renders it necessary or proper to do so. Such warrant may be directed to and be served by a private person named in said warrant, as well as by a qualified officer; and pending examination and hearing, such order may be made concerning the care, custody, or confinement of such alleged lunatic as the judge shall see fit.

Sec. 17. The judge may, in his discretion, issue a warrant to the sheriff, or his deputy, directing him to summon a jury of six lawful men, to hear and determine whether the alleged lunatic is insane.

In the State of *Rhode Island* patients may be committed upon the order of a justice's court or one of the justices of the Spureme Court, or by a guardian, or by relatives and friends, upon the certificate of two practicing physicians of good standing.

All have unrestricted communication with two commissioners appointed by the Legislature and are visited weekly by a committee of the trustees.

According to the *Connecticut* laws of 1869:

SECTION 1. Any lunatic or distracted person may be placed in a hospital, asylum, or retreat for the insane, or other suitable place of detention, either public or private, by his or her legal guardian, or relatives or friends in case of no guardian; but in no case without the certificate of one or more reputable physicians, after a personal examination made within one week of the date thereof, which certificate shall be duly acknowledged before some magistrate or other officer authorized to administer oaths, or to take the acknowledgment of deeds in the State where given, who shall certify to the genuineness of the signature, and to the respectability of the signer.

The laws of *New Jersey* regulating the protection and admission of the insane to asylums are quite numerous. Section 17 of the laws of 1875—76 is as follows:

And be it enacted, That no person shall be admitted into said asylum as a patient except upon an order of some court or judge authorized to send patients, without lodging with the superintendent first, a request, under the hand of the person by whose direction he is sent, stating his age and place of nativity, if known, his Christian name and surname, place of residence, occupation, and degrees of relationship or other circumstance of connection between him and the person requesting his admission; and, second, a certificate dated within one month, under oath signed by a respectable physician, of the fact of his being insane; each person signing such request or certificate shall annex his profession or occupation and the county and State of his residence, unless these facts appear on the face of the document.

In *Pennsylvania* the laws of 1869 thus provide for the incarceration of patients:

Insane persons may be placed in a hospital for the insane by their legal guardians, or by their relatives or friends in case they have no guardians, but never without the certificate of two or more reputable physicians, after a personal examination made within one week of the date thereof, and this certificate is to be duly acknowledged and sworn to, or affirmed, before some magistrate or judicial officer, who shall certify to the genuineness of the signatures and to the responsibility of the signers.

The law of *Virginia* requires, in order to commit a person to an asylum, that the suspected person must be brought before a commission in lunacy, called for the purpose, consisting of three (3) magistrates of the city or county in which he resides and that they shall summon the family physician and other witnesses, to make a thorough examination of the case. If after a careful investigation the person is adjudged insane, he is sent to an asylum, with a record of the examination, and the super-intendent is required to admit him, if there is a vacancy. The law does not contemplate the admission of the insane of other States.

In *Maryland* a patient may be committed upon the certificate of one physician. At the Mount Hope Asylum two certificates are required.

In *North Carolina*, sections 13 and 14 of the laws of 1881, thus provide:

SECTION 13.—The judges of the Superior Courts, in their respective districts, shall allow to be committed to the asylum, as a patient, any person who may be confined in jail on a criminal charge of any kind, or degree, or upon a peace warrant whenever the judge shall be satisfied, by a verdict of jury of inquisition, that the alleged criminal act was committed while such person was insane.

SEC. 14.—For admission into the asylum in other cases the following proceedings shall be had: Some respectable citizen, residing in the county of the alleged insane person, shall make before and file with a justice of the peace of the county an affidavit in writing.

In *Mississippi* the insane person may be committed by a "lunacy inquiry," requiring six jurors, or, as is usually the case, he may be received in an asylum, upon the certificate of two physicians, who shall swear before a justice or a county clerk.

In *Alabama* the indigent insane are received in the State Asylum on certificate of the Probate Judges of their respective counties, attested by one respectable physician and other witnesses, with or without a jury, as the judge may decide.

Paying patients are received on certificate of one respectable physician, accompanied by the usual bond to secure payment of board.

The laws of *Ohio* in relation to the care of the insane are quite voluminous. The Revised Statutes thus provide:

SECTION 702. For the admission of patients to any of the asylums for the insane, the following proceedings shall be had: Some resident citizen of the proper county shall file with the Probate Judge of such county an affidavit as follows:

THE STATE OF OHIO, ——— COUNTY, ss:

——— ———, the undersigned, a citizen of ——— County, Ohio, being sworn, says that he believes ——— ——— is insane (or, that, in consequence of his insanity, his being at large is dangerous to the community). He has a legal settlement in ——— township, in this County.

Dated this ——— day of ———, A. D. ———.

SEC. 703. When the affidavit is filed, the Probate Judge shall forthwith issue his warrant to some suitable person, commanding him to bring the person alleged to be insane before him, on a day therein named, which shall not be more than five days after the affidavit has been filed, and shall immediately issue subpœnas for such witnesses as he deems necessary (one of whom shall be a respectable physician), commanding the persons in such subpœnas named to appear before the judge on the return day of the warrant: and if any person disputes the insanity of the party charged, the Probate Judge shall issue subpœnas for such person or persons as are demanded on behalf of the person alleged to be insane; provided, that if, by reason of the character of the affliction or insanity of said person, it is deemed unsuitable or improper to bring the person into such Probate Court, then the Probate Judge shall personally visit said person and certify that he has so ascertained the condition of the person by actual inspection, and all proceedings as herein required may then be had in the absence of such person.

SEC. 704. At the time appointed (unless for good cause the investigation is adjourned) the judge shall proceed to examine the witnesses in attendance; and if, upon the hearing of the testimony, he is satisfied that the person so charged is insane, he shall cause a certificate to be made out by the medical witness in attendance, which shall set forth the following. (Here follows a list of questions relating to the patient's symptoms which are to be found in the certificate.)

SEC. 705. The Probate Judge, upon receiving the certificate of the medical witness, made out according to the provisions of the preceding section, shall forthwith apply to the superintendent of the asylum for the insane situated in the district in which such patient resides; he shall, at the same time, transmit copies, under his official seal, of the certificate of the medical witness, and of his finding in the case; upon receiving the application and certificate the superintendent shall immediately advise the Probate Judge whether the patient can be received, and, if so, at what time; the Probate Judge, when advised that the patient will be received, shall forthwith issue his warrant to the sheriff, or any other suitable person, commanding him to forthwith take charge of and convey such insane person to the asylum; if the Probate Judge is satisfied from proof that an assistant is necessary he may appoint one person as such assistant. The warrant of the Probate Judge shall be substantially as follows:

THE STATE OF OHIO, —— COUNTY, *ss.*
OFFICE OF THE PROBATE JUDGE OF SAID COUNTY.

To —— ——:

All the proceedings prescribed by law to entitle —— —— to be admitted into the asylum for the insane having been had, you are commanded forthwith to take charge of and convey said —— —— to the asylum for the insane at ——, and you are authorized to take ——

—— as assistant; after executing this warrant, you will make due return thereof to this office.

Witness my hand and official seal this —— day of ——, A. D.,——.

—— ——
Probate Judge.

Upon receiving such patient the superintendent shall indorse upon the warrant a receipt substantially as follows:

ASYLUM FOR THE INSANE, AT——,

—— ——, A. D., ——.

Received this day, of —— ——, the patient named in the within warrant.

—— ——,
Superintendent.

This warrant, with the receipt of the superintendent thereon, shall be returned to the Probate Judge who issued it, and shall be filed by him with the other papers relating to the case. In all cases the relatives of the insane person shall have a right, if they choose, to convey such insane person to the asylum for the insane, and in such case the warrant shall be directed to one of such relatives, directing him to take another of the relatives as his assistant. If the medical witness does not state in his certificate that the patient is free from all infectious diseases and from vermin, the Probate Judge shall refuse to make the application to the superintendent, as therein provided, until such certificate is furnished. The relatives of any person charged with insanity, or who is found to be insane, shall, in all cases, have the right to take charge of and keep such insane person charged with insanity, if they desire so to do; and in such case the Probate Judge before whom the inquest has been held shall deliver such insane person to them.

The insane of *Indiana* are committed by two magistrates, who are required to visit the alleged lunatic in person and report to the clerk of County Court, who subpœnas witnesses and sends a certified copy of proceedings to Superintendent of State Hospital, requesting admission.

Chapter 85, Revised Statutes of 1874 of the State of *Illinois,* contains the following provisions for the commitment of the insane:

That upon the petition of a near relative of the suspected person, or any respectable person in the county, made to a judge of the County Court, the latter may direct the clerk to issue a writ directed to the sheriff or person having in custody the alleged lunatic, to bring before him the person; and the clerk is furthermore directed to issue the necessary subpœnas for witnesses. A jury of six persons, one of whom shall be a physician, shall be empanelled to try the case. The case shall be tried in the presence of the alleged lunatic, who shall be entitled to the benefit of counsel. The jury shall return a written and signed verdict. If it be that the person is declared insane a committal is to be made out by the clerk, who shall confer with the superintendent, and a warrant shall be issued and directed to the sheriff, or in preference, the relatives of the insane person. The court may make an order to temporarily commit any person.

The law of *Michigan* which concerns the commitment of the insane is as follows:

SECTION 26. When a person in indigent circumstances, and not a pauper, becomes insane, application may be made in his behalf to the Probate Judge of the county where he resides; and said Probate Judge shall call two respectable physicians, and other credible witnesses, and also immediately notify the prosecuting attorney of his county and the supervisor of the township or ward in which such insane person resides, of the time and place of meeting, whose duty it shall be to attend the examination and act in behalf of said county; and said Probate Judge shall fully investigate the facts in the case, and either with or without the verdict of a jury, at his discretion, as to question of insanity, shall decide the case as to his indigence, but the decision as to indigence shall not be conclusive in such county; and if the Probate Judge certifies that satisfactory proof has been adduced, showing him insane, and his estate is insufficient to support him and his family, or, if he has no family, himself, under the visitation of insanity, on his certificate, under the seal of the Probate Court of said county, he shall be admitted into the asylum and supported there at the expense of the county to which he belongs until he shall be restored to soundness of mind, if effected in two years, and until otherwise ordered. The Judge of Probate in such cases shall have power to compel the attendance of witnesses and jurors, and shall file the certificates of physicians, taken under oath, and other papers, in his office, and enter the proper order in his (the) journal of the Probate Court in his office. The Judge of Probate shall report the result of his proceedings to the supervisors of his county, if such person belongs to that county, whose duty it shall be, at the next annual meeting thereafter, to raise money requisite to meet the expenses of support accordingly.

In *Kentucky* the insane are committed by the inquest of a jury and by order of court, their presence in open court being required, unless, upon the affidavit of two respectable physicians, it is shown that it would be dangerous to bring the supposed lunatic into court.

In *Iowa*, I am informed by Dr. Hill, the *modus operandi* of commitment is the following:

A practicing physician, a practicing lawyer, appointed by the Circuit

Judge, who usually continue in office during good behavior, and the clerk of the courts, constitute the commissioners of insanity. The physician on the commission, or the family physician, goes to the home of the patient, often without informing him why he is there, and obtains answers to questions in the " Return of Physician." Then the commissioners meet and decide whether the person is insane, and whether to send him to the hospital. " If they shall be of opinion, from such preliminary inquiries as they may make . . . that such a course would probably be injurious to such person or attended with no advantage, they may dispense with such presence." Two blanks are filed with the superintendent and one with the clerk of the court.

The insane has a right to appeal to court within ten days, a right to appeal to court once in six months thereafter, as well as the right to habeas corpus. They may be discharged from an asylum by the visiting committee.

The laws relating to the care of the Insane in *Wisconsin* are quite simple. Lunatics are committed only by the County Judge. The alleged insane person, or any person acting in his behalf, can request a jury trial, in which case it must be accorded. If tried by jury, the judge is authorized to clear the court of all persons except those immediately interested. If the person is found insane, he is regularly committed by order of the court and under its seal.

In this State, as well as in some others, the physicians who examine the patient are required to answer a long list of questions, relating to the circumstances of the patient, the history of the disease, his habits, heredity, etc.

In *Minnesota*, according to the laws of 1874, Sec. 134, and 1877, Sec. 75, p. 123:

Patients, how committed.—The Probate Judge, or in his absence the court commissioner of any county, upon information being filed before him that there is an insane person in his county needing care and treatment, shall thereupon make an order appointing some regular physician or physicians (not less than one, or more than three) to examine the said person, to ascertain the fact of insanity, a certified copy of which order shall be delivered to said physician or physicians, and shall proceed to the hearing of such information, and shall hear and examine the proofs of said information, and if the said person is found to be insane, he shall, upon the written certificate of the examining physician or physicians, " that the said person in his or their opinion is insane and a proper subject for hospital treatment," said certificate being verified by the oath of the physician or physicians, issue duplicate warrants committing the person so found insane to the care of the superintendent of the hospital, and shall place the warrant in the hands of some friend or other suitable person, whom he shall authorize to convey the said insane person to the hospital.

In *Missouri* there is a State law, and in St. Louis there is a separate local regulation governing the commitment of the insane.

The municipal law of 1882 is thus worded:

SECTION 1.—Ordinance number 11,668. It shall be the duty of the police of the city of St. Louis, if any lunatic, idiot, or person of unsound mind, who is a resident of the city of St. Louis, be found by them within the limits of the city of St. Louis, in such condition as to endanger the lives or property of themselves or of others, or who are unprotected by guardians or friends and without means, to take such person into custody and give notice thereof forthwith to the Chief of Police, who shall immediately notify the Health Commissioner that such person is in his custody, and in said notice he shall give the Health Commissioner the name, age, place of residence, length of residence in the city, occupation when known, the locality where person was arrested, circumstances causing the arrest, and all other information he may have or can obtain in relation to said person. The Health Commissioner, on receipt of such report from the Chief of Police, shall cause an examination to be made of such person by one or more physicians of the Health Department. If upon such examination such person is found to be of unsound mind and an unfit person to be at large, the physician making such examination shall certify such fact to the Health Commissioner, whose duty it shall then be to take charge of such lunatic, idiot, or insane person and place such person in the insane asylum of the city of St. Louis, and to report to the Board of Health his action thereon, and all facts and information regarding such lunatic, idiot, or insane person in his possession, or that may come into his possession; but if the physician making such examination shall certify to the Health Commissioner that the person or persons reported by the Chief of Police as lunatic, idiot, or insane person be not of unsound mind or an idiot, and in his opinion not a fit subject for treatment in an insane asylum, the Health Commissioner shall give notice of the fact to the Chief of Police, and shall not receive such person from his custody. If, however, the physician or physicians examining such person should certify to the Health Commissioner that such person be a fit subject for hospital treatment, then the Health Commissioner shall place such person in one of the hospitals of the city.

Whenever any lunatic, idiot, or person of unsound mind may be arrested by the police of the city of St. Louis, and is found to be a nonresident of the City of St. Louis, the Health Commissioner shall report the fact to the Mayor, who shall, if he thinks proper, order the Chief of Police to cause such persons to be returned to the locality to which they belong, and all expenses attending the return of such person shall be borne by the city of St. Louis, but if the Mayor is of the opinion that it is not practicable to return such person, then the person shall be disposed of as provided in Section 1 of this article."

In *Arkansas* the recent laws for the commitment of the insane provide:

SECTION 2. Whenever it shall appear that any person entitled to admission to the State Lunatic Asylum is insane, any reputable citizen of the State may file a written statement with the County and Probate Judge of the county in which such supposed insane person may reside.

SEC. 3. Any County and Probate Judge with whom a citizen's statement may have been filed, as set forth in Section 2 of this Act, shall appoint a time, as soon thereafter as may be practicable, for hearing, and at such time appointed shall proceed to hear the testimony of such competent witnesses as may be produced at such hearing, and in addition to the testimony of such witnesses, shall cause such alleged insane person to be examined by one or more regular practicing physicians of good standing, who shall present in writing to such County and Probate Judge a sworn statement of the result of his or their examination, including the following interrogatories, with their answers as part of the same.

(Here follow a number of interrogatories relating to the patient's antecedents, present condition, etc., which may be found in the blank certificate.)

In *Texas* the insane person is committed by a jury of six, a charge of lunacy having first been brought before the County Court. Witnesses are subpoenaed, and upon a verdict of lunacy the person is deprived of his liberty.

The law of *California* necessitates that application shall be made to a judge of the Supreme Court by the friends of the patient. That after the presentation of a certificate of examination, signed by at least two physicians of good standing, the judge shall sign an order of commitment, which also provides for the appointment of a guardian; as in some other States, a number of interrogatories are included in the medical certificate.

For the Napa State Asylum the following laws were passed in 1876:

SECTION 18. No case of idiocy, imbecility, harmless, chronic, mental unsoundness, or acute *mania à potu*, shall be committed to this asylum, and whenever in the opinion of the resident physician, after a careful examination of the case of any person committed, it shall be satisfactorily ascertained by him that the party has been unlawfully committed, and that he or she comes under the rule of exemptions provided for in this section, he shall have the authority to discharge such person so unlawfully committed, and return him or her to the county from which committed, at the expense of such county.

SEC. 19. The judge, shall inquire into the ability of insane persons committed by him to the asylum to bear the actual charges and expenses
24

for the time that such person may remain in the asylum. In case an insane person, committed to the asylum under the provisions of this Act, shall be possessed of real or personal property sufficient to pay such charges and expenses, the judge shall appoint a guardian for such person, who shall be subject to all the provisions of the general laws of this State in relation to guardians, as far as the same are applicable; and when there is not sufficient money in the hands of the guardian, the judge may order a sale of property of such insane person, or as much thereof as may be necessary, and from the proceeds of such sale the guardian shall pay to the Board of Trustees the sum fixed upon by them each month quarterly in advance for the maintenance of such ward; and he also shall, out of the proceeds of such sale, or such other fund as he may have belonging to such ward, pay for such clothing as the resident physician shall, from time to time, furnish such insane person; and he shall give a bond, with good and sufficient sureties, payable to the Board of Trustees, and approved by the judge, for the faithful performance of the duties required of him by this Act, as long as the property of his insane ward is sufficient for the purpose. Indigent insane persons having kindred of degree of husband or wife, father, mother, or children, living within this State of sufficient ability, said kindred shall support such insane persons to the extent prescribed for paying patients.

Sec. 20. Non-residents of this State, conveyed or coming herein while insane, shall not be committed to or supported in the Napa State Asylum for the Insane; but this prohibition shall not prevent the commitment to and temporary care in said asylum of persons stricken with insanity while traveling or temporarily sojourning in this State; or sailors attacked with insanity upon the high seas, and first arriving thereafter in some port within this State.

In *Oregon* a complaint must be made by two householders to the County Judge. The patient will be examined under supervision of the County Judge assisted by the District Attorney and two physicians, upon whose certificate (if found to be insane) the patient will be sent to the asylum. One copy of the commitment is sent to the asylum, one copy retained in county from which patient is sent, and one copy transmitted to Secretary of State.

I am indebted to Dr. J. M. Wallace for the following abstract of the *Canadian* laws:

There are two methods of committing insane persons to an asylum in Ontario. The ordinary process, and by warrant of the Lieutenant Governor of the province. The lunatic is committed to jail as a dangerous lunatic, and is kept there until he is examined by the County Judge and two physicians, who each certify that he is insane and dangerous to be at large. The certificates and other commitment papers are forwarded to the Provincial Secretary, who is a member of the Government; he advises the Lieutenant-Governor to issue a warrant for the transfer of the lunatic from the jail to an asylum. By the ordinary process, application for admission is made to the Medical Superintendent by the friends of the lunatic; blank forms are sent out to be filled, and when returned and found satisfactory, the Medical Superintendent sends an order for the admission of the patient.

PLATE I.

IDIOCY.

J. R——, aged forty-three, is a case well known in the literature of psychiatry. He weighs 72 pounds, is of short stature (4 feet 7½ inches), and his head is perhaps one of the smallest reported in this country. The circumference from a point in front one inch above the root of the nose, to one at the level of the occipital protuberance behind is 15 inches. The bi-aural arc measurement is 8 inches, and the antero-posterior arc is 8 inches. This is over a thick growth of short coarse hair, and at least an inch difference must exist between the true measurements and those made. His teeth are nearly all gone and he has a double cataract. His general health is good and he is well nourished. His intelligence is almost nil. He has been taught to swear, and can say a few words without any idea of their import. He is good-tempered and easily amused. His left . ear is the seat of old inflammatory contraction, and is much deformed.

PLATE II.

IMBECILITY.

A. B——, aged twenty-five. Received in the asylum two years before his death, which occurred last year (1882).

He had a very shrill voice, and was quite excitable, crying and laughing without cause. He was in the habit of collecting bits of paper and straw which he chewed, and kept a supply in his shoes and socks. He was sometimes so violent as to need restraint and was quite offensive in his habits. For some months previous to death he suffered from Bright's disease, from which he ultimately died.

PLATE III.

MELANCHOLIA ATTONITA.

C. C——, aged thirty-seven. Duration of insanity seven months. Cause unknown. Auditory hallucinations. She hears voices commanding her not to eat, and it is often necessary to feed her with the tube. She has delusions of persecution. Her movements are sluggish, and she assumes fixed attitudes. There is rarely any play of facial expression and she takes no notice of those about her.

PLATE IV.

CHRONIC MELANCHOLIA.

X—— has been melancholic for some years, and the disease is drifting into dementia.

PLATE V.

SUBACUTE MANIA.

E. E——, aged twenty-eight. Duration of insanity six years. Originally acute mania of a violent type. Cause unknown. Auditory hallucinations. She has communication with divine personages, and delusions of grandeur, believing that she is Queen of Ireland, and is the kinswoman of every one about her. She is remarkably obscene and alludes to her carnal relations, which are of a peculiar kind, and she is incoherent and loquacious. Her hair is coarse and becomes bristling and erect when she is excited.

PLATE VI.

CHRONIC MANIA.

J. B——, aged fifty-one, has been in the Ward's Island Asylum eleven years. No history of cause. He is incoherent and excitable, but quite tractable. Disclaims his proper name, and has delusions that his bones are all broken and his head mashed. Is clownish in his behavior, and sings at the top of his voice. He is fond of decorating himself with rubbish and dirty finery.

PLATE VII.

DEMENTIA.

A. W——, aged forty-four. Duration of insanity four years. Cause intemperance. Her dementia was the sequel of acute melancholia. She has had visual hallucinations, and has seen spectres and other frightful things. She has had suicidal tendencies, and has heard voices which told her to destroy herself. Her violence has been remarkable. She has now (March, 1883) lapsed into a condition of dementia attended by great restlessness and violence. It is necessary to keep her strapped in the chair. Her habits are of the filthiest kind, and she needs constant attention. There is a constant accumulation of saliva, which she ejects with violence, there being the automatic expectoration alluded to by some writers. She betrays no indication of the mental operations except in her appearance.

PLATE VIII.

DEMENTIA.

R. P. H——, aged thirty-six. There is a strong family history of insanity, five of his uncles being insane. He is profoundly demented, and is dirty, stupid, and careless. His disease has lasted nineteen years, and followed melancholia.

PLATE IX.

GENERAL PARESIS.

J. McK——, aged thirty-seven. He has been in the asylum two years. There is no known cause of the disease. He has had delusions of wealth, but is now demented and stupid.

INDEX.

www.ingramcontent.com/pod-product-compliance
Lightning Source LLC
Chambersburg PA
CBHW021350210326
41599CB00011B/825